GCSE
CHEMISTRY

B. EARL & L.D.R. WILFORD

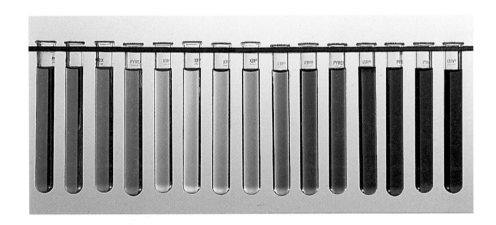

JOHN MURRAY

International hazard warning symbols

You will need to be familiar with these symbols when undertaking practical experiments in the laboratory.

 Corrosive

 Oxidising

 Explosive

 Radioactive

 Harmful or irritant

 Toxic

 Highly flammable

©B. Earl and L. D. R. Wilford, 1995

First published in 1995
by John Murray (Publishers) Ltd
50 Albemarle Street
London W1X 4BD

Reprinted 1996, 1997, 1998

Layouts by Eric Drewery.
Artwork by Philip Ford and Tom Cross.
Typeset in 11/13pt Garamond and News Gothic by Servis Filmsetting Ltd, Manchester.
Printed by Butler & Tanner Ltd, Frome, Somerset.

A catalogue entry for this title can be obtained from the British Library

ISBN 0-7195-5303-2

CONTENTS

CONTENTS

CONTENTS

Preface to the reader

This textbook has been written to help you in your study of chemistry to GCSE. Although you will be following a GCSE syllabus for only one particular examination group, this book contains the material needed by all the groups. For this reason it is not expected that you will need to study or learn everything in this textbook.

The different chapters in this book are split up into short topics. At the end of each of these topics are questions to test whether you have undertstood what you have read. At the end of each chapter there are larger study questions. Try to answer as many of the questions as you can as you come across them because asking and answering questions is at the heart of your study of chemistry.

A selection of examination questions, selected from examination papers published by the different examination groups, is included at the end of the book. In many cases they are designed to test your ability to apply your chemical knowledge. The questions may provide certain facts and ask you to make an interpretation of them. In such cases, the factual information may not be covered in the text.

To help draw attention to the more important words, scientific terms are printed in bold the first time they are used. Checklists also are presented at the end of each chapter summarising the important points covered.

This textbook will provide you with the information you need for your particular syllabus. We hope you enjoy using this book.

Acknowledgements

The authors would like to thank Irene, Katharine and Barbara for their never-ending patience and encouragement through the production of this textbook. In addition great thanks is given to Mr Dennis Richards, Headteacher, St Aidan's Church of England High School, Harrogate for his help and support.

B. Earl
L. D. R. Wilford

We use coloured strips at the edges of pages to define different areas of chemistry:
- ■ 'starter' chapters – basic principles
- ■ inorganic chemistry
- ■ physical chemistry
- organic chemistry and the living world.

1 MATTER

Chemistry is about what **matter** is like and how it behaves and our explanations and predictions of its behaviour. What is **matter**? This word is used to cover all the substances and materials from which the physical universe is composed. There are many millions of different substances known, and all of them can be categorised as solids, liquids or gases (Figure 1.1). These are what we call the **three states of matter**.

Figure 1.1 Water in three different states

Solids, liquids and gases

A **solid**, at a given temperature, has a definite volume and shape which may be affected by changes in temperature. Solids usually increase slightly in size when heated (**expansion**) and usually decrease in size if cooled (**contraction**).

A **liquid**, at a given temperature, has a fixed volume and will take up the shape of any container into which it is poured. Like a solid, a liquid's volume is slightly affected by changes in temperature.

A **gas**, at a given temperature, has neither a definite shape nor volume. It will take up the shape of any container into which it is placed and will spread out evenly within it. Unlike those of solids and liquids, the volumes of gases are affected quite markedly by changes in temperature.

Liquids and gases, unlike solids, are relatively **compressible**. This means that the volume can be reduced by the application of pressure. Gases are much more compressible than liquids.

The kinetic theory of matter

Figure 1.2a Model of a chrome alum crystal

The **kinetic theory** helps to explain the way in which matter behaves. The evidence is consistent with the idea that all matter is made up of tiny **particles**. This theory explains the physical properties of matter in terms of the movement of its constituent particles.

The main points of the theory are:

- all matter is made up of tiny, moving particles, invisible to the naked eye. Different substances have different types of particles (atoms, molecules or ions) which have different sizes
- the particles move all the time. The higher the temperature, the faster they move on average
- heavier particles move more slowly than lighter ones at a given temperature.

The kinetic theory can be used as a scientific model to explain how the arrangement of particles relates to the properties of the three states of matter.

Explaining the states of matter

In a solid the particles attract one another. There are bonds between the particles which hold them close together. The particles have little freedom of movement and are only able to vibrate about a fixed position. They are arranged in a regular manner, which explains why many solids form crystals.

It is possible to model such crystals by using spheres to represent the particles (Figure 1.2a). If the spheres are built up in a regular way then the shape compares very closely with that of a part of a chrome alum crystal (Figure 1.2b).

Figure 1.2b A chrome alum crystal

Studies using X-ray crystallography (p. 47) have confirmed how the particles are arranged in crystal structures. When crystals of a pure substance form under a given set of conditions, the particles present are always packed in the same way. However, the particles may be packed in different ways in crystals of different substances. For example, common salt (sodium chloride) has its particles arranged to give cubic crystals as shown in Figure 1.3.

In a liquid the particles are still quite close together but they move around in a random way and often collide with one another. The forces of attraction between the particles in a liquid are weaker than those in a solid. Particles in the liquid form of a substance have more energy on average than the particles in the solid form of the same substance.

In a gas the particles are relatively far apart. They are free to move anywhere within the container in which they are held. They move randomly at very high velocities, much more rapidly than those in a

Figure 1.3 Sodium chloride crystals

liquid. They collide with each other, but less often than in a liquid, and they also collide with the walls of the container. They exert virtually no forces of attraction on each other because they are relatively far apart. Such forces, however, are very significant. If they did not exist we could not have solids or liquids (see Changes of state).

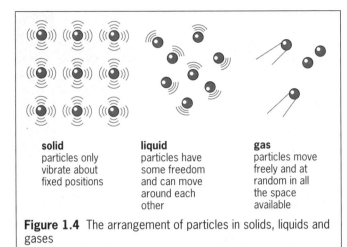

solid
particles only vibrate about fixed positions

liquid
particles have some freedom and can move around each other

gas
particles move freely and at random in all the space available

Figure 1.4 The arrangement of particles in solids, liquids and gases

The arrangement of particles in solids, liquids and gases is shown in Figure 1.4.

QUESTIONS

a Use your research skills to find out about the technique of X-ray crystallography and how this technique can be used to determine the crystalline structure of sodium chloride.

b Liquids and solids cannot be compressed as much as gases can. Explain why this should be the case.

Changes of state

The kinetic theory model can be used to explain how a substance changes from one state to another. If a solid is heated the particles vibrate faster as they gain energy. This makes them 'push' their neighbouring particles further away from themselves. This causes an increase in the volume of the solid and the solid expands. Expansion has taken place.

Eventually, the particles break free from one another as the heat energy overcomes the forces of attraction and the bonds are broken. The regular pattern of the structure breaks down. The particles can now move around each other. The solid has melted. The temperature at which this takes place is called the **melting point** of the substance. The temperature of a pure melting solid will not rise until it has all melted. When the substance has become a liquid there are still some forces of attraction between the particles, which is why it is a liquid and not a gas.

Solids which have high melting points have stronger forces of attraction between their particles than those which have low melting points. A list of some substances with their corresponding melting and boiling points is shown in Table 1.1.

If the liquid is heated the particles will move around even faster as their average energy increases.

Table 1.1

Substance	Melting point (°C)	Boiling point (°C)
Sodium chloride	801	1413
Water	0	100
Magnesium oxide	2827	3627
Methane	−182	−164
Aluminium	661	2467
Oxygen	−218	−183
Ethanol	−117	79
Sulphur	113	445

Some particles at the surface of the liquid have enough energy to overcome the forces of attraction between themselves and the other particles in the liquid and they escape to form a gas. The liquid begins to **evaporate** as a gas is formed.

Eventually, a temperature is reached at which the particles are trying to escape from the liquid so quickly that bubbles of gas actually start to form inside the bulk of the liquid. This temperature is called the **boiling point** of the substance. At the boiling point the pressure of the gas created above the liquid equals **atmospheric pressure**.

Liquids with high boiling points have stronger forces between their particles than liquids with low boiling points.

When a gas is cooled the average energy of the particles decreases and the particles move closer together. The forces of attraction between the particles now become significant and cause the gas to **condense** into a liquid. When a liquid is cooled it **freezes** to form a solid. In each of these changes energy is given out.

Changes of state are examples of **physical changes**. Whenever a physical change of state occurs, the temperature remains constant during the change (see Heating and cooling curves, p. 4). During a physical change no new substance is formed.

An unusual state of matter

Liquid crystals are an unusual state of matter (Figure 1.5). These substances look like liquids, flow

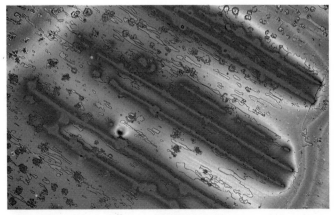

Figure 1.5 A polarised light micrograph of liquid crystals

Figure 1.6 Liquid crystals are used in this calculator's display

Figure 1.7 Dry ice (solid carbon dioxide) sublimes on heating

like liquids but have some order in the arrangement of the particles, and so in some ways they behave like crystals.

Liquid crystals are now part of our everyday life. They are widely used in displays for digital watches and calculators (Figure 1.6), and in portable televisions. They are also used in thermometers because liquid crystals change colour as the temperature changes.

An unusual change of state

There are a few substances which change directly from a solid to a gas without becoming a liquid when they are heated. This rapid spreading out of the particles is called **sublimation**. Cooling causes a change from a gas directly back to a solid. Examples of substances that behave in this way are carbon dioxide (Figure 1.7) and iodine.

$$\text{solid} \underset{\text{sublimation}}{\overset{\text{sublimation}}{\rightleftharpoons}} \text{gas}$$

Carbon dioxide is a white solid called dry ice at temperatures below -78 °C. When heated to just above -78 °C it changes into carbon dioxide gas. The changes of state are summarised in Figure 1.8.

Heating and cooling curves

The graph shown in Figure 1.9 was drawn by plotting the temperature of water as it was heated steadily from -5 °C to 100 °C. You can see from the curve that changes of state have taken place. When the temperature was first measured only ice was present.

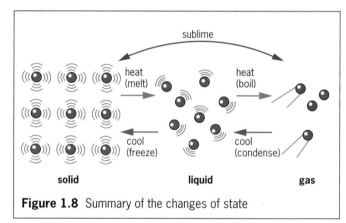

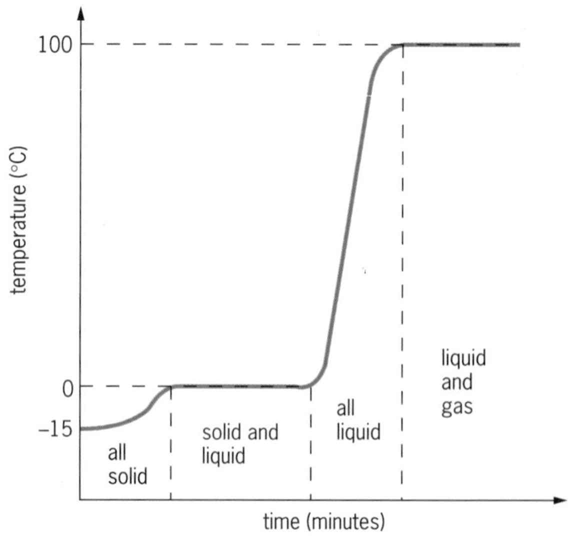

Figure 1.8 Summary of the changes of state

Figure 1.9 Graph of temperature against time for the change from ice to water to steam

After 2 minutes the curve flattens, showing that even though heat energy is being put in the temperature remains constant.

In ice the particles of water are close together and are attracted to one another. For ice to melt the particles must obtain sufficient energy to overcome the forces of attraction between the water particles. This is where the heat energy is going.

The temperature will only begin to rise again after all the ice has melted. Generally, the heating curve for a pure solid always stops rising at its melting point and gives rise to a sharp melting point. The addition or presence of impurities lowers the melting point. You can try to find the melting point of a substance using the apparatus shown in Figure 1.10.

In the same way, if you want to boil a liquid such as water you have to give it some extra energy. This can be seen on the graph where the curve levels out at the boiling point of water.

The reverse processes of condensing and freezing occur on cooling. This time, however, energy is given out when the gas condenses to the liquid and the liquid freezes to give the solid.

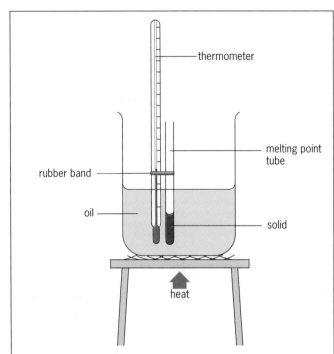

Figure 1.10 Apparatus shown here if heated slowly can be used to find the melting point of a substance

QUESTIONS

a Write down as many uses as you can for liquid crystals.
b Why do gases expand more than solids for the same increase in temperature?
c Ice on a car windscreen will disappear as you drive along, even without the heater on. Explain why this happens.
d When salt is placed on ice the ice melts. Explain why this happens.
e Draw and label the graph you would expect to produce if water at 100 °C was allowed to cool to a temperature of –5 °C.

Diffusion—evidence for moving particles

When you walk past a cosmetics counter in a department store you can usually smell the perfumes. For this to happen gas particles must be leaving the open perfume bottles and be spreading out through the air in the store. This spreading out of a gas is called **diffusion** and it takes place in a haphazard and random way.

All gases diffuse to fill the space available to them. As you can see from Figure 1.11, after a day the brown-red fumes of gaseous bromine have spread evenly throughout both gas jars from the liquid present in the lower gas jar.

Gases diffuse at different rates. If one piece of cotton wool is soaked in concentrated ammonia solution and another is soaked in concentrated hydrochloric

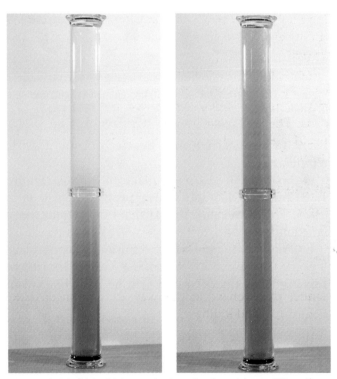

Figure 1.11 After 24 hours the bromine fumes have diffused throughout both gas jars

acid and these are put at opposite ends of a dry glass tube, then after a few minutes a white cloud of ammonium chloride appears (Figure 1.12). This shows the position at which the two gases meet and react. The white cloud forms in the position shown because the ammonia particles are lighter than the hydrogen chloride particles (released from the hydrochloric acid) and so move faster. Generally, light

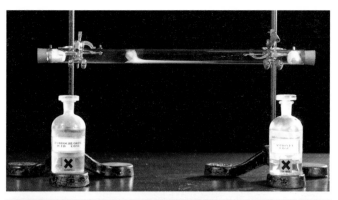

Figure 1.12 Hydrochloric acid (left) and ammonia (right) diffuse at different rates

Figure 1.13 Diffusion within copper(II) sulphate solution can take days to reach the stage shown right

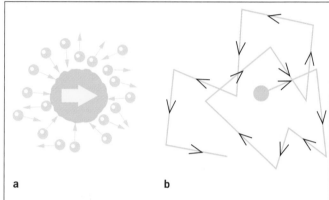

a   **b**

Figure 1.14 (**a**, left) Pollen particles being bombarded by water molecules and (**b**, right) Brownian motion

particles move faster than heavier ones at a given temperature.

Diffusion will also take place in liquids (Figure 1.13) but it is a much slower process than it is in gases. This is because the particles of a liquid move much more slowly.

In liquids and gases where diffusion is taking place we are dealing with a process called **intimate mixing**. The kinetic theory can be used to explain this process. It states that collisions are taking place between particles in a liquid or a gas and that there is sufficient space between the particles of one substance for the particles of the other substance to move into.

Brownian motion

Evidence for the movement of particles in liquids came to light in 1827 when a botanist, Robert Brown, observed that fine pollen grains on the surface of water were not stationary. Through his microscope he noticed that the grains were moving about in a random way. It was 36 years later, in 1863, that another scientist called Wiener explained what Brown had observed. He said that the pollen grains were moving because the much smaller and faster-

moving water particles were constantly colliding with them (Figure 1.14a).

This random motion of visible particles (pollen grains) caused by much smaller, invisible ones (water particles) is called **Brownian motion**, after the scientist who first observed this phenomenon (Figure 1.14b).

QUESTIONS

a Explain why bromine would diffuse faster in evacuated gas jars.
b Describe, with the aid of a diagram, the diffusion of copper(II) sulphate solution.
c Explain why diffusion in liquids is slower than in gases.

Gas laws

What do you think has caused the difference between the balloons in Figure 1.15?

The **pressure** inside a balloon is caused by the gas particles striking the inside surface of the balloon (Figure 1.16). There is an increased pressure inside the balloon at a higher temperature due to

Figure 1.15 The balloon is partly inflated at room temperature (left) and becomes fully expanded at a higher temperature

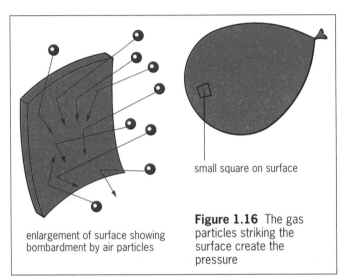

enlargement of surface showing bombardment by air particles

small square on surface

Figure 1.16 The gas particles striking the surface create the pressure

the gas particles having more energy, moving around faster and so striking the inside surface of the balloon more frequently; this in turn leads to an increase in pressure. Since the balloon is an elastic envelope, the increased pressure causes the skin to stretch and the volume to increase. This increase in volume with increased temperature is a property of all gases. In 1781, from this sort of observation, a French scientist called Jacques Charles concluded that when the temperature of a gas increased the volume also increased, at a constant pressure.

Investigating Charles' Law

Charles' Law applies to a fixed mass of gas. The air column in the glass tube shown in Figure 1.17a has been trapped by a drop of concentrated acid, which moves up the tube as the gas expands. The length of the air column can be taken as a measure of the volume of air that is trapped. Readings are taken of the temperature and the length of the air column as the water bath is heated. If these data are plotted, then a volume against temperature graph like the one shown in Figure 1.17b is produced.

The graph shows that at −273 °C the volume of the gas should contract to zero! This temperature is called **absolute zero** and it is the lowest possible temperature. At this temperature, theoretically, particles have no motion and therefore possess no energy.

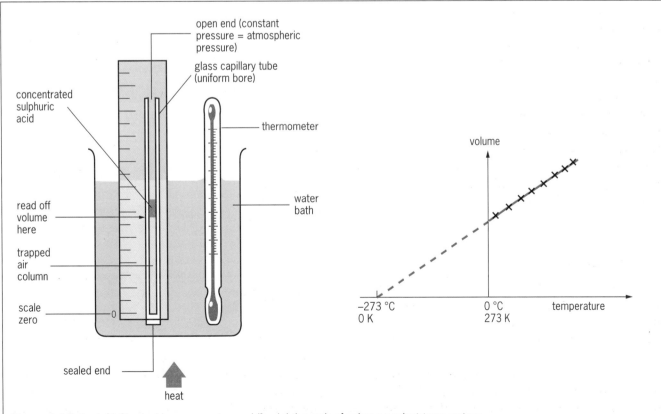

Figure 1.17 (**a**, left) Charles' Law apparatus and (**b**, right) graph of volume against temperature

We can define a new temperature scale, which was proposed by Lord Kelvin in 1854, called the Kelvin scale of temperature, which has 0 K at absolute zero. Kelvins are the same size as degrees on the Celsius scale. On the Kelvin scale, water freezes at 273 K and boils at 373 K. Note that we write 273 K without a ° (degree) sign. In general, to convert a Celsius temperature to a Kelvin temperature add 273.

$$K = {}^{\circ}C + 273$$

Charles' Law states the volume, V, of a fixed mass of gas is directly proportional to its absolute temperature, T, if the pressure is kept constant. Later, it was found that a more accurate representation of Charles' Law required the temperature to be measured on the absolute temperature scale. Mathematically,

$$V \propto T$$

or

$$V = \text{constant} \times T$$

or

$$\frac{V}{T} = \text{constant}$$

Boyle's Law

You can feel the increased pressure of the gas on your finger by pushing in the piston of a bicycle pump. When you do this you push the same number of particles into a smaller volume (Figure 1.18). This means they hit the walls of the pump more often so increasing the pressure.

In 1662 a scientist called Robert Boyle deduced from experiments he carried out on various gases that when the pressure was increased, the volume of the gas was reduced (Figure 1.19).

Investigating Boyle's Law

A car foot pump is used to increase the pressure on a fixed mass of air trapped by oil in a strong glass tube (Figure 1.20). The length of the air column is measured at different pressures.

A graph plotted of volume against $1/p$ gives a straight line through the origin. This demonstrates Boyle's Law, which states that the volume, V, of a fixed mass of gas is inversely proportional to its pressure, p, if the temperature is constant. Mathematically,

$$V \propto \frac{1}{p}$$

or

$$V = \frac{\text{constant}}{p}$$

or

$$pV = \text{constant}$$

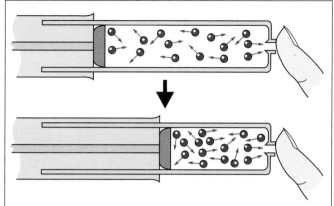

Figure 1.18 A higher pressure is created by pushing in the piston

Figure 1.19 Robert Boyle (1627–1691)

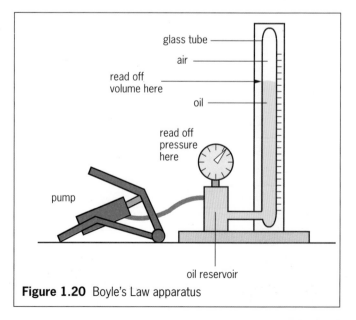

Figure 1.20 Boyle's Law apparatus

Combining the gas laws

By combining Boyle's and Charles' Laws it is possible to show the relationship between the pressure, p, the volume, V, and the temperature, T, of a fixed mass of gas as:

$$\frac{pV}{T} = \text{constant}$$

or

$$\frac{p_1 V_1}{T_1} = \frac{p_2 V_2}{T_2}$$

where p_1, V_1 and T_1 and p_2, V_2 and T_2 are the pressure, volume and temperature (in Kelvin) in two different situations.

An example of the use of this relationship is given below.

If the volume of a gas collected at $40\,^\circ\text{C}$ and a pressure of $1 \times 10^5\,\text{Pa}$ was $100\,\text{cm}^3$, what would be the volume of the gas at a temperature of $10\,^\circ\text{C}$ and a pressure of $2 \times 10^5\,\text{Pa}$?

Using

$$\frac{p_1 V_1}{T_1} = \frac{p_2 V_2}{T_2}$$

$p_1 = 1 \times 10^5\,\text{Pa}$ $p_2 = 2 \times 10^5\,\text{Pa}$
$V_1 = 100\,\text{cm}^3$ $V_2 = ?$
$T_1 = (40 + 273) = 313\,\text{K}$ $T_2 = (10 + 273)\,\text{K} = 283\,\text{K}$

$$\frac{(1 \times 10^5) \times 100}{313} = \frac{(2 \times 10^5) \times V_2}{283}$$

$$V_2 = \frac{(1 \times 10^5) \times 100 \times 283}{313 \times (2 \times 10^5)}$$

$$V_2 = 45.21\,\text{cm}^3$$

QUESTIONS

a When a gas is cooled the particles move more slowly. Explain what will happen to the volume of the cooled gas if the pressure is kept constant.

b A bubble of methane gas rises from the bottom of the North Sea.
  (i) What will happen to the size of the bubble as it rises to the surface?
  (ii) Explain your answer to (i).

c A gas syringe contains $50\,\text{cm}^3$ of oxygen gas at $20\,^\circ\text{C}$. If the temperature was increased to $45\,^\circ\text{C}$, what would be the volume occupied by this gas, assuming constant pressure throughout?

d A bicycle pump contains $50\,\text{cm}^3$ of air at a pressure of $1 \times 10^5\,\text{Pa}$. What would be the volume of the air if the pressure was increased to $2.1 \times 10^5\,\text{Pa}$ at constant temperature?

e If the volume of a gas collected at $60\,^\circ\text{C}$ and $1 \times 10^5\,\text{Pa}$ pressure was $70\,\text{cm}^3$, what would be the volume at a temperature of $0\,^\circ\text{C}$ and a pressure of $4 \times 10^5\,\text{Pa}$?

Checklist

After studying Chapter 1 you should know and understand the following terms.

Matter   Anything which occupies space and has a mass.

Solids, liquids and gases   The three states of matter to which all substances belong.

Kinetic theory   A theory which accounts for the bulk properties of matter in terms of the constituent particles.

Melting point   The temperature at which a solid begins to liquefy. Pure substances have a sharp melting point.

Boiling point   The temperature at which the pressure of the gas created above the liquid equals atmospheric pressure.

Atmospheric pressure   The pressure exerted by the atmosphere on the surface of the Earth due to the weight of the air.

Condensation   The change of a vapour or a gas into a liquid. This process is accompanied by the evolution of heat.

Evaporation   A process occurring at the surface of a liquid involving the change of state of a liquid into a vapour at a temperature below the boiling point.

Sublimation   The direct change of state from solid to gas and the reverse process.

Diffusion   The process by which different substances mix as a result of the random motions of their particles.

Charles' Law   At constant pressure the volume of a given mass of gas is directly proportional to the absolute temperature.

$$V \propto T$$

Boyle's Law   At a constant temperature the volume of a given mass of gas is inversely proportional to the pressure.

$$V \propto \frac{1}{p}$$

Absolute temperature   A temperature measured with respect to absolute zero on the Kelvin scale. Absolute zero is the lowest possible temperature for all substances. The Kelvin scale is usually denoted by T.

$$T\,\text{K} = {}^\circ\text{C} + 273$$

QUESTIONS

1 a Explain how particles move in:
   (i) a solid
   (ii) a liquid
   (iii) a gas.

b Explain what happens to the particles in a gas as it cools down.

2 Explain the meaning of each of the following terms. In each case include in your answer an example to help with your explanation.
   a Diffusion.
   b Expansion.
   c Contraction.
   d Brownian motion.
   e Random movement.
   f Sublimation.
   g Intimate mixing.
   h Physical change.

3 a Why is heat given out when a vapour condenses to a liquid?
   b Why is evaporation accompanied by cooling?

4 Use the kinetic theory to explain the following.
   a The pressure in a car tyre will be higher at the end of a long journey.
   b You can smell hot coffee several metres away, yet you have to be near cold coffee to smell it.
   c A football is blown up until it is hard on a hot summer's day. In the evening the football feels softer.
   d When a person wearing perfume enters a room it takes several minutes for the smell to reach the back of the room.
   e Washing will dry better on a windy day than on a still day.
   f When a balloon filled with hydrogen gas at ground level was released it burst when it reached a height of 1 km.

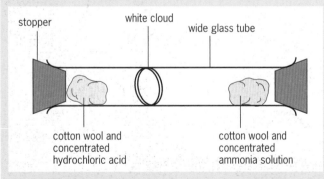

stopper   white cloud   wide glass tube

cotton wool and concentrated hydrochloric acid   cotton wool and concentrated ammonia solution

5 The apparatus shown above was set up. Give explanations for the following observations.
   a The formation of a white cloud.
   b It took a few minutes before the white cloud formed.
   c The white cloud formed nearer to the cotton wool soaked in the hydrochloric acid.
   d Warming the concentrated ammonia and concentrated hydrochloric acid before carrying out the experiment reduced the time taken for the white cloud to form.

6 The following diagram shows the three states of matter and how they can be interchanged.
   a Name the changes **A** to **E**.
   b Name a substance which will undergo change **E**.
   c Name a substance which will undergo changes from solid to liquid to gas between 273 K and 373 K.

d Describe what happens to the particles of the solid during change **E**.
   e Which of the changes **A** to **E** will involve:
   (i) an input of heat energy?
   (ii) an output of heat energy?
   f Draw diagrams to show the arrangement of particles in:
   (i) solids
   (ii) liquids
   (iii) gases.

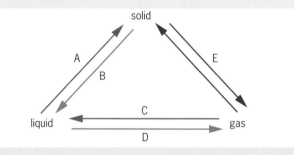

solid

A   B   E

liquid   C   gas

D

7 Some copper(II) sulphate solution was carefully placed in the bottom of a beaker of water. The beaker was then covered and left for several hours.

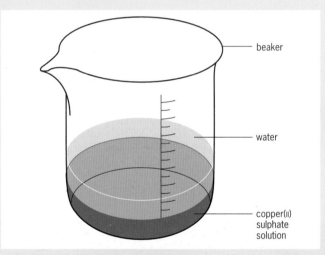

beaker

water

copper(II) sulphate solution

a Describe what you would see after:
   (i) a few minutes
   (ii) several hours.
   b Explain your answer to (a) using your ideas of the kinetic theory of particles.
   c What is the name of the physical process that takes place in this experiment?

8 A bicycle pump contains $50\,cm^3$ of air when the piston is fully drawn out. The air in the pump is initially at a pressure of 1 atmosphere at $17\,°C$. Calculate the pressure of the air as it is forced into a tyre if compression of the piston reduces the volume of the pump to $15\,cm^3$ and at the same time the temperature in the pump is raised to $27\,°C$. (1 atmosphere = $1 \times 10^5\,Pa$)

9 A balloon made of an elasticated fabric which carries meteorological instruments contains $30\,m^3$ of hydrogen gas at atmospheric pressure and at a temperature of $20\,°C$.
   a What will be the volume of the hydrogen gas when the balloon has reached a height of $11\,000\,m$, where the pressure is now one-third that at sea level and the temperature is $-25\,°C$? (1 atmosphere = $1 \times 10^5\,Pa$)
   b Give an explanation of the answer you obtained in (a).

2 ELEMENTS, COMPOUNDS AND MIXTURES

The universe is made up of a very large number of substances. Our own world is no exception (Figure 2.1).

If this vast array of substances is examined more closely, it is found that they are made up of some basic substances which were given the name **elements** in 1661 by Robert Boyle.

In 1803, John Dalton (Figure 2.2) suggested that each element was composed of its own kind of particles, which he called **atoms**. Atoms are much too small to be seen. We now know that about 20×10^6 of them would only stretch over a length of 1 cm.

Figure 2.1 Planet Earth from space

Figure 2.2 John Dalton (1766–1844)

Elements

Robert Boyle used the name element for any substance that cannot be broken down into a simpler substance. This definition can be extended to include the fact that each element is made up of only one kind of atom. The word atom comes from the Greek word for 'unsplittable'.

For example, aluminium is an element which is made up of only aluminium atoms. It is not possible to obtain a simpler substance chemically from the aluminium atoms. You can only make more complicated substances from it, such as aluminium oxide, aluminium nitrate or aluminium sulphate.

There are 112 elements which have now been identified. Twenty-one of these do not occur in nature and have been made artificially by scientists. They include elements such as plutonium, curium and unnilpentium. Ninety-one of the elements occur naturally and range from some very reactive gases, such as fluorine and chlorine, to gold and platinum, which are unreactive elements.

All elements can be classified according to their various properties. A simple way to do this is to classify them as **metals** or **non-metals** (Figures 2.3 and 2.4). Table 2.1 shows the physical data for some common metallic and non-metallic elements.

You will notice that many metals have high densities, high melting points and high boiling points, and that most non-metals have low densities, low melting points and low boiling points. Table 2.2 gives a summary of the different properties of metals and non-metals.

A discussion of the chemical properties of metals is given in Chapters 3, 4 and 9. The chemical properties of certain non-metals are discussed in Chapters 3, 4, 15 and 16.

Figure 2.3 Some metals: copper, mercury and magnesium

Figure 2.4 Some non-metals: bromine liquid, sticks of sulphur and neon used in advertising signs

Table 2.1 Physical data for some metallic and non-metallic elements at room temperature and pressure

Element	Metal/non-metal	Density (g cm^{-3})	Melting point (°C)	Boiling point (°C)
Nickel	Metal	8.90	1453	2732
Aluminium	Metal	2.70	660	2580
Iron	Metal	7.87	1535	2750
Copper	Metal	8.92	1083	2567
Gold	Metal	19.29	1065	2807
Magnesium	Metal	1.74	649	1107
Lead	Metal	11.34	328	1740
Zinc	Metal	7.14	420	907
Silver	Metal	10.50	962	2212
Hydrogen	Non-metal	0.07[a]	−259	−253
Carbon	Non-metal	2.25	2652	Sublimes
Oxygen	Non-metal	1.15[b]	−218	−183
Sulphur	Non-metal	2.07	113	445
Nitrogen	Non-metal	0.88[c]	−210	−196

Source: Earl B., Wllford L.D.R. Chemistry data book. Nelson Blackie, 1991

[a] At −254 °C
[b] At −184 °C
[c] At −197 °C

Table 2.2 Different properties of metals and non-metals

Property	Metal	Non-metal
Physical state at room temperature	Usually solid (occasionally liquid)	Solid, liquid or gas
Malleability	Good	No—usually soft or brittle
Ductility	Good	when solid
Appearance (solids)	Shiny (lustrous)	Dull
Melting point	Usually high	Usually low
Boiling point	Usually high	Usually low
Density	Usually high	Usually low
Conductivity (thermal and electrical)	Good	Very poor

Atoms—the smallest particles

Everything is made up of billions of atoms. The atoms of all elements are extremely small; in fact they are too small to be seen. The smallest atom known is hydrogen, with each atom being represented as a sphere having a diameter of 0.00000007 mm (or 7×10^{-8} mm) (Table 2.3). Atoms of different elements have different diameters as well as different masses. How many atoms of hydrogen would have to be placed side by side along the edge of your ruler to fill a 1 mm division?

Chemists use shorthand symbols to label the elements and their atoms. The symbol consists of one or two letters, the first of which must be a capital. Where several elements have the same initial letter, a second letter of the name or subsequent letter is added. For example, **C** is used for **carbon**, **Ca** for **calcium** and **Cl** for **chlorine**. Some symbols seem to have no relationship to the name of the element, for example **Na** for **sodium** and **Pb** for **lead**. These symbols come from their Latin names, natrium for sodium and plumbum for lead. A list of some common elements and their symbols is given in Table 2.4.

Table 2.3 Sizes of atoms

Atom	Diameter of atom (mm)
H	7×10^{-8}
O	12×10^{-8}
S	20.8×10^{-8}

Table 2.4 Some common elements and their symbols

Element	Symbol	Physical state at room temperature and pressure
Aluminium	Al	Solid
Argon	Ar	Gas
Barium	Ba	Solid
Boron	B	Solid
Bromine	Br	Liquid
Calcium	Ca	Solid
Carbon	C	Solid
Chlorine	Cl	Gas
Chromium	Cr	Solid
Copper (Cuprum)	Cu	Solid
Fluorine	F	Gas
Germanium	Ge	Solid
Gold (Aurum)	Au	Solid
Helium	He	Gas
Hydrogen	H	Gas
Iodine	I	Solid
Iron (Ferrum)	Fe	Solid

Element	Symbol	Physical state at room temperature and pressure
Lead (Plumbum)	Pb	Solid
Magnesium	Mg	Solid
Mercury (Hydragyrum)	Hg	Liquid
Neon	Ne	Gas
Nitrogen	N	Gas
Oxygen	O	Gas
Phosphorous	P	Solid
Potassium (Kalium)	K	Solid
Silicon	Si	Solid
Silver (Argentum)	Ag	Solid
Sodium (Natrium)	Na	Solid
Sulphur	S	Solid
Tin (Stannum)	Sn	Solid
Zinc	Zn	Solid

The complete list of the elements with their corresponding symbols is shown in the periodic table on page 35.

Molecules

The atoms of some elements are joined together in small groups. These small groups of atoms are called **molecules**. For example, the atoms of the elements hydrogen, oxygen, nitrogen, fluorine, chlorine, bromine and iodine are each joined in pairs and they are known as **diatomic** molecules. In the case of phosphorus and sulphur the atoms are joined in larger numbers, four and eight respectively. In chemical shorthand the molecule of iodine shown in Figure 2.5 is written as I_2.

The gaseous elements helium, neon, argon, krypton, xenon and radon are composed of separate and individual atoms. When an element exists as separate atoms, then the molecules are said to be **monatomic**. In chemical shorthand these monatomic molecules are written as He, Ne, Ar, Kr, Xe and Rn respectively.

Molecules are not always formed by atoms of the same type joining together. For example, water exists as molecules containing oxygen and hydrogen atoms.

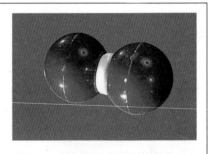

I —— I

Figure 2.5 An iodine molecule—as a letter-and-stick and a space-filling model

QUESTIONS

a How would you use a similar chemical shorthand to write a representation of the molecules of oxygen and nitrogen?

b Using the periodic table on page 35 write down the symbols for the following elements and give their physical states at room temperature.
  (i) nickel
  (ii) xenon
  (iii) platinum.

Compounds

Compounds are pure substances which are formed when two or more elements chemically combine together. Water is a simple compound formed from the elements hydrogen and oxygen (Figure 2.6). This combining of the elements can be represented by a word equation:

hydrogen + oxygen → water

Water molecules contain two atoms of hydrogen and one atom of oxygen, and hence water has the **chemical formula** H_2O. Elements other than hydrogen will also react with oxygen to form compounds called oxides. For example, magnesium reacts violently with oxygen gas to form the white powder magnesium oxide. This reaction is accompanied by a release of energy as new chemical bonds are formed (Figure 2.7).

When a new substance is formed during a chemical reaction, we say that a **chemical change** has taken place.

magnesium + oxygen → magnesium oxide

When substances such as hydrogen and magnesium combine with oxygen in this way they are said to have been **oxidised**. The chemical process is known as **oxidation**.

Reduction is the opposite of oxidation. In this process oxygen is removed instead of being added. For example, the oxygen has to be removed in the extraction of iron from iron(III) oxide. This can be done in a blast furnace with carbon monoxide. The iron(III) oxide loses oxygen to the carbon monoxide. Iron(III) oxide is reduced to iron while carbon monoxide is oxidised to carbon dioxide. You will deal in more detail with this extraction process in Chapter 9.

iron(III) + carbon → iron + carbon
oxide   monoxide   dioxide

Both **red**uction and **ox**idation have taken place in this chemical process, and so this is known as a **redox** reaction. A further discussion of oxidation and reduction takes place in Chapter 6.

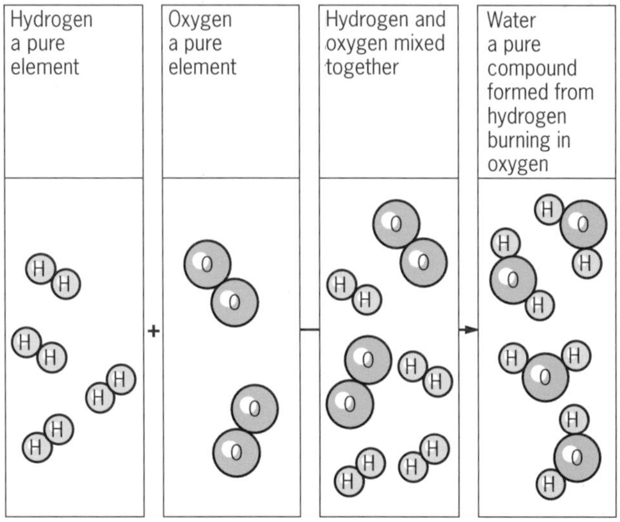

Hydrogen a pure element	Oxygen a pure element	Hydrogen and oxygen mixed together	Water a pure compound formed from hydrogen burning in oxygen

Figure 2.6 The element hydrogen reacts with the element oxygen to produce the compound water

Figure 2.7 Magnesium burns brightly in oxygen to produce magnesium oxide

More about formulae

The formula of a compound is made up from the symbols of the elements present and numbers to show the ratio in which the different atoms are present. Carbon dioxide has the formula CO_2. This tells you that it contains one carbon atom for every two oxygen atoms. The 2 in the formula tells you that there are two oxygen atoms present in each molecule of carbon dioxide.

The following table shows the names and formulae of some common compounds which you will meet in your study of chemistry.

Table 2.5 Names and formulae of some common compounds

Compound	Formula
Carbon monoxide	CO
Ethanol	C_2H_5OH
Sulphuric acid	H_2SO_4
Ammonia	NH_3
Sodium hydroxide	NaOH
Glucose	$C_6H_{12}O_6$
Calcium hydroxide	$Ca(OH)_2$

The ratio of atoms within a chemical compound is usually constant. Compounds are made up of fixed proportions of their elements: they have a fixed composition. Chemists call this the **Law of constant composition**.

Balancing chemical equations

Word equations are a useful way of representing chemical reactions but a better and more useful method is to produce a **balanced chemical equation**. This type of equation gives the formulae of the reactants and the products as well as showing the relative numbers of each particle involved. Balanced equations often include the physical state symbols:

(s) = solid, (l) = liquid, (g) = gas, (aq) = aqueous solution

The word equation to represent the reaction between iron and sulphur is:

$$iron + sulphur \xrightarrow{heat} iron(\text{II}) \ sulphide$$

When we replace the words with symbols for the reactants and the products and include their physical state symbols, we obtain:

$$Fe(s) + S(s) \xrightarrow{heat} FeS(s)$$

Since there is the same number of each type of atom on both sides of the equation this is a **balanced** chemical equation.

In the case of magnesium reacting with oxygen, the word equation was:

$$magnesium + oxygen \xrightarrow{heat} magnesium \ oxide$$

When we replace the words with symbols for the reactants and the products and include their physical state symbols, it is important to remember that oxygen is a diatomic molecule:

$$Mg(s) + O_2(g) \xrightarrow{heat} MgO(s)$$

In the equation there are two oxygen atoms on the left-hand side (O_2) but only one on the right (MgO).

We cannot change the formula of magnesium oxide, so to produce the necessary two oxygen atoms on the right-hand side we will need 2MgO—this means $2 \times MgO$. The equation now becomes:

$$Mg(s) + O_2(g) \xrightarrow{heat} 2MgO(s)$$

There are now two atoms of magnesium on the right-hand side and only one on the left. By placing a '2' in front of the magnesium, we obtain a balanced chemical equation.

$$2Mg(s) + O_2(g) \xrightarrow{heat} 2MgO(s)$$

This balanced chemical equation now shows us that two atoms of magnesium react with one molecule of oxygen gas when heated to produce two units of magnesium oxide.

QUESTIONS

a Write the word and balanced chemical equations for the reactions which take place between:
  (i) calcium and oxygen
  (ii) zinc and oxygen.
b Write down the ratio of the atoms present in each formula for each of the compounds shown in Table 2.5.
c Iron is extracted from iron(III) oxide in a blast furnace by a redox reaction. What does the term 'redox reaction' mean?

Mixtures

Many everyday things are not pure substances, they are mixtures. A mixture contains more than one substance (elements and/or compounds). An example of a common mixture is sea water (Figure 2.8).

Figure 2.8 Sea water is a common mixture

What is the difference between mixtures and compounds?

There are differences between compounds and mixtures. This can be shown by considering the reaction between iron filings and sulphur. A mixture of iron filings and sulphur looks different from the individual elements (Figure 2.9). This mixture has the properties of both iron and sulphur; for example, a magnet can be used to separate the iron filings from the sulphur (Figure 2.10).

Substances in a mixture have not undergone a chemical reaction and it is possible to separate them provided that there is a suitable difference in their physical properties (p. 22). If the mixture of iron and sulphur is heated a chemical reaction occurs and a new substance is formed called iron(II) sulphide (Figure 2.9). The word equation for this reaction is:

$$\text{iron} + \text{sulphur} \xrightarrow{\text{heat}} \text{iron(II) sulphide}$$

During the reaction heat energy is given out as new chemical bonds are formed. This is called an **exothermic** reaction and accompanies a chemical change (Chapter 13, p. 166). The iron(II) sulphide formed has totally different properties to the mixture of iron and sulphur (Table 2.6). Iron(II) sulphide, for example, would not be attracted towards a magnet.

In iron(II) sulphide, FeS, one atom of iron has combined with one atom of sulphur. No such ratio exists in a mixture of iron and sulphur, because the atoms have not chemically combined. Table 2.7 summarises the major differences between mixtures and compounds.

Figure 2.9 The elements (left to right) iron and sulphur, and below a mixture of iron and sulphur and black iron(II) sulphide

Figure 2.10 A magnet will separate the iron from the mixture

Table 2.6 Different properties of iron, sulphur, an iron/sulphur mixture and iron(II) sulphide

Substance	Appearance	Effect of a magnet	Effect of dilute hydrochloric acid
Iron	Dark grey powder	Attracted to it	Very little action when cold. When warm, a gas is produced with a lot of bubbling (effervescence)
Sulphur	Yellow powder	None	No effect when hot or cold
Iron/sulphur mixture	Dirty yellow powder	Iron powder attracted to it	Iron powder reacts as above
Iron(II) sulphide	Dark grey solid	No effect	A foul-smelling gas is produced with some effervescence

Table 2.7 The major differences between mixtures and compounds

Mixture	Compound
It contains two or more substances	It is a single substance
The composition can vary	The composition is always the same
No chemical change takes place when a mixture is formed	When the new substance is formed it involves chemical change
The properties are those of the individual elements	The properties are very different to those of the component elements
The components may be separated quite easily by physical means	The components can only be separated by one or more chemical reactions

a Make a list of some other common mixtures, stating what they are mixtures of.

Separating mixtures

Many mixtures contain useful substances mixed with unwanted material. In order to obtain these useful substances, chemists often have to separate them from the impurities. Chemists have developed many different methods of separation, particularly for separating compounds from complex mixtures. How they separate them depends on what is in the mixture and the properties of the substances present. It also depends on whether the substances to be separated are solids, liquids or gases.

Separating solid/liquid mixtures

If a solid substance is added to a liquid it may **dissolve** to form a **solution**. In this case the solid is said to be **soluble** and is called the **solute**. The liquid it has dissolved in is called the **solvent**. An example of this type of process is when sugar is added to tea or coffee. Can you think of other examples where this type of process takes place?

Sometimes the solid does not dissolve in the liquid. The solid is said to be **insoluble**. For example, tea leaves do not dissolve in boiling water when tea is made from them.

Filtration

When a cup of tea is poured through a tea strainer you are carrying out a **filtering** process. **Filtration** is a common separation technique used in chemistry laboratories throughout the world. It is used when a solid needs to be separated from a liquid. For example, sand can be separated from a mixture with

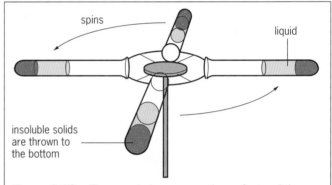

Figure 2.12a The sample is spun round very fast and the solid is flung to the bottom of the tube

Figure 2.11 It is important when filtering not to overfill the filter paper

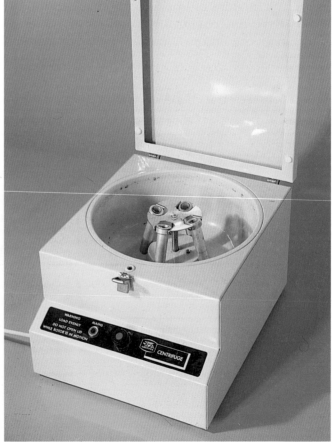

Figure 2.12b An open centrifuge

water by filtering through filter paper as shown in Figure 2.11.

The filter paper contains holes which are large enough to allow the molecules of water through but not the sand particles. It acts like a sieve. The sand gets trapped in the filter paper and the water passes through it. The sand is called the **residue** and the water is called the **filtrate**.

Decanting

Carrots do not dissolve in water. When you have boiled some carrots it is easy to separate them from the water by pouring it off. This process is called **decanting**. This technique is used quite often to separate an insoluble solid, that has settled at the bottom of a flask, from a liquid.

Centrifuging

Another way to separate a solid from a liquid is to use a **centrifuge**. This technique is sometimes used instead of filtration. It is usually used when the solid particles are so small that they spread out (disperse) throughout the liquid and remain in **suspension**. They do not settle to the bottom of the container, as heavier particles would do, under the force of gravity. The technique of **centrifuging** or **centrifugation** involves the suspension being spun round very fast in a centrifuge so that the solid gets flung to the bottom of the tube (Figure 2.12a and b).

The pure liquid can be decanted after the solid has been forced to the bottom of the tube. This method of separation is used extensively to separate blood cells from blood plasma (Figure 2.13). In this case, the solid particles (the blood cells) are flung to the bottom of the tubes, allowing the liquid plasma to be decanted.

Evaporation

If the solid has dissolved in the liquid it cannot be separated by filtering or centrifuging. Instead, the solution can be heated so that the liquid evaporates completely and leaves the solid behind. The simplest way to obtain salt from its solution is by slow evaporation as shown in Figure 2.14.

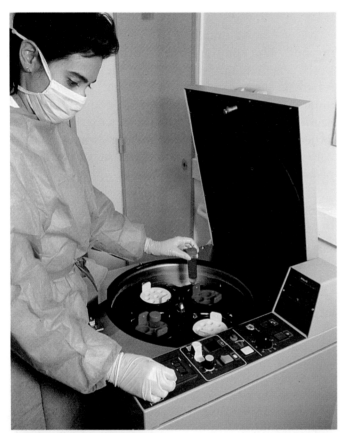

Figure 2.13 Whole blood (top) is separated by centrifuging into blood cells and plasma (bottom)

Figure 2.14 Apparatus used to slowly evaporate a solvent

Figure 2.15 Salt is obtained in Rio de Janeiro, Brazil, by evaporation of sea water

Crystallisation

In many parts of the world salt is obtained from sea water on a vast scale. This is done by using the heat of the sun to evaporate the water to leave a saturated solution of salt known as brine. A **saturated solution** is defined as one that contains as much solute as can be dissolved at a particular temperature. When the solution is saturated the salt begins to **crystallise**, and it is removed using large scoops (Figure 2.15).

Simple distillation

If we want to obtain the solvent from a solution, then the process of **distillation** can be carried out. The apparatus used in this process is shown in Figure 2.16.

Water can be obtained from salt water using this method. The solution is heated in the flask until it boils. The steam rises into the Liebig condenser, where it condenses back into water. The salt is left behind in the flask. In hot and arid countries such as Saudi Arabia this sort of technique is used on a much larger scale to obtain pure water for drinking (Figure 2.17). This process is carried out in a desalination plant.

Separating liquid/liquid mixtures

In recent years there have been many oil tanker disasters, just like the one shown in Figure 2.18. These have resulted in millions of litres of oil being washed into the sea. Oil and water do not mix easily. They are said to be **immiscible**. When cleaning up disasters of this type, a range of chemicals can be added to the oil to make it more soluble. This results in the oil and water mixing with each other. They are now said to be **miscible**. The following techniques can be used to separate mixtures of liquids.

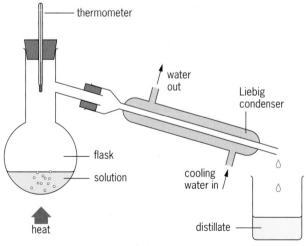

Figure 2.16 Water can be obtained from salt water by distillation

Figure 2.17 This plant produces large quantities of drinking water in Bahrain

Figure 2.18 Millions of litres of oil are spilt in such disasters and the cleaning-up operation is a slow and costly process

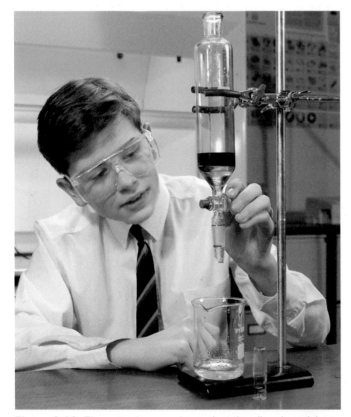

Figure 2.19 The water is more dense than the oil and so sinks to the bottom of the separating funnel. When the tap is opened the water can be run off

Liquids which are immiscible

Using a separating funnel

If two liquids are immiscible they can be separated using a **separating funnel**. The mixture is poured into the funnel and the layers allowed to separate. The lower layer can then be run off by opening the tap as shown in Figure 2.19.

Liquids which are miscible

If miscible liquids are to be separated, then this can be done by **fractional distillation**. The apparatus used for this process is shown in Figure 2.20. This apparatus could be used to separate a mixture of ethanol and water.

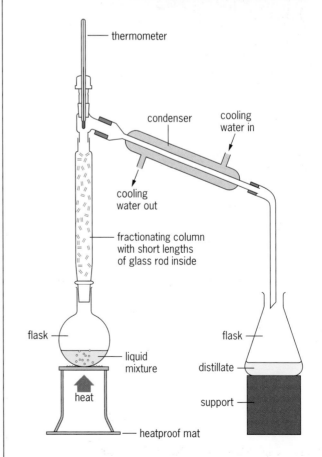

Figure 2.20 Typical fractional distillation apparatus

- thermometer
- condenser
- cooling water in
- cooling water out
- fractionating column with short lengths of glass rod inside
- flask
- liquid mixture
- heat
- heatproof mat
- flask
- distillate
- support

Fractional distillation relies upon the liquids having different boiling points. When the mixture is heated the vapours of ethanol and water boil off at different temperatures and can be condensed and collected separately.

Ethanol boils at 78 °C whereas water boils at 100 °C. When the mixture is heated the vapour produced is mainly ethanol with some steam. Because water has the higher boiling point, it condenses out from the mixture with ethanol. This is what takes place in the fractionating column. The water condenses and drips back into the flask while the ethanol vapour moves up the column and into the condenser, where it condenses into liquid ethanol and is collected in the receiver. When all the ethanol has distilled over, the temperature reading on the thermometer rises steadily to 100 °C, showing that the steam is now entering the condenser. At this point the receiver can be changed and the now condensing water can be collected.

The technique is used extensively to separate the individual gases from the air as well as to separate crude oil (Figure 2.21).

Separating solid/solid mixtures

You saw earlier in this chapter (p. 17) that it was possible to separate iron from sulphur using a magnet. In that case we were using one of the physical properties of iron, that is, the fact that it is magnetic. In a similar way, it is possible to separate scrap iron from other metals by using a large electromagnet like the one shown in Figure 2.22.

It is essential that when separating solid/solid mixtures you pay particular attention to the individual physical properties of the components. If,

Figure 2.21 (**a**, top) Fractional distillation unit for crude oil. (**b**, bottom) Gases from the air are extracted in this fractional distillation plant

Figure 2.22 Magnetic separation of iron-containing materials

Figure 2.23 Apparatus used to separate an iodine/salt mixture. The iodine sublimes on heating

for example, you wish to separate two solids, one of which sublimes, then this property should dictate the method you employ.

In the case of an iodine/salt mixture the iodine sublimes but salt does not. Iodine can be separated by heating the mixture in a fume cupboard as shown in Figure 2.23. The iodine sublimes and re-forms on the cool inverted funnel.

Chromatography

What happens if you have to separate two or more solids that are soluble? This type of problem is encountered when you have mixtures of coloured materials such as inks and dyes. A technique called **chromatography** is widely used to separate these materials so that they can be identified.

There are several types of chromatography; however, they all follow the same basic principles. The simplest kind is paper chromatography. To separate the different-coloured dyes in a sample of black ink, a spot of the ink is put onto a piece of chromatography paper. This paper is then set in a suitable solvent as shown in Figure 2.24.

As the solvent moves up the paper, the dyes are carried with it and begin to separate. They separate because the substances have different solubilities in the solvent and are absorbed to different degrees by the chromatography paper. As a result, they are separated gradually as the solvent moves up the paper. The **chromatogram** in Figure 2.24b shows how the ink contains three dyes, P, Q and R. Numerical measurements known as R_f **values** can be obtained from chromatograms. An R_f value is defined as the ratio of the distance travelled by the solute (for example P, Q or R) to the distance travelled by the solvent.

Chromatography and electrophoresis (separation according to charge) are used extensively in hospital, health and forensic science laboratories to separate a variety of mixtures (Figure 2.25).

The substances to be separated do not have to be coloured. Colourless substances can be made visible by spraying the chromatogram with a **locating agent**. The locating agent will react with the colourless substances to form a coloured product. In other situations the position of the substances on the chromatogram may be located using ultraviolet light.

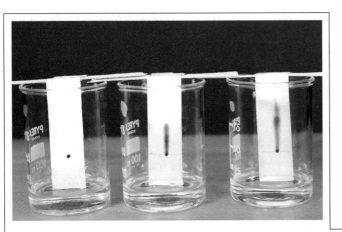

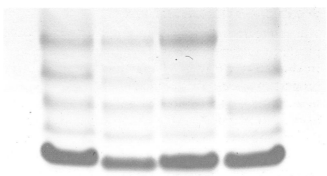

Figure 2.25 Protein samples are separated by electrophoresis in hospital laboratories

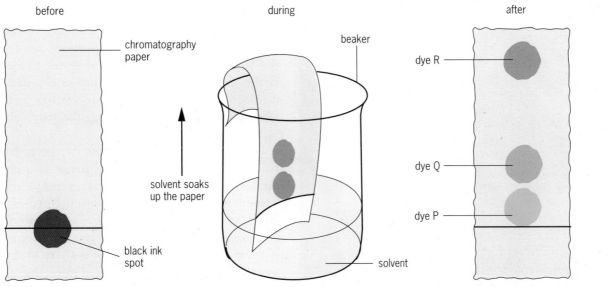

before   during   after

chromatography paper

beaker

dye R

solvent soaks up the paper

dye Q

black ink spot

dye P

solvent

Figure 2.24 (**a**, top) Chromatographic separation of black ink. (**b**, bottom) The black ink separates into three dyes: P, Q and R

Solvent extraction

Sugar can be obtained from crushed sugar cane by adding water. The water dissolves the sugar from the sugar cane (Figure 2.26). This is an example of **solvent extraction**. In a similar way some of the green substances can be removed from ground-up grass using ethanol. The substances are extracted from a mixture by using a solvent which dissolves only those substances required.

Drugs have to be manufactured to a very high degree of purity (Figure 2.27). To ensure that the highest possible purity is obtained, the drugs are dissolved in a suitable solvent and subjected to fractional crystallisation.

Figure 2.26 Sugar can be extracted from sugar cane by using a suitable solvent

Figure 2.27 Drugs are manufactured to a high degree of purity by fractional crystallisation

---

QUESTIONS

a Write down as many examples as you can think of in which a centrifuge is used.

b What is the difference between fractional distillation and simple distillation?

c Describe how you would use chromatography to show that grass contains a mixture of pigments.

d Explain the following terms:
   (i) miscible
   (ii) immiscible
   (iii) evaporation
   (iv) condensation
   (v) solvent extraction.

e Devise a method for obtaining salt from sea water in the school laboratory.

Gels, foams and emulsions

Gels, foams and **emulsions** are all examples of mixtures which are formed by mixing two substances (or phases) which cannot mix. This may seem a strange statement to make until you look closely at some food substances shown in Figure 2.28.

When you mix a solid with a liquid you sometimes get a **gel**. A gel is a semi-solid which can move around but not as freely as a liquid. Within a gel the solid makes a kind of network which traps the liquid and makes it unable to flow freely (Figure 2.29).

A gelatine gel is made with warm water. Gelatine is a protein. Proteins are natural polymers (Chapter 14, p. 185) and the molecules of proteins are very large. The large molecules disperse in water to form a gel. As the gelatine in water mixture cools, the gelatine molecules are attracted to each other and form a continuous network. In this way the jelly you eat as a pudding is formed.

When you pour out a glass of fizzy drink, the frothy part at the top of the drink is a gas/liquid mixture called a **foam**. The gas, carbon dioxide, has formed tiny bubbles in the liquid but has not dissolved in it. If left to stand, foams like this one collapse as the tiny bubbles in the foam join together to form bigger bubbles which then escape. It is possible to form solid foams where the gases are trapped in a solid structure. This happens in foam rubber and bread (Figure 2.30).

Figure 2.28 (**a**, top) These are mixtures of solid and liquid. (**b**, middle) These foams have been formed by trapping bubbles of gas in liquids or solids. (**c**, bottom) These have been formed by mixing immiscible liquids

Figure 2.30 Examples of solid foams

network of gelatine molecules

water molecules trapped in a network of gelatine

Figure 2.29 Network within a gel

Emulsions are mixtures of liquids which are immiscible. Earlier in this chapter you found out that when two liquids are immiscible they do not mix but form two different layers. Oil and water are like this but if you shake the mixture it becomes cloudy.

The apparent mixing that you see is due to the fact that one of the liquids has been broken into tiny droplets which float suspended in the other liquid. If the mixture of oil and water is now left to stand the two layers will re-form. To make emulsions, such as mayonnaise, an **emulsifier** is used to stop the droplets joining back together again to form a separate layer. The emulsifier used when making mayonnaise is egg yolk. If you examine the ingredients on the side of many packets found in your kitchen cupboard you will find that emulsifiers have 'E' numbers in the range E322 to E494. For example, ammonium phosphatide is used as the emulsifier in cocoa and chocolate.

It is worth noting that gels, foams and emulsions are all examples of different kinds of solutions. In true solutions the two phases completely mix together but in these systems the two phases are separate. They are all examples of **colloids**.

Mixtures for strength

Composite materials

Composite materials are those that combine the properties of two constituents in order to get the exact properties needed for a particular job.

Glass-reinforced plastic (GRP) is an example of a composite material combining the properties of two different materials. It is made by embedding short fibres of glass in a matrix of plastic. The glass fibres give the plastic extra strength so that it does not break when it is bent or moulded into shape. The finished product has the lightness of plastic as well as the strength and flexibility of the glass fibres (Figure 2.31).

Many composite materials can be found in nature. Bone is a composite material formed from strands of protein and the mineral calcium phosphate (Figure 2.32). The calcium phosphate is hard and therefore gives strength to the bone. Another example is wood. Wood consists of cellulose fibres mixed with lignin, which is largely responsible for the strength of the wood.

Figure 2.31 The glass-reinforced plastic used to make boats like this is a composite material

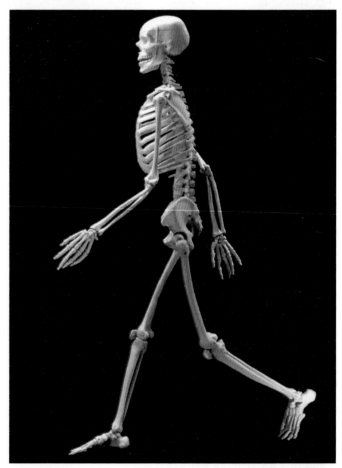

Figure 2.32 Bone is a composite material

Checklist

After studying Chapter 2 you should know and understand the following terms.

Element   A substance which cannot be further divided into simpler substances by chemical methods.

Compound   A substance formed by the combination of two or more elements in fixed proportions.

Mixture   A system of two or more substances that can be separated by physical means.

Metals   A class of chemical elements which have a characteristic lustrous appearance and which are good conductors of heat and electricity.

Non-metals   A class of chemical elements that are typically poor conductors of heat and electricity.

Atom   The smallest part of an element that can exist as a stable entity.

Molecule   A group of atoms chemically bonded together.

Monatomic molecule   A molecule which consists of only one atom, for example neon and argon.

Diatomic molecule   A molecule containing two atoms, for example hydrogen, H_2, and oxygen, O_2.

Chemical change   A permanent change in which a new substance is formed.

Oxidation   The process of combining with oxygen.

Reduction   The process of removing oxygen.

Redox reaction   A reaction which involves the two processes of reduction and oxidation.

Chemical formula   A shorthand method of representing chemical elements and compounds.

Exothermic reaction   A chemical reaction in which heat energy is produced.

Law of constant composition   Compounds always have the same elements joined together in the same proportions.

Solution   This is formed when a substance (solute) disappears (dissolves) into another substance (solvent).

$$\text{solute} + \text{solvent} \xrightarrow{\text{dissolves}} \text{solution}$$

Saturated solution   This is a solution which contains as much dissolved solute as it can at a particular temperature.

Soluble   If the solute dissolves in the solvent it is said to be soluble.

Insoluble   If the solute does not dissolve in the solvent it is said to be insoluble.

Filtration   The process of separating a solid from a liquid, using a fine filter paper which does not allow the solid to pass through.

Residue   The solid left behind in the filter paper after filtration has taken place.

Filtrate   The liquid which passes through the filter paper during filtration.

Centrifuging   The separation of the components of a mixture by rapid spinning. The denser particles are flung to the bottom of the containing tubes. The liquid can then be decanted off.

Decanting   The process of removing a liquid from a solid which has settled or from an immiscible heavier liquid by pouring.

Distillation   The process of boiling a liquid and then condensing the vapour produced back into a liquid. It is used to purify liquids and to separate mixtures of liquids.

Crystallisation   The process of forming crystals from a liquid.

Miscible   When two liquids form a homogeneous layer when mixed together, they are said to be miscible.

Immiscible   When two liquids form two layers when mixed together, they are said to be immiscible.

Chromatography   A technique employed for the separation of mixtures of dissolved substances.

R_f value   This is the ratio of the distance travelled by the solute to the distance travelled by the solvent in chromatography.

Gel   A mixture formed between a solid and a liquid in which the solid forms a network which traps the liquid so that it cannot flow freely.

Foam   A mixture formed between a gas and a liquid. The gas forms tiny bubbles in the liquid but has not dissolved in it.

Emulsion   The apparent mixing of two immiscible liquids by the use of an emulsifier which breaks down one of the liquids into tiny droplets. The droplets of this liquid float suspended in the other liquid so that they do not separate out into different layers.

Composite materials   These are materials which combine the properties of two substances in order to get the exact properties required for a particular job.

QUESTIONS

1 Define the following terms using specific examples to help with your explanation:
  a element
  b metal
  c non-metal
  d compound
  e molecule
  f mixture
  g atom
  h gel
  i foam
  j emulsion.

2 Which of the substances listed below are:
  a metallic elements?
  b non-metallic elements?
  c compounds?
  d mixtures?
Phosphorus, sea water, potassium, argon, water, air, carbon monoxide, iron, sodium chloride, diamond, brass, zinc, dilute sulphuric acid, sulphur.

3 At room temperature and pressure (rtp), which of the substances listed below is:
  a a solid element?
  b a liquid element?
  c a gaseous mixture?
  d a solid mixture?
  e a liquid compound?
  f a solid compound?
Mercury, salt, carbon dioxide, helium, solder, air, oil, copper, water, sand, tin, bronze.

4 A student heated a mixture of sulphur and iron filings quite strongly. She saw a red glow spread through the mixture. At the end of the experiment a black solid had been formed.
  a Explain what the red glow indicates.
  b Give the chemical name of the black solid.
  c Write a word equation and a balanced chemical equation to represent the reaction which has taken place.
  d The black solid is a compound. Explain the difference between the mixture of sulphur and iron and the compound formed by the chemical reaction between them.

5 Name the method which is most suitable for separating the following:
  a ethanol (boiling point 78 °C) from a mixture of ethanol and water
  b the sediment formed at the bottom of a wine bottle
  c nitrogen from liquid air
  d red blood cells from plasma
  e petrol and kerosene from crude oil
  f coffee grains from coffee solution
  g pieces of steel from engine oil
  h a complex mixture of proteins.

6 The table below shows the melting points, boiling points and densities of substances **A** to **D**.

Substance	Melting point (°C)	Boiling point (°C)	Density (g cm^{-3})
A	1009	2506	8.9
B	−256	−248	0.08
C	44	97	1.6
D	−10	63	0.9

  a Which substance is a gas at room temperature?
  b Which substance is a liquid at room temperature?
  c Which substances are solids at room temperature?
  d Which substance is most likely to be a metal?
  e Which substance will be a liquid at −250 °C?
  f What is the melting point of the least dense non-metal?
  g Which substance is a gas at 72 °C?

7 a How many atoms of the different elements are there in the formulae of the compounds given below?
  (i) nitric acid, HNO_3
  (ii) methane, CH_4
  (iii) copper nitrate, $Cu(NO_3)_2$
  (iv) ethanoic acid, CH_3COOH
  (v) sugar, $C_{12}H_{22}O_{11}$
  (vi) phenol, C_6H_5OH
  (vii) ammonium carbonate, $(NH_4)_2CO_3$
  b Balance the following equations:
  (i) $Cu(s) + O_2(g) \longrightarrow CuO(s)$
  (ii) $Fe(s) + Cl_2(g) \longrightarrow FeCl_3(s)$
  (iii) $Na(s) + O_2(g) \longrightarrow Na_2O(s)$
  (iv) $H_2(g) + O_2(g) \longrightarrow H_2O(g)$
  (v) $Mg(s) + CO_2(g) \longrightarrow MgO(s) + C(s)$

8 Carbon-fibre-reinforced plastic (CRP) is used in the manufacture of golf clubs and tennis rackets.
  a What are composite materials?
  b Which two substances are used to manufacture this composite material?
Consider the data below.

Material	Strength (GPa)	Stiffness (GPa)	Density (g cm^{-3})	Relative cost
Aluminium	0.2	75	2.7	low
Steel	1.1	200	7.8	low
CRP	1.8	195	1.6	high

  c Discuss the advantages and disadvantages of using the three materials above in the manufacture of golf clubs.

3 ATOMIC STRUCTURE AND THE PERIODIC TABLE

We have already seen in Chapter 2 that everything you see around you is made out of tiny particles called atoms (Figure 3.1). When John Dalton developed his atomic theory, about 200 years ago, he stated that the atoms of any one element were identical and that each atom was 'indivisible'. Scientists in those days believed that atoms were solid particles like marbles.

However, in the last hundred years or so it has been proved by great scientists, such as Bohr, Einstein, Moseley, Thomson, Rutherford and Chadwick, that atoms are in fact made up of even smaller 'sub-atomic' particles. The most important of these are **electrons**, **protons** and **neutrons**, although 70 sub-atomic particles have now been discovered.

Inside atoms

The three sub-atomic particles are found in distinct and separate regions. The protons and neutrons are found in the centre of the atom, which is called the **nucleus**. The neutrons have no charge and protons are positively charged. The nucleus occupies only a very small volume of the atom but is very dense.

The rest of the atom surrounding the nucleus is where electrons are most likely to be found. The electrons are negatively charged and move around very quickly in **electron shells** or **energy levels**. The electrons are held within the atom by an **electrostatic force of attraction** between themselves and the positive charge of the protons in the nucleus (Figure 3.2).

About 1837 electrons are equal in mass to the mass of one proton or one neutron. A summary of each type of particle, its mass and relative charge is shown

Figure 3.1 Atoms. A field ion micrograph of atoms of iridium. The tiny dots are the locations of individual atoms, and the ring-like patterns are facets of a single crystal

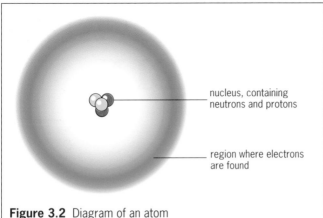

Figure 3.2 Diagram of an atom

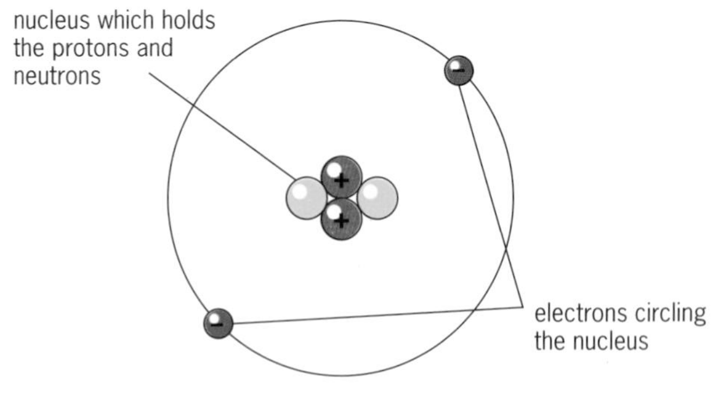

Figure 3.3 An atom of helium has 2 protons, 2 electrons and 2 neutrons

nucleus which holds the protons and neutrons

electrons circling the nucleus

Table 3.1 Characteristics of a proton, a neutron and an electron

Particle	Symbol	Relative mass (amu)	Relative charge
Proton	p	1	+1
Neutron	n	1	0
Electron	e	1/1837	−1

amu = atomic mass unit

in Table 3.1. You will notice that the masses of all these particles are measured in **atomic mass units (amu)**. This is because they are so light that their masses cannot be measured usefully in grams.

Although atoms contain electrically charged particles, the atoms themselves are electrically neutral (they have no overall electric charge). This is because atoms contain equal numbers of electrons and protons. For example, the diagram in Figure 3.3 represents the atom of the non-metallic element helium. The atom of helium possesses two protons, two neutrons and two electrons. The electrical charge of the protons in the nucleus is, therefore, balanced by the opposite charge of the two electrons.

Atomic number and mass number

The number of protons in the nucleus of an atom is called the **atomic number** (or proton number) and is given the symbol Z. Hence in the diagram shown in Figure 3.3, the helium atom has an atomic number of 2, since it has two protons in its nucleus. Each element has its own atomic number and no two different elements have the same atomic number. For example, a different element, lithium, has an atomic number of 3, since it has three protons in its nucleus.

Neutrons and protons have a similar mass. Electrons possess very little mass. So the mass of any atom depends on the number of neutrons and protons in its nucleus. The total number of protons and neutrons found in the nucleus of an atom is called the **mass number** (or nucleon number) and is given the symbol A.

$$\text{mass number} = \text{atomic number} + \text{number of}$$
$$(A) \qquad\qquad (Z) \qquad\qquad \text{neutrons}$$

Hence, in the example shown in Figure 3.3 the helium atom has a mass number of 4, since it has two protons and two neutrons in its nucleus. If we consider the metallic element lithium, it has three protons and four neutrons in its nucleus. It therefore has a mass number of 7.

The atomic number and mass number of an element are usually written in the following shorthand way:

mass number (A) ⟍
$\qquad\qquad {}^{4}_{2}\text{He}\leftarrow$ symbol of the element
atomic number (Z) ⟋

The number of neutrons present can be calculated by rearranging the relationship between the atomic number, mass number and number of neutrons to give:

$$\text{number of} = \text{mass number} - \text{atomic number}$$
$$\text{neutrons} \qquad (A) \qquad\qquad (Z)$$

For example, the number of neutrons in one atom of ${}^{24}_{12}\text{Mg}$ is:

$$24 - 12 = 12$$
$$(A)\ (Z)$$

and the number of neutrons in one atom of ${}^{207}_{82}\text{Pb}$ is:

$$207 - 82 = 125$$
$$(A)\ (Z)$$

Table 3.2 shows the number of protons, neutrons and electrons in the atoms of some common elements.

Ions

An ion is an electrically charged particle. When an atom loses one or more electrons it becomes a positively charged ion. For example, during the chemical reactions of potassium, each atom loses an electron to form a positive ion, K^+.

Table 3.2 Number of protons, neutrons and electrons in some elements

Element	Symbol	Atomic number	Number of electrons	Number of protons	Number of neutrons	Mass number
Hydrogen	H	1	1	1	0	1
Helium	He	2	2	2	2	4
Carbon	C	6	6	6	6	12
Nitrogen	N	7	7	7	7	14
Oxygen	O	8	8	8	8	16
Fluorine	F	9	9	9	10	19
Neon	Ne	10	10	10	10	20
Sodium	Na	11	11	11	12	23
Magnesium	Mg	12	12	12	12	24
Sulphur	S	16	16	16	16	32
Potassium	K	19	19	19	20	39
Calcium	Ca	20	20	20	20	40
Iron	Fe	26	26	26	30	56
Copper	Cu	29	29	29	35	64

$$_{19}K^+$$

19 protons	= 19+
18 electrons	= 18–
Overall charge	= 1+

When an atom gains one or more electrons it becomes a negatively charged ion. For example, during some of the chemical reactions of chlorine it gains an electron to form a negative ion, Cl^-.

$$_{17}Cl^-$$

17 protons	= 17+
18 electrons	= 18–
Overall charge	= 1–

Table 3.3 shows some common ions. You will notice from Table 3.3 that:

- some ions contain more than one type of atom, for example NO_3^-
- an ion may possess more than one unit of charge (either negative or positive), for example Al^{3+}, O^{2-}, SO_4^{2-}.

Table 3.3 Some common ions

Name	Formula
Lithium ion	Li^+
Sodium ion	Na^+
Potassium ion	K^+
Magnesium ion	Mg^{2+}
Aluminium ion	Al^{3+}
Zinc ion	Zn^{2+}
Ammonium ion	NH_4^+
Fluoride ion	F^-
Chloride ion	Cl^-
Bromide ion	Br^-
Oxide ion	O^{2-}
Sulphide ion	S^{2-}
Nitrate ion	NO_3^-
Sulphate ion	SO_4^{2-}

Isotopes

Not all of the atoms in a sample of chlorine, for example, will be identical. Some atoms of the same element can contain different numbers of neutrons. Atoms of the same element which have different numbers of neutrons are called **isotopes**.

The two isotopes of chlorine are shown in Figure 3.4. Generally, isotopes behave in the same way during chemical reactions. The only effect of the extra neutrons is to alter the mass of the atom and properties which depend on it, such as density. Some other examples of atoms with isotopes are shown in Table 3.4.

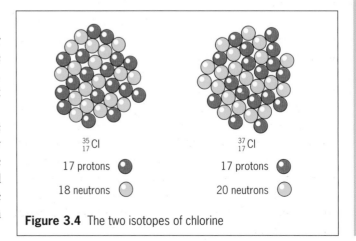

$$^{35}_{17}Cl$$

17 protons ●

18 neutrons ○

$$^{37}_{17}Cl$$

17 protons ●

20 neutrons ○

Figure 3.4 The two isotopes of chlorine

Table 3.4 Some atoms and their isotopes

Element	^A_Z Symbol	Particles present
Hydrogen	^1_1H	1 e, 1 p, 0 n
	^2_1H	1 e, 1 p, 1 n
	^3_1H	1 e, 1 p, 2 n
Carbon	$^{12}_6C$	6 e, 6 p, 6 n
	$^{13}_6C$	6 e, 6 p, 7 n
	$^{14}_6C$	6 e, 6 p, 8 n
Oxygen	$^{16}_8O$	8 e, 8 p, 8 n
	$^{17}_8O$	8 e, 8 p, 9 n
	$^{18}_8O$	8 e, 8 p, 10 n
Strontium	$^{86}_{38}Sr$	38 e, 38 p, 48 n
	$^{88}_{38}Sr$	38 e, 38 p, 50 n
	$^{90}_{38}Sr$	38 e, 38 p, 52 n
Uranium	$^{235}_{92}U$	92 e, 92 p, 143 n
	$^{238}_{92}U$	92 e, 92 p, 146 n

Some of the atoms of certain isotopes are unstable because of the extra number of neutrons, and they are said to be radioactive. The best known elements which have radioactive isotopes are uranium and carbon. The major isotopes of these elements are shown in Table 3.4. Radioactivity is discussed in more detail in Chapter 17.

The mass spectrometer

How do we know isotopes exist? They were first discovered by scientists using apparatus called a mass spectrometer (Figure 3.5). The first mass spectrometer was built by Francis Aston in 1919. This enabled scientists to compare the relative masses of atoms accurately for the first time.

A vacuum exists inside a mass spectrometer. A sample of the vapour of the element is injected into a chamber where it is bombarded by electrons. The subsequent collisions that take place cause the atoms in the vapour to lose one of their electrons and so form positive ions. The beam of positive ions is accelerated by an electric field and then deflected by a magnetic field. The amount of deflection depends on the different masses of the positive ions. The lighter ions, which were formed from the lighter isotopes, are deflected more than the heavier ones. In this way particles with different masses can be separated and identified. A detector counts the number of each of the ions that fall upon it and so a measure of the percentage abundance of each isotope is obtained. A typical mass spectrum for chlorine is shown in Figure 3.6.

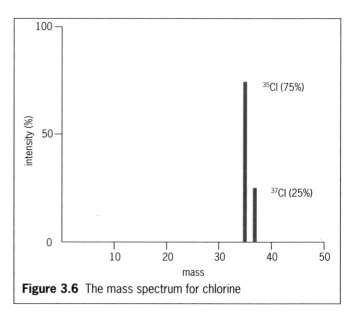

Figure 3.6 The mass spectrum for chlorine

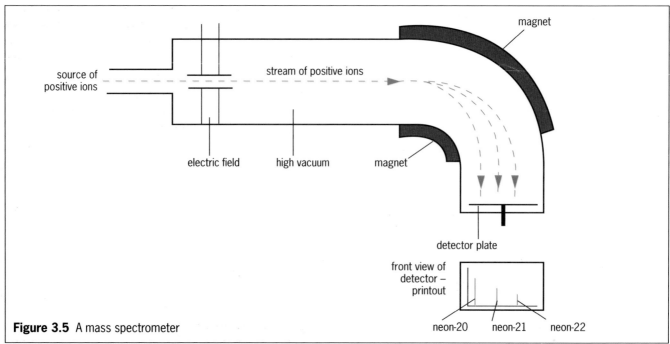

Figure 3.5 A mass spectrometer

Relative atomic mass

The average mass of a large number of atoms of an element is called its **relative atomic mass** (symbol A_r). This quantity takes into account the percentage abundance of all the isotopes of an element which exist.

In 1961 the International Union of Pure and Applied Chemistry (IUPAC) recommended that the standard used for the A_r scale was carbon-12. An atom of carbon-12 was taken to have a mass of 12 amu. The A_r of an element is now defined as the average mass of its isotopes compared to one-twelfth the mass of one atom of carbon-12:

$$A_r = \frac{\text{average mass of isotopes of the element}}{\frac{1}{12} \times \text{mass of 1 atom of carbon-12}}$$

Note: $\frac{1}{12}$ of the mass of one carbon-12 atom = 1 amu.

For example, in the case of chlorine, which has two isotopes:

	$^{35}_{17}Cl$	$^{37}_{17}Cl$
% Abundance	75	25
Ratio of atoms	3	1

Hence the 'average mass' of a chlorine atom is:

$$\frac{(3 \times 35) + (1 \times 37)}{4} = 35.5$$

$$A_r = \frac{35.5}{1}$$

$$= 35.5 \text{ amu}$$

QUESTIONS

a Calculate the number of neutrons in the following atoms:

  (i) $^{27}_{13}Al$

  (ii) $^{31}_{15}P$

  (iii) $^{262}_{107}Uns$

  (iv) $^{190}_{76}Os$

b Given that the percentage abundance of $^{20}_{10}Ne$ is 90% and that of $^{22}_{10}Ne$ is 10%, calculate the A_r of neon.

The arrangement of electrons in atoms

The nucleus of an atom contains the heavier sub-atomic particles, the protons and the neutrons. The electrons, the lightest of the sub-atomic particles, move around the nucleus at great distances from the nucleus relative to their size. They move very fast in electron energy levels very much as the planets orbit the Sun.

It is not possible to give the exact position of an electron in an energy level. However, we can state that electrons can only occupy certain, definite energy levels and that they cannot exist between them. Each of the electron energy levels can only hold a certain number of electrons.

- First energy level holds up to 2 electrons.
- Second energy level holds up to 8 electrons.
- Third energy level holds up to 18 electrons.

There are further energy levels which contain increasing numbers of electrons.

The third energy level can be occupied by a maximum of 18 electrons. However, when eight electrons have occupied this level a certain stability is given to the atom and the next two electrons go into the fourth energy level, and then the remaining ten electrons complete the third energy level.

The electrons fill the energy levels starting from the energy level nearest to the nucleus, which has the lowest energy. When this is full (with two electrons) the next electron goes into the second energy level. When this energy level is full with eight electrons, then the electrons begin to fill the third and fourth energy levels as stated above.

For example, a $^{12}_{6}C$ atom has an atomic number of 6 and therefore has six electrons. Two of the six electrons enter the first energy level, leaving four to occupy the second energy level, as shown in Figure 3.7. The electron configuration for carbon can be written in a shorthand way as 2,4.

There are 112 elements, and Table 3.5 shows the way in which the electrons are arranged in the first

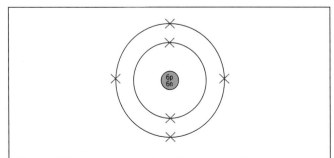

Figure 3.7 *Arrangement of electrons in a carbon atom*

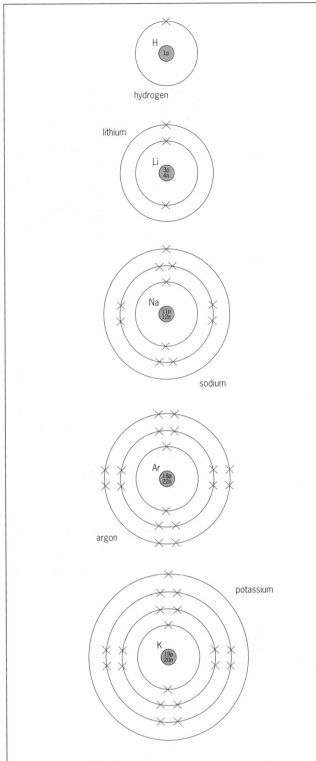

Figure 3.8 Electron arrangements of hydrogen, lithium, sodium, argon and potassium

Table 3.5 Electron arrangement in the first 20 elements

Element	Symbol	Atomic number	Number of electrons	Electron structure
Hydrogen	H	1	1	1
Helium	He	2	2	2
Lithium	Li	3	3	2,1
Beryllium	Be	4	4	2,2
Boron	B	5	5	2,3
Carbon	C	6	6	2,4
Nitrogen	N	7	7	2,5
Oxygen	O	8	8	2,6
Fluorine	F	9	9	2,7
Neon	Ne	10	10	2,8
Sodium	Na	11	11	2,8,1
Magnesium	Mg	12	12	2,8,2
Aluminium	Al	13	13	2,8,3
Silicon	Si	14	14	2,8,4
Phosphorus	P	15	15	2,8,5
Sulphur	S	16	16	2,8,6
Chlorine	Cl	17	17	2,8,7
Argon	Ar	18	18	2,8,8
Potassium	K	19	19	2,8,8,1
Calcium	Ca	20	20	2,8,8,2

20 of these elements. The way the electrons are arranged is called the **electron structure** or **electron configuration**. Figure 3.8 shows how the electrons are arranged in the electron energy levels of a selection of atoms.

QUESTIONS

a How many electrons may be accommodated in the first three energy levels?

b What is the same about the electron structures of:
  (i) lithium, sodium and potassium?
  (ii) beryllium, magnesium and calcium?

The periodic table

The periodic table was devised in 1869 by the Russian Dmitri Mendeléev, who was the Professor of Chemistry at St Petersburg University (Figure 3.9). His periodic table was based on the chemical and physical properties of the 60 elements that had been discovered at that time. Other scientists, including Döbereiner, Newlands and Meyer, had attempted to categorise the known elements but Mendeléev's classification proved to be the most successful. As a scientific idea it was tested by making predictions about elements which were unknown at the time. Subsequently, as elements were discovered they were found to fit neatly into the classification. For example, Mendeléev predicted the properties of the missing element he called 'Eka-Silicon'. He predicted the colour, density and melting point as well as its relative atomic mass. In 1886 the element we now know as germanium was discovered in Germany by Clemens Winkler, and its properties were almost exactly those Mendeléev had predicted.

Mendeléev's periodic table has been modified in the light of work carried out by Rutherford and Moseley. Discoveries about sub-atomic particles led them to realise that the elements should be arranged by atomic number. In the modern periodic table the 112 known elements are arranged in order of increasing atomic number (Figure 3.10).

Those elements with similar chemical properties are found in the same columns or **groups**. There are

Figure 3.9 Dmitri Mendeléev (1834–1907)

eight groups of elements. The first column is called group 1; the second group 2; and so on up to group 7. The final column in the periodic table is called group 0. Some of the groups have been given names.

Group 1: The alkali metals
Group 2: The alkaline earth metals
Group 7: The halogens
Group 0: Inert gases or noble gases

The horizontal rows are called **periods** and these are numbered 1–7 going down the periodic table.

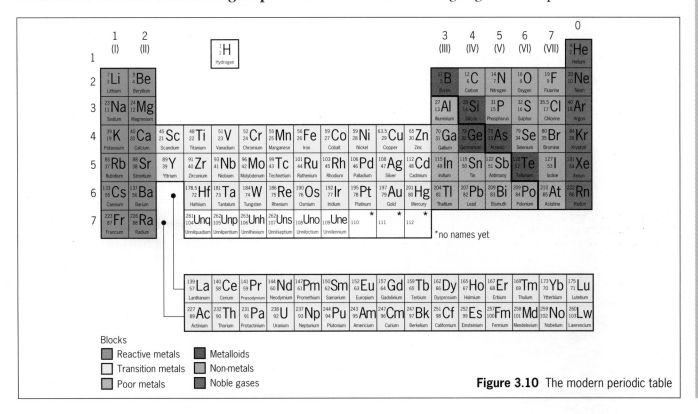

Blocks
- Reactive metals
- Transition metals
- Poor metals
- Metalloids
- Non-metals
- Noble gases

Figure 3.10 The modern periodic table

Figure 3.11 Some transition elements (left to right): copper, cobalt and nickel

Between groups 2 and 3 is the block of elements known as the transition elements (Figure 3.11).

The periodic table can be divided into two as shown by the bold line (see p. 35). The elements to the left of this line are metals and those on the right are non-metals. The elements which lie on this dividing line are known as metalloids (Figure 3.12). These elements behave in some ways as metals and in others as non-metals.

Figure 3.12 The metalloid silicon. It is used to make silicon 'chips'

Electron configuration and the periodic table

Now that the number of electrons in the outer energy level has been established, it can be seen that it corresponds with the number of the group in the periodic table in which the element is found. For example, the elements shown in Table 3.6 have one electron in their outer energy level and they are all found in group 1. The elements in group 0, however, are an exception to this rule, as they have two or eight electrons in their outer energy level. The outer electrons are mainly responsible for the chemical properties of any element, and, therefore, elements in the same group have similar chemical properties (Tables 3.7 and 3.8).

Table 3.6 Electron configuration of the first three elements of group 1

Element	Symbol	Atomic number	Electron configuration
Lithium	Li	3	2,1
Sodium	Na	11	2,8,1
Potassium	K	19	2,8,8,1

Table 3.7 Electron configuration of the first three elements of group 2

Element	Symbol	Atomic number	Electron configuration
Beryllium	Be	4	2,2
Magnesium	Mg	12	2,8,2
Calcium	Ca	20	2,8,8,2

Table 3.8 Electron configuration of the first three elements in group 7

Element	Symbol	Atomic number	Electron configuration
Fluorine	F	9	2,7
Chlorine	Cl	17	2,8,7
Bromine	Br	35	2,8,18,7

Group 1 — the alkali metals

Group 1 consists of the five metals lithium, sodium, potassium, rubidium and caesium, and the radioactive element francium. Lithium, sodium and potassium are commonly available for use in school. They are all very reactive metals and they are stored under oil to prevent them coming into contact with water or air. These three metals have the following properties.

- They are good conductors of electricity and heat.
- They are soft metals.
- They are metals with low densities.

Figure 3.13 Freshly cut sodium

- They have shiny surfaces when freshly cut with a knife (Figure 3.13).
- They burn in oxygen or air, with characteristic flame colours, to form white, solid oxides. For example, lithium reacts with the oxygen in air to form white lithium oxide, according to the equation:

$$\text{lithium} + \text{oxygen} \rightarrow \text{lithium oxide}$$
$$4\text{Li}(s) + \text{O}_2(g) \rightarrow 2\text{Li}_2\text{O}(s)$$

These group 1 oxides all dissolve in water to form alkaline solutions of the metal hydroxide.

$$\text{lithium oxide} + \text{water} \rightarrow \text{lithium hydroxide}$$
$$\text{Li}_2\text{O}(s) + \text{H}_2\text{O}(l) \rightarrow 2\text{LiOH}(aq)$$

- They react vigorously with water to give an alkaline solution of the metal oxide as well as hydrogen gas.

For example:

$$\text{potassium} + \text{water} \rightarrow \text{potassium} + \text{hydrogen gas}$$
$$\text{hydroxide}$$
$$2\text{K}(s) + 2\text{H}_2\text{O}(l) \rightarrow 2\text{KOH}(aq) + \text{H}_2(g)$$

Of these three metals, potassium is the most reactive towards water (Figure 3.14), followed by sodium and then lithium. Such gradual changes we call **trends**. Trends are useful to chemists as they allow predictions to be made about elements we have not observed in action.

Considering the group as a whole, the further down the group you go the more reactive the metals become. Francium is, therefore, the most reactive of the group 1 metals.

Table 3.6 shows the electron configuration of the first three elements of group 1. You will notice in each case that the outer energy level contains only

Figure 3.14 (**a**, top) Potassium reacts very vigorously with cold water. (**b**, bottom) An alkaline solution is produced when potassium reacts with water

one electron. When these elements react they lose this outer electron, and in doing so become more stable, because they obtain the electron configuration of a noble gas. You will learn more about the stable nature of these gases later in this chapter (p. 40).

When, for example, the element sodium reacts it loses its outer electron. This requires energy to overcome the electrostatic attractive forces between the outer electron and the positive nucleus (Figure 3.15).

Look at Figure 3.16. Why do you think potassium is more reactive than lithium or sodium?

Potassium is more reactive because less energy is required to remove the outer electron from its atom than for lithium or sodium. This is because as you go down the group the size of the atoms increases and the outer electron gets further away from the nucleus and becomes easier to remove.

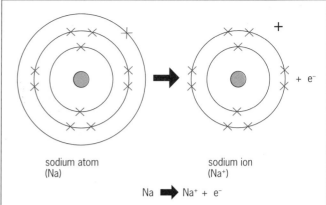

Figure 3.15 Sodium atom losing an electron to give a sodium ion

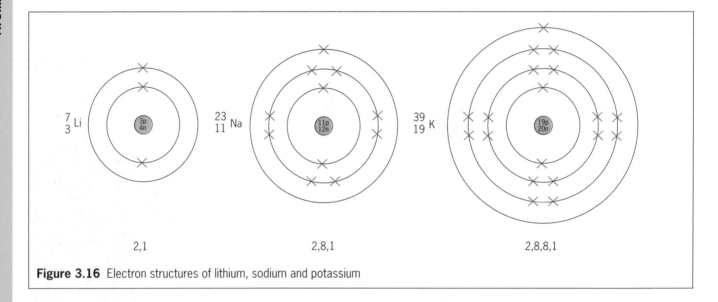

Figure 3.16 Electron structures of lithium, sodium and potassium

Group 2 — the alkaline earth metals

Group 2 consists of the five metals beryllium, magnesium, calcium, strontium and barium, and the radioactive element radium. Magnesium and calcium are generally available for use in school. These metals have the following properties.

- They are harder than those in group 1.
- They are silvery-grey in colour when pure and clean. They tarnish quickly, however, when left in air due to the formation of a metal oxide (Figure 3.17).
- They are good conductors of heat and electricity.
- They burn in oxygen or air with characteristic flame colours to form solid white oxides. For example:

    magnesium + oxygen → magnesium oxide
    $$2Mg(s) + O_2(g) \rightarrow 2MgO(s)$$

- They react with water, but they do so much less vigorously than the elements in group 1. For example:

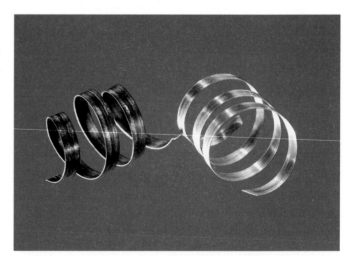

Figure 3.17 Tarnished (left) and cleaned-up magnesium

calcium + water → calcium hydroxide + hydrogen gas
$$Ca(s) + 2H_2O(l) \rightarrow Ca(OH)_2(aq) + H_2(g)$$

Considering the group as a whole, the further down the group you go, the more reactive the metals become.

Flame colours

If a clean nichrome wire is dipped into a metal compound and then held in the hot part of a Bunsen flame, the flame can become coloured (Figure 3.18). Certain metal ions may be detected in their compounds by observing their characteristic flame colours (Table 3.9).

A flame colour is obtained as a result of the electrons in the particular ions being excited when they absorb energy from the flame which is then emitted as visible light. The different electron configurations of the different ions, therefore, give rise to the different colours (Figure 3.19).

Table 3.9 Characteristic flame colours shown by some metal ions

	Metal	Flame colour
Group 1	Lithium	Crimson
	Sodium	Golden yellow
	Potassium	Lilac
	Rubidium	Red
	Caesium	Blue
Group 2	Calcium	Brick red
	Strontium	Crimson
	Barium	Apple green
Others	Lead	Blue–white
	Copper	Green

Figure 3.18 The golden yellow colour is characteristic of sodium

Figure 3.19 Different metal ions are used to produce the colours seen in fireworks

Group 7—the halogens

Group 7 consists of the four elements fluorine, chlorine, bromine and iodine, and the radioactive element astatine. Of these five elements, chlorine, bromine and iodine are generally available for use in school. These elements:

- are coloured (Table 3.10)
- exist as diatomic molecules, for example Cl_2, Br_2 and I_2
- show a gradual change from a gas (Cl_2) through liquid (Br_2) to solid (I_2) (Figure 3.20).

For details of the extraction of bromine, see Chapter 10, p. 133.

Table 3.10 Colours of some halogens

Halogen	Colour
Chlorine	Pale green
Bromine	Red–brown
Iodine	Purple–black

Figure 3.20 Chlorine, bromine and iodine

Displacement reactions

If chlorine is bubbled into a solution of potassium bromide the less reactive halogen, bromine, is displaced by the more reactive halogen, chlorine, as you can see from Figure 3.21:

potassium + chlorine → potassium + bromine
bromide   chloride

$$2KBr(aq) + Cl_2(g) \rightarrow 2KCl(aq) + Br_2(aq)$$

The observed order of reactivity of the halogens, confirmed by similar displacement reactions, is:

Decreasing reactivity

chlorine   bromine   iodine

You will notice that, unlike the elements of groups 1 and 2, the order of reactivity decreases on going down the group.

Table 3.11 shows the electron configuration for chlorine and bromine. In each case the outer energy level contains seven electrons. When these elements react they gain one electron per atom to gain the stable electron configuration of a noble gas. You will learn more about the stable nature of these gases in the next section. For example, when chlorine reacts it gains a single electron and forms a negative ion (Figure 3.22).

Chlorine is more reactive than bromine because the incoming electron is being more strongly attracted into the outer energy level of the smaller atom. The attractive force on it will be greater than in the case of bromine, since the outer energy level of chlorine is closer to the nucleus. As you go down the group this outermost, extra electron is further from the nucleus. It will, therefore, be held less securely, and the resulting reactivity of the elements in group 7 will decrease down the group.

Group 0—the noble gases

Helium, neon, argon, krypton, xenon and the radio-active element radon make up a most unusual group of non-metals. They were all discovered after Mendeléev had published his periodic table. They:

- are colourless gases
- exist as monatomic molecules, for example He, Ne
- are very unreactive.

No compounds of helium, neon or argon have ever been found. However, more recently a number of compounds of xenon and krypton with fluorine and oxygen have been produced.

These gases are chemically unreactive because they have electron configurations which are stable and very difficult to change (Table 3.12). They are so stable that other elements attempt to attain these

Figure 3.21 Bromine being displaced

Table 3.11 Electron configuration of chlorine and bromine

Element	Symbol	Atomic number	Electron configuration
Chlorine	Cl	17	2,8,7
Bromine	Br	35	2,8,18,7

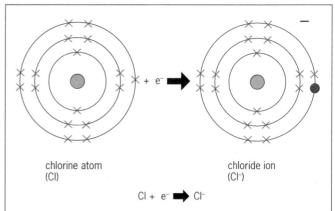

chlorine atom (Cl)   chloride ion (Cl⁻)

$$Cl + e^- \rightarrow Cl^-$$

Figure 3.22 A chlorine atom gaining an electron to form a chloride ion

Table 3.12 Electron configuration of helium, neon and argon

Element	Symbol	Atomic number	Electron configuration
Helium	He	2	2
Neon	Ne	10	2,8
Argon	Ar	18	2,8,8

electron configurations during chemical reactions (Chapter 4, p. 44). You have probably seen this in your study of the elements of groups 1, 2 and 7.

Although unreactive, they have many uses. Argon, for example, is the gas used to fill light bulbs to prevent the tungsten filament reacting with air. Neon is used extensively in advertising signs and in lasers. Further uses of these gases are discussed in Chapter 10, p. 131.

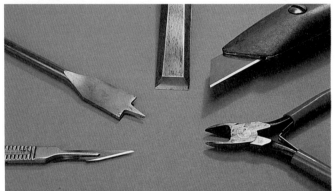

Figure 3.23 Everyday uses of transition metals

Helium is separated from natural gas by the liquefaction of the other gases. The other noble gases are obtained in large quantities by the fractional distillation of liquid air (Chapter 10, p. 130).

Transition elements

This block of metals includes many you will be familiar with, for example copper, iron, nickel, zinc and chromium (Figure 3.23).

- They are harder and stronger than the metals in groups 1 and 2.
- They have much higher densities than the metals in groups 1 and 2.
- They are less reactive metals.
- They form a range of brightly coloured compounds (Figure 3.24).
- They are good conductors of heat and electricity.
- They show catalytic activity (Chapter 11, p. 144) as elements and compounds. For example, iron is used in the industrial production of ammonia gas (Haber process, Chapter 15, p. 192).
- They form more than one simple ion. For example, copper forms Cu^+ and Cu^{2+} and iron forms Fe^{2+} and Fe^{3+}.

The position of hydrogen

Hydrogen is often placed by itself in the periodic table. This is because the properties of hydrogen are unique. However, profitable comparisons can be

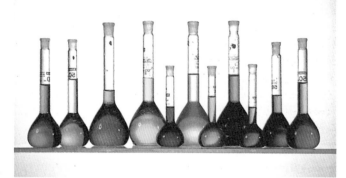

Figure 3.24 Some solutions of coloured transition metal compounds

made with the other elements. It is often shown at the top of either group 1 or group 7, but it cannot fit easily into the trends shown by either group, as shown in Table 3.13.

Table 3.13 Comparison of hydrogen with lithium and fluorine

Lithium	Hydrogen	Fluorine
Solid	Gas	Gas
Forms a positive ion	Forms positive or negative ions	Forms a negative ion
1 electron in outer energy level	1 electron in outer energy level	1 electron short of a full outer energy level
Loses 1 electron to form a noble gas configuration	Needs 1 electron to form a noble gas configuration	Needs 1 electron to form a noble gas configuration

QUESTIONS

a Write word and balanced chemical equations for the reactions between:
  (i) sodium and oxygen
  (ii) sodium and water.
b Write word and balanced chemical equations for the reactions between:
  (i) magnesium and water
  (ii) calcium and oxygen.
c Write word and balanced chemical equations for the reactions between:
  (i) bromine and potassium iodide solution
  (ii) bromine and potassium chloride solution.
  If no reaction will take place, write 'no reaction' and explain why.
d Write down the names and symbols for the noble gases not given in Table 3.12 and use your research skills to find a use for each.
e Look at the photographs in Figure 3.23 and decide which properties are important when considering the particular use the metal is being put to.
f Which groups in the periodic table contain:
  (i) only metals?
  (ii) only non-metals?
  (iii) both metals and non-metals?
g Account for the fact that calcium is more reactive than magnesium.

Checklist

After studying Chapter 3 you should know and understand the following terms.

Electron A fundamental sub-atomic particle with a negative charge present in all atoms within energy levels around the nucleus.

Proton A fundamental sub-atomic particle which has a positive charge equal in magnitude to that of an electron. The proton occurs in all nuclei.

Neutron A fundamental, uncharged sub-atomic particle present in the nuclei of atoms.

Electrostatic force of attraction A strong force of attraction between opposite charges.

Electron energy levels The allowed energies of electrons in atoms.

Atomic mass unit Exactly $\frac{1}{12}$ of the mass of one atom of the most abundant isotope of carbon-12.

Atomic number (proton number) Symbol Z. The number of protons in the nucleus of an atom. The number of electrons present in an atom. The order of the element within the periodic table.

Mass number (nucleon number) Symbol A. The total number of protons and neutrons found in the nucleus of an atom.

Isotopes Atoms of the same element which possess different numbers of neutrons. They differ in mass number (nucleon number).

Mass spectrometer A device in which atoms or molecules are ionised and then accelerated. Ions are separated according to their mass.

Relative atomic mass Symbol A_r.
$$A_r = \frac{\text{average mass of isotopes of the element}}{\frac{1}{12} \times \text{mass of a carbon-12 atom}}$$

Electron configuration A shorthand method of describing the arrangement of electrons within the energy levels of an atom.

Periodic table A table of elements arranged in order of increasing atomic number to show the similarities of the chemical elements with related electron configurations.

Group A vertical column of the periodic table containing elements with similar properties with the same number of electrons in their outer energy levels. They have an increasing number of inner energy levels as you descend the group.

Periods Horizontal rows of the periodic table. Within a period the atoms of all the elements have the same number of occupied energy levels but have an increasing number of electrons in the outer energy level.

Metalloid (semi-metal) Any of the class of chemical elements intermediate in properties between metals and non-metals, for example boron and silicon.

Ion An atom or group of atoms which has either lost one or more electrons, making it positively charged, or gained one or more electrons, making it negatively charged.

Displacement reaction A reaction in which a more reactive element displaces a less reactive element from solution.

QUESTIONS

1 An atom **X** has an atomic number of 19 and relative atomic mass of 39.
  a How many electrons, protons and neutrons are there in an atom of **X**?
  b How many electrons will there be in the outer energy level (shell) of an atom of **X**?
  c Write down the symbol for the ion **X** will form.
  d Which group of the periodic table would **X** be in?
  e (i) How would you expect **X** to react with water?
  (ii) Write a word and balanced chemical equation for this reaction.

2 The atomic number of barium (Ba) is 56. It is in group 2 of the periodic table.
  a How many electrons would you expect a barium atom to contain in its outer energy level?
  b How would you expect barium to react with chlorine? Write a word and balanced chemical equation for this reaction.
  c How would you expect barium to react with water? Write a word and balanced chemical equation for this reaction.
  d Write down the formulae of the bromide and sulphate of barium.

3 Find the element germanium (Ge) in the periodic table.
  a Which group of the periodic table is this element in?
  b How many electrons will it have in its outer energy level (shell)?
  c Is germanium a metal or a non-metal?
  d What is the formula of the chloride of germanium?
  e Name and give the symbols of the other elements in this group.

4 Three members of the halogens are: $^{35.5}_{17}Cl$, $^{80}_{35}Br$ and $^{127}_{53}I$.
  a (i) Write down the electron structure of an atom of chlorine.
  (ii) Why is the relative atomic mass of chlorine not a whole number?
  (iii) How many protons are there in an atom of iodine?
  (iv) How many neutrons are there in an atom of bromine?
  (v) State and account for the order of reactivity of these elements.
  b When potassium is allowed to burn in a gas jar of chlorine, in a fume cupboard, clouds of white smoke are produced.
  (i) What does the white smoke consist of?
  (ii) Write a word and balanced chemical equation for this reaction.
  (iii) Describe what you would expect to see when potassium is allowed to burn safely in a gas jar of bromine vapour. Write a word and balanced chemical equation for this reaction.

5 'By using displacement reactions it is possible to deduce the order of reactivity of the halogens.' Discuss this statement with reference to the elements bromine, iodine and chlorine only.

6 Use the information given in the table below to answer the questions below concerning the elements **Q, R, S, T** and **X**.

Element	Atomic number	Mass number	Electron structure
Q	3	7	2,1
R	12	24	2,8,2
S	18	40	2,8,8
T	8	18	2,6
X	19	39	2,8,8,1

  a Which element has 20 neutrons in each atom?
  b Which element is a noble gas?
  c Which two elements form ions with the same electron structure as neon?
  d Which two elements are in the same group of the periodic table and which group is this?
  e Place the elements in the table above into the periods in which they belong.
  f Which is the most reactive metal element shown in the table?
  g (i) Which of the above elements is lithium?
  (ii) What colour flame would compounds containing lithium produce?

7 a $^{238}_{92}U$ and $^{235}_{92}U$ are isotopes of uranium.
  With reference to this example, explain what you understand by the term isotope.
  b A sample of gallium contains 60% of atoms of $^{69}_{31}Ga$ and 40% of atoms of $^{71}_{31}Ga$. Calculate the relative atomic mass of this sample of gallium.

8 Copy and complete the following table with reference to the periodic table on p. 35.

Element name	Symbol	Atomic number	Mass number	Number of neutrons	$^A_Z X$
		5	11		
			40	22	
		14	28		
			20	10	
		26		30	
			84	48	
		52		76	

4 BONDING AND STRUCTURE

In Chapter 3 we have seen that the noble gases are stable or unreactive because they have full electron energy levels (Figure 4.1). It would seem that when elements react to form compounds they do so to achieve full electron energy levels. This idea forms the basis of the electronic theory of chemical bonding.

Figure 4.1 The unreactive noble gas neon is used in advertising lights such as those of Piccadilly Circus in London

Ionic bonding

Ionic bonds are usually found in compounds that contain metals combined with non-metals. When this type of bond is formed, electrons are **transferred** from the metal atoms to the non-metal atoms during the chemical reaction. In doing this, the atoms become more stable by getting full outer energy levels.

For example, consider what happens when sodium and chlorine combine to make sodium chloride.

$$\text{sodium} + \text{chlorine} \rightarrow \text{sodium chloride}$$

Sodium has just one electron in its outer energy level ($_{11}$Na 2,8,1). Chlorine has seven electrons in its outer energy level ($_{17}$Cl 2,8,7). When these two elements react, the outer electron of each sodium atom is transferred to the outer energy level of a chlorine atom (Figure 4.2). In this way both the atoms obtain full outer energy levels and become 'like' the nearest noble gas. The sodium atom has become a sodium ion with an electron configuration like neon, while the chlorine atom has become a chloride ion with an electron configuration like argon.

Only the outer electrons are important in bonding, so we can simplify the diagrams by missing out the inner energy levels (Figure 4.3).

The charges on the sodium and chloride ions are equal but opposite. They balance each other and the resulting formula for sodium chloride is NaCl. These oppositely charged ions attract each other and are pulled, or **bonded**, to one another by strong electrostatic forces. This type of bonding is called **ionic bonding**. The alternative name, **electrovalent bonding**, is derived from the fact that there are electrical charges on the atoms involved in the bonding.

Figure 4.4 shows the electron transfers that take place during the formation of magnesium oxide.

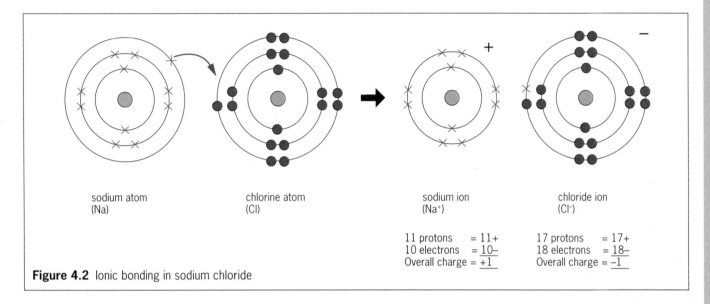

sodium atom (Na)   chlorine atom (Cl)   sodium ion (Na⁺)   chloride ion (Cl⁻)

11 protons	$= 11+$	17 protons	$= 17+$
10 electrons	$= \underline{10-}$	18 electrons	$= \underline{18-}$
Overall charge	$= \underline{+1}$	Overall charge	$= \underline{-1}$

Figure 4.2 Ionic bonding in sodium chloride

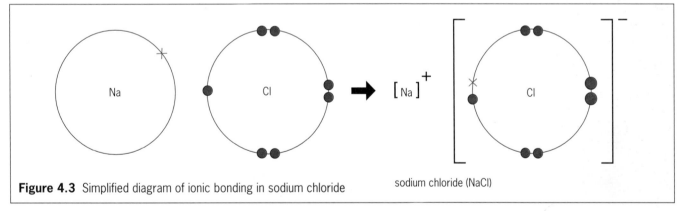

Figure 4.3 Simplified diagram of ionic bonding in sodium chloride

sodium chloride (NaCl)

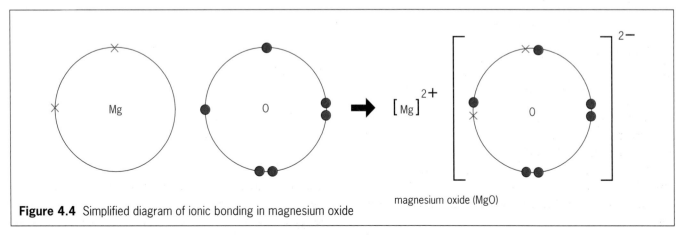

magnesium oxide (MgO)

Figure 4.4 Simplified diagram of ionic bonding in magnesium oxide

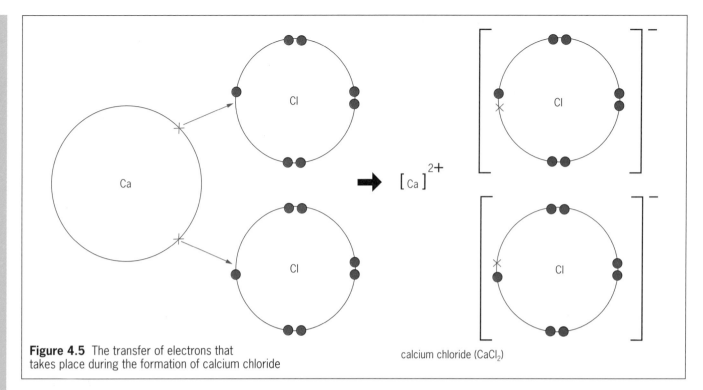

Figure 4.5 The transfer of electrons that takes place during the formation of calcium chloride

calcium chloride (CaCl$_2$)

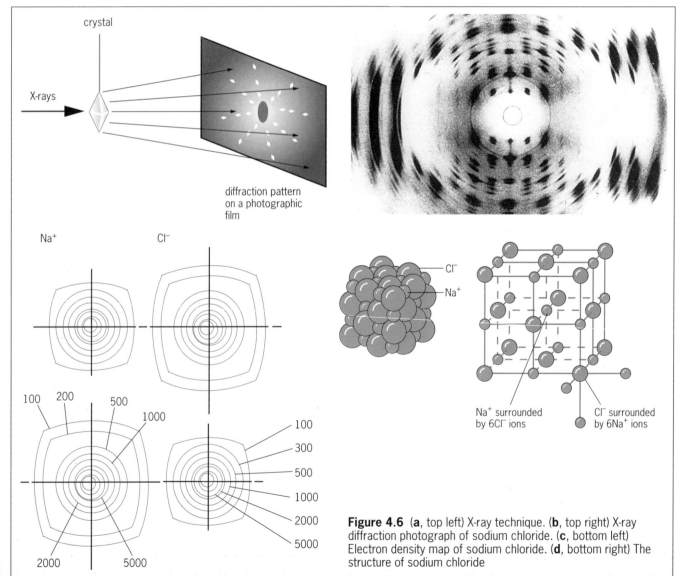

Figure 4.6 (**a**, top left) X-ray technique. (**b**, top right) X-ray diffraction photograph of sodium chloride. (**c**, bottom left) Electron density map of sodium chloride. (**d**, bottom right) The structure of sodium chloride

Magnesium obtains a full outer energy level by losing two electrons. These are transferred to the oxygen atom. In magnesium oxide, the Mg^{2+} and O^{2-} are oppositely charged and are attracted to one another. The formula for magnesium oxide is MgO.

Figure 4.5 shows the electron transfers that take place during the formation of calcium chloride. When these two elements react, the calcium atom gives each of two chlorine atoms one electron. In this case, a compound is formed containing two chloride ions (Cl^-) for each calcium ion (Ca^{2+}). Its formula is $CaCl_2$.

QUESTIONS

a Draw diagrams to represent the bonding in each of the following ionic compounds:
   (i) calcium oxide (CaO)
   (ii) magnesium chloride ($MgCl_2$)
   (iii) lithium fluoride (LiF)
   (iv) potassium chloride (KCl).

Ionic structures

Ionic structures are solids at room temperature and have high melting and boiling points. The ions are packed together in a regular arrangement called a **lattice**. Within the lattice, oppositely charged ions attract one another strongly. Scientists, using **X-ray diffraction** (Figure 4.6a), have obtained photographs which indicate the way in which the ions are arranged (Figure 4.6b). The electron density map of sodium chloride is shown in Figure 4.6c.

Figure 4.6d shows the structure of sodium chloride as determined by the X-ray diffraction technique. The study of crystals using X-ray diffraction was pioneered by Sir William Bragg and his son Sir Laurence Bragg in 1912. X-rays are a form of electromagnetic radiation. They have a much shorter wavelength than light therefore it is possible to use them to investigate extremely small structures.

When X-rays are passed through a crystal of sodium chloride, for example, you get a pattern of spots called a diffraction pattern (Figure 4.6b). This pattern can be recorded on photographic film and used to work out how the ions or atoms are arranged in the crystal. Crystals give particular diffraction patterns depending on their structure, and this makes X-ray diffraction a particularly powerful technique in the investigation of crystal structures.

Figure 4.6d shows only a tiny part of a small crystal of sodium chloride. Many millions of sodium ions and chloride ions would be arranged in this way in a crystal of sodium chloride to make up the **giant ionic structure**. Each sodium ion in the lattice is

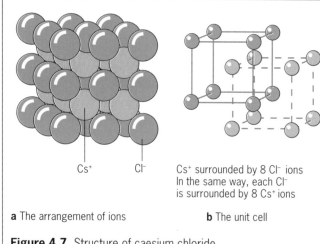

a The arrangement of ions   **b** The unit cell

Cs^+ surrounded by 8 Cl^- ions In the same way, each Cl^- is surrounded by 8 Cs^+ ions

Figure 4.7 Structure of caesium chloride

surrounded by six chloride ions, and each chloride ion is surrounded by six sodium ions.

Not all ionic substances form the same structures. Caesium chloride (CsCl), for example, forms a different structure due to the larger size of the caesium ion compared with that of the sodium ion. This gives rise to the structure shown in Figure 4.7, which is called a body-centred cubic structure. Each caesium ion is surrounded by eight chloride ions and, in turn, each chloride ion is surrounded by eight caesium ions.

Properties of ionic compounds

Ionic compounds have the following properties.

- They are usually solids at room temperature, with high melting points. This is due to the strong electrostatic forces holding the crystal lattice together. A lot of energy is therefore needed to separate the ions and melt the substance.
- They are usually hard substances.
- They usually cannot conduct electricity when solid, because the ions are not free to move.
- They mainly dissolve in water. This is because water molecules are able to bond with both the positive and the negative ions, which breaks up the lattice and keeps the ions apart.
- They usually conduct electricity when in the molten state or in aqueous solution. The forces of attraction between the ions are broken and the ions are free to move. This allows an electric current to be passed through the molten sodium chloride. For a further discussion of this process see Chapter 6 (p. 71).

Formulae of ionic substances

On p. 45 we saw that ionic compounds contain positive and negative ions, whose charges balance. For example, sodium chloride contains one Na^+ ion

Table 4.1 Some common valencies

	Valency					
	1		**2**		**3**	
Metals	Lithium	(Li^+)	Magnesium	(Mg^{2+})	Aluminium	(Al^{3+})
	Sodium	(Na^+)	Calcium	(Ca^{2+})	Iron	(Fe^{3+})
	Potassium	(K^+)	Copper	(Cu^{2+})		
	Silver	(Ag^+)	Zinc	(Zn^{2+})		
	Copper	(Cu^+)	Iron	(Fe^{2+})		
			Lead	(Pb^{2+})		
			Barium	(Ba^{2+})		
Non-metals	Fluoride	(F^-)	Oxide	(O^{2-})		
	Chloride	(Cl^-)	Sulphide	(S^{2-})		
	Bromide	(Br^-)				
	Hydrogen	(H^+)				
Groups of atoms	Hydroxide	(OH^-)	Carbonate	(CO_3^{2-})	Phosphate	(PO_4^{3-})
	Nitrate	(NO_3^-)	Sulphate	(SO_4^{2-})		
	Ammonium	(NH_4^+)				
	Hydrogencarbonate	(HCO_3^-)				

for every Cl^- ion, giving rise to the formula NaCl. This method can be used to write down formulae which show the ratio of the number of ions present in any ionic compound.

The formula of magnesium chloride is $MgCl_2$. This formula is arrived at by each Mg^{2+} ion combining with two Cl^- ions, and once again the charges balance. The size of the charge on an ion is a measure of its **valency** or combining power. Na^+ has a valency of 1, but Mg^{2+} has a valency of 2. Na^+ can bond (combine) with only one Cl^- ion, whereas Mg^{2+} can bond with two Cl^- ions.

Some elements, such as copper and iron, possess two ions with different valencies. Copper can form the Cu^+ ion and the Cu^{2+} ion, and therefore it can form two different compounds with chlorine, CuCl (copper(I) chloride) and $CuCl_2$ (copper(II) chloride). Iron forms the Fe^{2+} and Fe^{3+} ions.

Table 4.1 shows the valencies of a series of ions you will normally meet in your study of chemistry.

You will notice that Table 4.1 includes groups of atoms which have net charges. For example, the nitrate ion is a single unit composed of one nitrogen atom and three oxygen atoms and has one single negative charge. The formula, therefore, of magnesium nitrate would be $Mg(NO_3)_2$. You will notice that the NO_3 has been placed in brackets with a 2 outside the bracket. This indicates that there are two nitrate ions present for every magnesium ion. The ratio of the atoms present is therefore:

$$Mg \ (N \ \ O_3)_2$$
$$1Mg:2N:6O$$

QUESTIONS

a Using the information in Table 4.1, write the formulae of:
  (i) iron(II) chloride
  (ii) calcium phosphate
  (iii) iron(III) chloride
  (iv) silver oxide.
b Using the formulae in your answer to (a), write down the ratio of atoms present for each of the compounds.

Covalent bonding

Another way in which atoms can gain the stability of the noble gas electron configuration is by sharing the electrons in their outer energy levels. This occurs between non-metal atoms, and the bond formed is called a **covalent bond**. The simplest example of this type of bonding can be seen by considering the hydrogen molecule, H_2.

Each hydrogen atom in the molecule has one electron. In order to obtain a full outer energy level and gain the electron configuration of the noble gas, helium, each of the hydrogen atoms must have two electrons. To do this, the two hydrogen atoms allow their outer energy levels to overlap (Figure 4.8). A molecule of hydrogen is formed, with two hydrogen atoms sharing a pair of electrons (Figure 4.9). This shared pair of electrons is known as a **single covalent bond** and is represented by a single line as in dihydrogen:

$$H—H$$

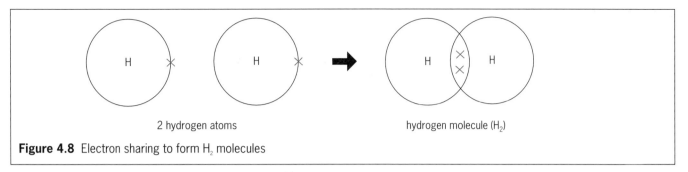

2 hydrogen atoms   hydrogen molecule (H_2)

Figure 4.8 Electron sharing to form H_2 molecules

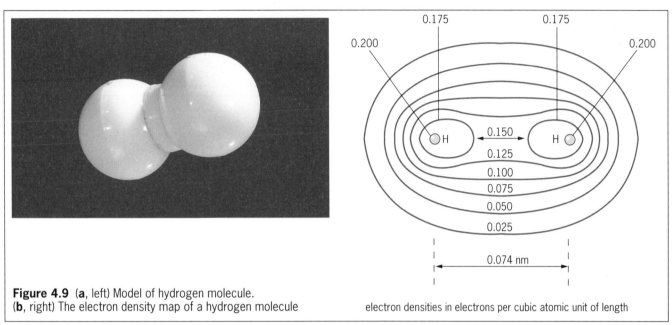

0.074 nm

Figure 4.9 (**a**, left) Model of hydrogen molecule.
(**b**, right) The electron density map of a hydrogen molecule

electron densities in electrons per cubic atomic unit of length

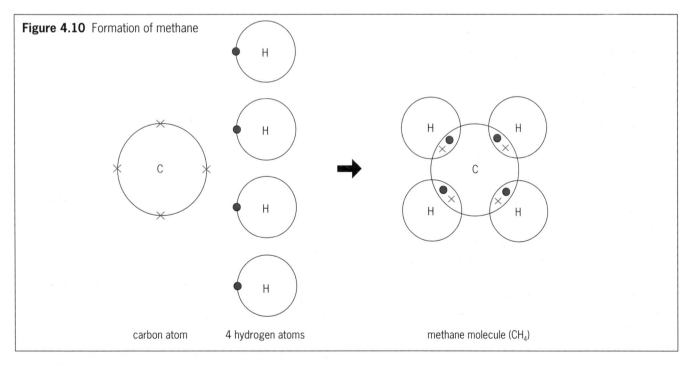

Figure 4.10 Formation of methane

carbon atom   4 hydrogen atoms   methane molecule (CH_4)

Other covalent compounds

Methane (natural gas) is a gas whose molecules contain atoms of carbon and hydrogen. The electron structures are:

$$_6C\ 2,4 \qquad\qquad _1H\ 1$$

The carbon atom needs four more electrons to attain the electron configuration of the noble gas neon. Each hydrogen atom only needs one electron to form the electron configuration of helium. Figure 4.10 shows how the atoms gain these electron configurations by

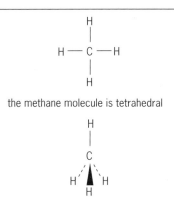

the methane molecule is tetrahedral

Figure 4.11 (**a**, left) Methane molecule. (**b**, right) Model of the methane molecule. The tetrahedral shape is caused by the repulsion of the C–H bonding pairs of the electrons

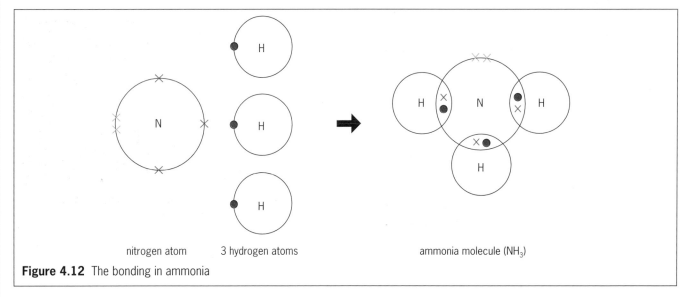

nitrogen atom   3 hydrogen atoms   ammonia molecule (NH_3)

Figure 4.12 The bonding in ammonia

the sharing of electrons. You will note that only the outer electron energy levels are shown. Figure 4.11 shows the shape of the methane molecule.

Ammonia is a gas containing the elements nitrogen and hydrogen. It is used in large amounts to make fertilisers. The electron configuration of the two elements is:

$$_7N\ 2,5 \qquad _1H\ 1$$

The nitrogen atom needs three more electrons to obtain the noble gas structure of neon. Each hydrogen requires only one electron to form the noble gas structure of helium. The nitrogen and hydrogen atoms share electrons, forming three single covalent bonds (Figure 4.12). Unlike methane the shape of an ammonia molecule is pyramidal (Figure 4.13).

Water is a liquid containing the elements hydrogen and oxygen. The electron configuration of the two elements is:

$$_8O\ 2,6 \qquad _1H\ 1$$

The oxygen atom needs two electrons to gain the electron configuration of neon. Each hydrogen

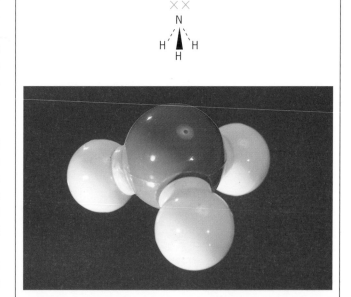

Figure 4.13 (**a**, top) Ammonia molecule. (**b**, bottom) Model of the ammonia molecule. The pyramidal shape is caused by the repulsion between the bonding pairs of electrons as well as the lone pair (or non-bonding pair) of electrons

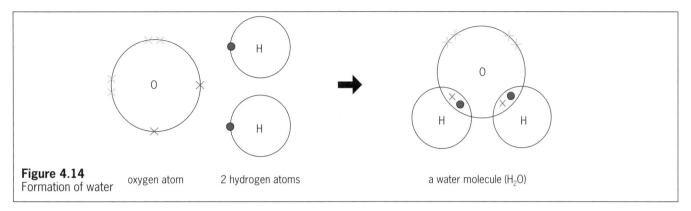

Figure 4.14
Formation of water

oxygen atom   2 hydrogen atoms   a water molecule (H₂O)

requires one more electron to gain the electron configuration of helium. The oxygen and hydrogen atoms share electrons, forming a water molecule with two single covalent bonds as shown in Figure 4.14. A water molecule is V-shaped (Figure 4.15).

Carbon dioxide is a gas containing the elements carbon and oxygen. The electron configurations of the two elements are:

$$_6C\ 2,4 \qquad _8O\ 2,6$$

In this case each carbon atom needs to share four electrons to gain the electron configuration of neon. Each oxygen needs to share two electrons to gain the electron configuration of neon. This is achieved by forming two **double covalent bonds** in which two pairs of electrons are shared in each case, as shown in Figure 4.16. Carbon dioxide is a linear molecule (Figure 4.17).

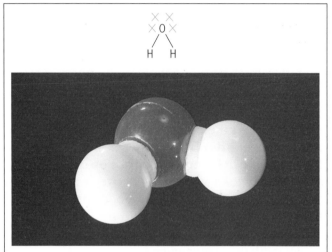

Figure 4.15 (**a**, top) Water molecule. (**b**, bottom) Model of a water molecule. This is a V-shaped molecule

Figure 4.16 Formation of carbon dioxide

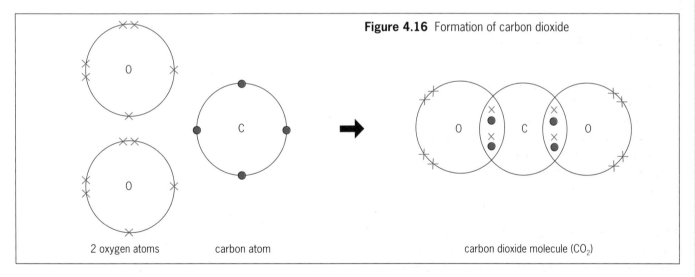

2 oxygen atoms   carbon atom   carbon dioxide molecule (CO₂)

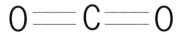

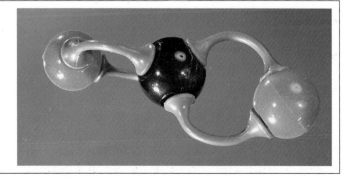

Figure 4.17 (**a**, left) Carbon dioxide molecule. Note the double covalent bond is represented by a double line.
(**b**, right) Model of the linear carbon dioxide molecule

QUESTIONS

a Draw diagrams to represent the bonding in each of the following covalent compounds:

(i) tetrafluoromethane (CF_4)
(ii) nitrogen gas (N_2)
(iii) hydrogen sulphide (H_2S)
(iv) hydrogen chloride (HCl)
(v) ethene (C_2H_4)
(vi) methanol (CH_3OH).

b Explain why the water molecule in Figure 4.15 is V-shaped.

Covalent structures

Compounds containing covalent bonds have molecules whose structures can be classified as either **simple molecular** or **giant molecular**.

Simple molecular structures are simple, formed from only a few atoms. They have strong covalent bonds between the atoms within a molecule (**intramolecular bonds**) but have weak bonds between the molecules (**intermolecular bonds**) (Figure 4.18). One type of weak bond between molecules is known as the **van der Waals'** bond (or force), and these forces increase steadily with the increasing size of the molecule. Examples of simple molecules are iodine, methane, water and ethanol.

Giant molecular structures contain many hundreds of thousands of atoms joined by strong covalent bonds. Examples of substances showing this type of structure are diamond, graphite, silicon(IV) oxide (Figure 4.19) and plastics (polymers, Chapter 14, p. 176) such as polyethene. Plastics are a tangled mass of very long molecules in which the atoms are joined together by strong covalent bonds to form long chains. Molten plastics can be made into fibres by being forced through hundreds of tiny holes in a 'spinneret' (Figure 4.20). This process aligns the long chains of atoms along the length of the fibre. For a further discussion of plastics, see Chapter 14, p. 173.

Properties of covalent compounds

Covalent compounds have the following properties.

• As simple molecular substances, they are usually gases, liquids or solids with low melting and boiling points. The melting points are low because of the weak intermolecular forces of attraction which exist between simple molecules. Giant molecular substances have higher melting points, because the whole structure is held together by strong covalent bonds within the giant molecule.

• Generally, they do not conduct electricity when molten or dissolved in water. This is because they do not contain ions. However, some molecules actually react with water to form ions. For example,

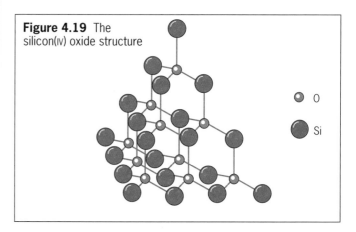

Figure 4.18 Strong covalent and weak intermolecular forces in iodine

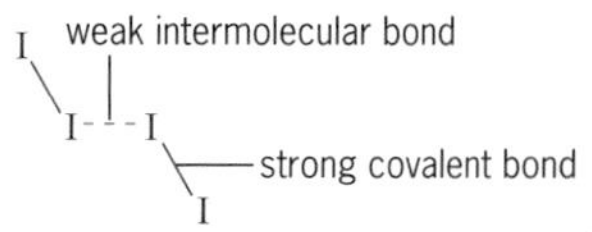

weak intermolecular bond

strong covalent bond

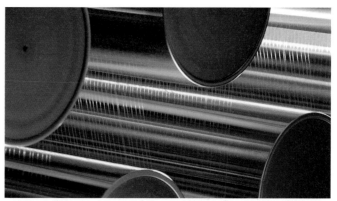

Figure 4.19 The silicon(IV) oxide structure

O
Si

Figure 4.20 These are magnified nylon fibres formed by forcing molten plastic through hundreds of tiny holes

hydrogen chloride gas produces aqueous hydrogen ions and chloride ions when it dissolves in water:

$$HCl(g) \xrightarrow{\text{water}} H^+(aq) + Cl^-(aq)$$

• Generally, they do not dissolve in water. However, water is an excellent solvent and can interact with and dissolve some covalent molecules better than others.

Allotropy

When an element can exist in more than one physical form in the same state it is said to exhibit **allotropy** (or polymorphism). Each of the different physical forms is called an **allotrope**. Allotropy is actually quite a common feature of the elements in the periodic table. Some examples of elements which show allotropy are sulphur, tin, iron and carbon.

Allotropes of carbon

Carbon is a non-metallic element which exists in more than one solid structural form. Its allotropes are called **graphite** and **diamond**. Each of the allotropes has a different structure (Figures 4.21 and 4.22) and

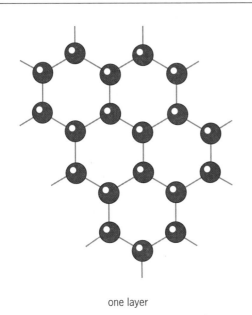

one layer

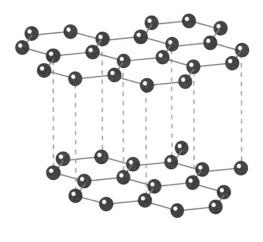

showing how the layers fit together

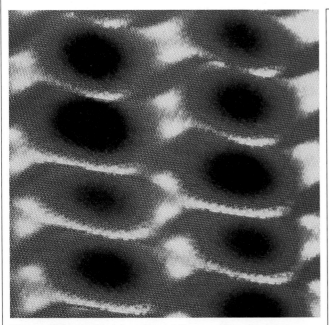

Figure 4.21 (**a**, top) A portion of the graphite structure. (**b**, bottom) A piece of graphite as seen through a scanning tunnelling microscope

Table 4.2 Physical properties of graphite and diamond

Property	Graphite	Diamond
Appearance	A dark grey, shiny solid	A colourless transparent crystal which sparkles in light
Electrical conductivity	Conducts electricity	Does not conduct electricity
Hardness	A soft material with a slippery feel	A very hard substance
Density (g cm^{-3})	2.25	3.51

Table 4.3 Uses of graphite and diamond

Graphite	Diamond
Pencils	Jewellery
Electrodes	Glass cutters
Lubricant	Diamond-studded saws
	Drill bits
	Polishers

so the allotropes exhibit different physical properties (Table 4.2). The different physical properties that they exhibit lead to the allotropes being used in different ways (Table 4.3).

Graphite

Figure 4.21a shows the structure of graphite. This is a layer structure. Within each layer each carbon atom is bonded to three others by strong covalent bonds. Each layer is therefore a giant molecule. Between these layers there are weak forces of attraction (van der Waals' forces) and so the layers will pass over each other easily.

With only three covalent bonds formed between carbon atoms within the layers, an unbonded electron is present on each carbon atom. These 'spare' (or **delocalised**) electrons form electron clouds between the layers and it is because of these spare electrons that graphite conducts electricity.

In recent years a set of interesting compounds known as **graphitic compounds** have been developed. In these compounds different atoms have been fitted in between the layers of carbon atoms to produce a substance with a greater electrical conductivity than pure graphite. Graphite is used as a component in certain sports equipment.

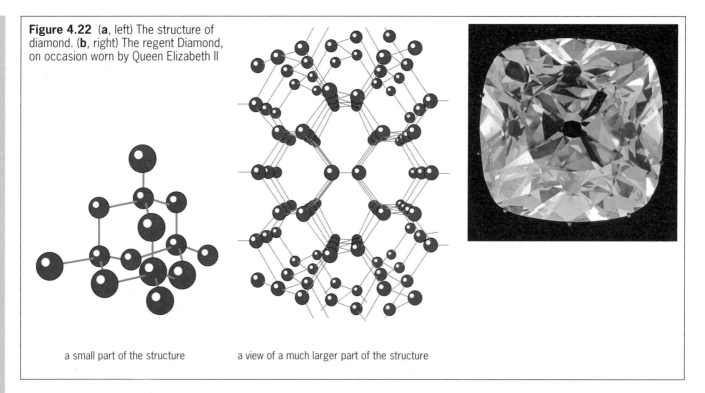

Figure 4.22 (**a**, left) The structure of diamond. (**b**, right) The regent Diamond, on occasion worn by Queen Elizabeth II

a small part of the structure   a view of a much larger part of the structure

Figure 4.23 Uses of graphite (as a pencil, in a squash racket) and diamond (as a toothed saw to cut marble, on a dentist's drill)

Diamond

Figure 4.22 shows the diamond structure. Each of the carbon atoms in the giant structure is covalently bonded to four others. They form a tetrahedral arrangement. This bonding scheme gives rise to a very rigid, three-dimensional structure and accounts for the extreme hardness of the substance. All the outer energy level electrons of the carbon atoms are used to form covalent bonds, so there are no electrons available to enable diamond to conduct electricity.

It is possible to manufacture both allotropes of carbon. Diamond is made by heating graphite to about 300 °C at very high pressures. Diamond made by this method is known as industrial diamond. Graphite can be made by heating a mixture of coke and sand at a very high temperature in an electric arc furnace for about 24 hours.

The various uses of graphite and diamond result from their differing properties (Figure 4.23).

Buckminsterfullerene — an unusual form of carbon

In 1985 a new form of carbon was obtained. It was formed by the action of a laser beam on a sample of graphite. The structure of buckminsterfullerene can be seen in Figure 4.24.

This spherical structure is composed of 60 carbon atoms covalently bonded together. The discovery of further spherical forms of carbon, 'bucky balls', has led to a whole new branch of inorganic carbon chemistry. It is thought that this type of molecule exists in chimney soot. Chemists have suggested that due to the large surface area of the bucky balls they may have uses as catalysts (Chapter 11, p. 144).

Buckminsterfullerene is named after an American architect, Buckminster Fuller, who built complex geometrical structures (Figure 4.25).

QUESTION

a Make a summary table of the properties of substances with covalent structures. Your table should include examples of simple molecular and giant molecular substances.

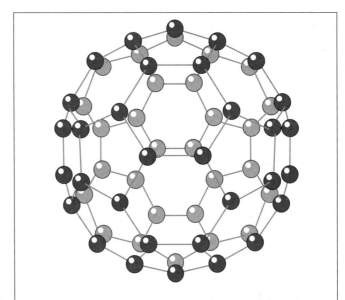

Figure 4.24 Buckminsterfullerene — a 'bucky ball' (C_{60})

Figure 4.25 C_{60} has a structure similar to a football and to the structure of the dome at the Expo in Montreal, Canada

Glasses and ceramics

Glasses

Glasses are irregular giant molecular structures held together by strong covalent bonds. Glass can be made by heating silicon(IV) oxide with other substances until a thick viscous liquid is formed. As this liquid cools, the atoms present cannot move freely enough to return to their arrangement within the pure silicon(IV) oxide structure. Instead they are forced to form a disordered arrangement as shown in Figure 4.26. Glass is called a **supercooled liquid**.

The glass used in bottles and windows is **soda glass**. This type of glass is made by heating a mixture of sand (silicon(IV) oxide), soda (sodium carbonate) and lime (calcium oxide). Pyrex is a borosilicate glass. It is made by incorporating some boron oxide into the silicon(IV) oxide structure so that silicon atoms are replaced by boron atoms. This type of glass is tougher than soda glass and more resistant to temperature changes. It is, therefore, used in the manufacture of cooking utensils and laboratory glassware (Figure 4.27).

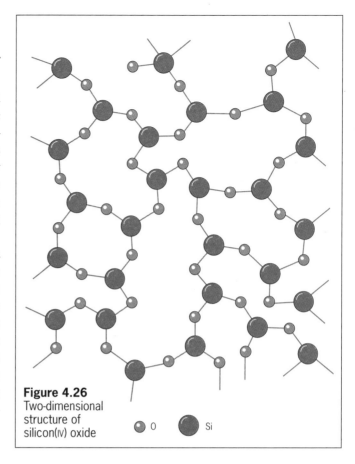

Figure 4.26
Two-dimensional structure of silicon(IV) oxide

○ O ● Si

Ceramics

The word **ceramic** comes from the Greek word meaning pottery or 'burnt stuff'. Clay dug from the ground contains a mixture of several materials. The main one is a mineral called kaolinite, $Al_2Si_2O_5(OH)_4$, in which the atoms are arranged in layers in a giant structure. While wet, the clay can be moulded because the kaolinite crystals move over one another. However, when it is dry the clay becomes rigid because the crystals stick together.

During firing in a furnace, the clay is heated to a temperature of 1000 °C (Figure 4.28). A complicated series of chemical changes take place, new minerals are formed and some of the substances in the clay react to form a type of glass. The material produced at the end of the firing, the ceramic, consists of many minute mineral crystals bonded together with glass.

Figure 4.27 This glassware is made from Pyrex

Figure 4.28 During firing the clay is heated to 1000 °C

QUESTION

a Draw up a table to summarise the properties of the different types of substances you have met in this chapter. Your table should include examples from ionic substances, covalent substances (simple and giant), ceramics and glasses.

Metallic bonding

Another way in which atoms obtain a more stable electron structure is found in metals. The electrons in the outer energy level of the atom of a metal move

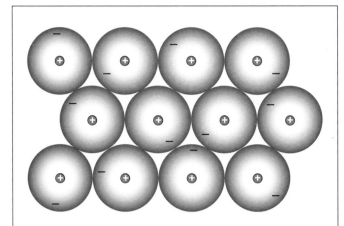

Figure 4.29 Metals consist of positive ions surrounded by a 'sea' of electrons

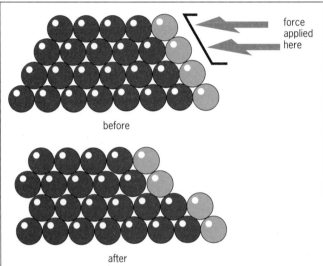

Figure 4.30 The positions of the positive ions in a metal before and after a force has been applied

Figure 4.31 Arrangement of ions in the crystal lattice of a metal

freely throughout the structure (they are **delocalised** forming a mobile 'sea' of electrons (Figure 4.29)). When the metal atoms lose these electrons, they become positive ions. Therefore, metals consist of positive ions embedded in moving electrons. The negatively charged electrons attract all the positive metal ions and bond them together with strong electrostatic forces of attraction as a single unit. This is the **metallic bond**.

Properties of metals

Metals have the following properties.

- They usually have high melting and boiling points due to the strong attraction between the positive metal ions and the mobile 'sea' of electrons.
- They conduct electricity due to the mobile 'sea' of electrons within the metal structure. When a metal is connected in a circuit, the mobile 'sea' of electrons moves towards the positive terminal while at the same time electrons are fed into the other end of the metal from the negative terminal.
- They are malleable and ductile. Unlike those in diamond, the bonds are not rigid but are still strong. If a force is applied to a metal, rows of ions can slide over one another. They reposition themselves and the strong bonds reform as shown in Figure 4.30. Malleable means that metals can be hammered into different shapes. Ductile means that the metals can be pulled out into thin wires.
- They have high densities. This arises because the atoms are very closely packed in a regular manner as can be seen in Figure 4.31. Different metals show different types of packing and in doing so they produce the arrangements of ions shown in Figure 4.32.

QUESTIONS

a Explain the terms:
   (i) malleable **(ii)** ductile.
b Explain why metals are able to conduct heat and electricity.
c Explain why the melting point of magnesium (649 °C) is much higher than the melting point of sodium (97.9 °C).

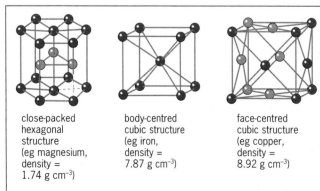

close-packed hexagonal structure (eg magnesium, density = 1.74 g cm⁻³)

body-centred cubic structure (eg iron, density = 7.87 g cm⁻³)

face-centred cubic structure (eg copper, density = 8.92 g cm⁻³)

Figure 4.32 Relating different structures to the density of metal

Checklist

After studying Chapter 4 you should know and understand the following terms.

Ionic (electrovalent) bond   A strong electrostatic force of attraction between oppositely charged ions.

Covalent bond   A chemical bond formed by the sharing of one or more pairs of electrons between two atoms.

Lattice   A regular three-dimensional arrangement of atoms/ions in a crystalline solid.

X-ray diffraction   A technique often used to study crystal structures.

Giant ionic structure   A lattice held together by the electrostatic forces of attraction between ions.

Valency (combining power)   The combining power of an atom or group of atoms. The valency of an ion is equal to its charge.

Simple molecular substance   These substances possess between one and a few hundred atoms per molecule.

Giant molecular substance   A substance containing thousands of atoms per molecule.

Intramolecular bonds   Forces which act within a molecule, for example covalent bonds.

Intermolecular bonds   Attractive forces which act between molecules, for example van der Waals' forces.

Allotropy   The existence of an element in two or more different forms in the same physical state.

Metallic bond   An electrostatic force of attraction between the mobile 'sea' of electrons and the regular array of positive metal ions within the solid metal.

Supercooled liquid   One which has cooled below its freezing point without solidification.

Glass   A supercooled liquid which forms a hard, brittle substance that is usually transparent and resistant to chemical attack.

Ceramics   Materials such as pottery made from inorganic chemicals by high-temperature processing.

QUESTIONS

1 Draw diagrams to show the bonding in each of the following compounds:
  a calcium fluoride (CaF_2)
  b oxygen (O_2)
  c magnesium chloride ($MgCl_2$)
  d tetrachloromethane (CCl_4).

2 Use the information given in Table 4.1 to work out the formula for:
  a silver oxide
  b zinc chloride
  c potassium sulphate
  d calcium nitrate
  e iron(II) nitrate
  f copper(II) carbonate
  g iron(III) hydroxide
  h aluminium fluoride.

3 Atoms of elements **X**, **Y**, and **Z** have 8, 9 and 11 electrons, respectively. Atoms of neon have 10 electrons.
  a Determine the formulae of the compounds formed by the combination of the atoms of the elements:
   (i) **X** and **Z**
   (ii) **Y** and **Z**
   (iii) **X** with themselves.
  b In each of the cases shown in (a)(i)–(iii) above, name the type of chemical bond formed.
  c Give two properties you would expect to be shown by the compounds formed in (a)(ii) and (a)(iii).

4 The diagram shows the arrangement of the outer electrons only in a molecule of ethanoic acid.

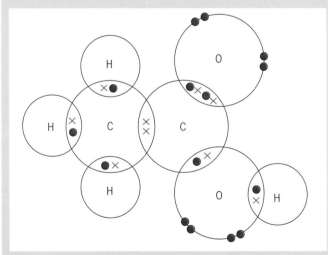

  a Name the different elements found in this compound.
  b What is the total number of atoms present in this molecule?
  c Between which two atoms is there a double covalent bond?
  d How many single covalent bonds does each carbon atom have?
  e Write a paragraph explaining the sorts of properties you would expect this sort of substance to have.

5 With the aid of an example in each case, explain the following words and phrases:
  a lattice
  b valency
  c allotropy
  d glass
  e ceramic material
  f intermolecular bond.

6 The elements sodium and chlorine react together to form the compound sodium chloride, which has a giant ionic lattice structure.
  a What type of structure do the elements (i) sodium and (ii) chlorine have?
  b Draw a diagram to represent how the ions are arranged in the crystal lattice of sodium chloride.
  c Explain how the ions are held together in this crystal lattice.
  d Draw diagrams to show how the electrons are arranged in a sodium ion and a chloride ion. (The atomic numbers of sodium and chlorine are 11 and 17, respectively.)
  e Make a table showing the properties of the three substances sodium, chlorine and sodium chloride. Include in your table:
   (i) the physical state at room temperature
   (ii) solubility in (or reaction with) water
   (iii) colour
   (iv) electrical conductivity.

7 Explain the following:
  a Metals are good conductors of electricity.
  b Metals generally have high densities.
  c The melting point of sodium chloride (NaCl) is 801 °C but that of magnesium oxide (MgO) is 2640 °C.
  d Solid sodium chloride does not conduct electricity but molten sodium chloride conducts very well.
  e Methane is a gas at room temperature.

8 Explain the difference between covalent and ionic bonding. Discuss in what ways the noble gases are important in both of these theories of bonding.

5 CHEMICAL CALCULATIONS

Chemists often need to know how much of a substance has been formed or used up during a chemical reaction (Figure 5.1). This is particularly important in the chemical industry, where the substances being reacted (**the reactants**) and the substances being produced (the **products**) are worth thousands of pounds. Waste costs money!

To solve this problem we need a way of counting atoms, ions or molecules. Atoms, ions and molecules are very tiny particles and it is impossible to measure out a dozen or even a hundred of them. Instead, chemists weigh out a very large number of particles. This number is 6×10^{23} atoms, ions or molecules and is called **Avogadro's constant** after the famous Italian scientist Amedeo Avogadro (1776-1856). An amount of substance containing 6×10^{23} particles is called a **mole** (often abbreviated to mol).

So, a mole of the element aluminium is 6×10^{23} atoms of aluminium and a mole of the element iron is 6×10^{23} atoms of iron (Figure 5.2).

Figure 5.1 Calcium oxide production at Swindon quarry—Tilcon chemists need to know how much calcium oxide is going to be produced

Figure 5.2 (a, top) A mole of aluminium. (**b**, bottom) A mole of iron on a balance

60

Calculating moles

You have already seen in Chapter 3 how we can compare the masses of all the other atoms with the mass of carbon atoms. This is the basis of the **relative atomic mass scale**. Chemists have found by experiment that if you take the relative atomic mass of an element in grams, it always contains 6×10^{23} or one mole of its atoms.

Moles and elements

For example, the relative atomic mass (A_r) of iron is 56, so one mole of iron is 56g. Therefore, 56g of iron contains 6×10^{23} atoms.

The A_r for aluminium is 27. In 27g of aluminium it is found that there are 6×10^{23} atoms. Therefore, 27g of aluminium is one mole of aluminium atoms.

The mass of a substance present in any number of moles can be calculated using the relationship:

$$\begin{array}{ll} \text{mass} & = \text{number} \times \text{mass of 1 mole} \\ \text{(in grams)} & \text{of moles} \quad \text{of the element} \end{array}$$

Example
Calculate the mass of (a) 2 moles and (b) 0.25 moles of iron (A_r: Fe = 56).

a mass of 2 moles of iron
= number of moles × relative atomic mass (A_r)
= 2×56
= 112g

b mass of 0.25 moles of iron
= number of moles × relative atomic mass (A_r)
= 0.25×56
= 14g

If we know the mass of the element then it is possible to calculate the number of moles of that element using:

$$\frac{\text{number of}}{\text{moles}} = \frac{\text{mass of the element}}{\text{mass of 1 mole of that element}}$$

Example
Calculate the number of moles of aluminium present in (a) 108g and (b) 13.5g of the element (A_r: Al = 27).

a number of moles of aluminium
$= \dfrac{\text{mass of aluminium}}{\text{mass of 1 mole of aluminium}}$
$= \dfrac{108}{27}$
$= 4$

b number of moles of aluminium
$= \dfrac{\text{mass of aluminium}}{\text{mass of 1 mole of aluminium}}$
$= \dfrac{13.5}{27}$
$= 0.5$

Moles and compounds

The idea of the mole has been used so far only with elements and atoms. However, it can also be used with compounds (Figure 5.3).

We cannot discuss the atomic mass of a molecule or of a compound because more than one type of atom is involved. Instead, we have to discuss the **relative formula mass** (RFM). This is the sum of the relative atomic masses of all those elements shown in the formula of the substance.

What is the mass of 1 mole of water (H_2O) molecules? (A_r: H = 1; O = 16)

From the formula of water, H_2O, you will see that 1 mole of water molecules contains 2 moles of hydrogen (H) atoms and 1 mole of oxygen (O) atoms. The mass of 1 mole of water molecules is therefore:

$$(2 \times 1) + (1 \times 16) = 18\text{g}$$

The mass of 1 mole of a compound is called its molar mass. If you write the molar mass of a compound without any units then it is the relative formula mass, often called the **relative molecular mass** (M_r). So the relative formula mass of water is 18.

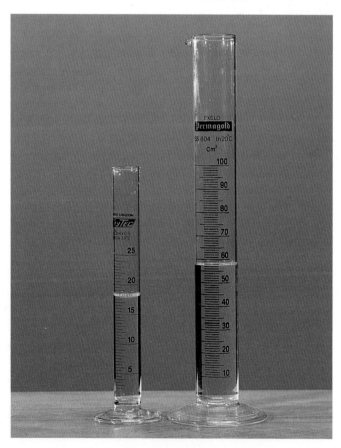

Figure 5.3 One mole of water (left) and one mole of ethanol in separate measuring cylinders

Examples

1   What is (a) the mass of 1 mole and (b) the relative formula mass (RFM) of ethanol, C_2H_5OH? (A_r: C=12; H=1; O=16)

    a   One mole of C_2H_5OH contains 2 moles of carbon atoms, 6 moles of hydrogen atoms and 1 mole of oxygen atoms. Therefore,

        mass of 1 mole of ethanol
        $= (2 \times 12) + (6 \times 1) + (1 \times 16)$
        $= 46g$

    b   The RFM of ethanol is 46.

2   What is (a) the mass of 1 mole and (b) the RFM of nitrogen gas, N_2? (A_r: N=14)

    a   Nitrogen is a diatomic gas. Each nitrogen molecule contains two atoms of nitrogen. Therefore:

        mass of 1 mole of N_2
        $= 2 \times 14$
        $= 28g$

    b   The RFM of N_2 is 28.

The mass of a compound found in any number of moles can be calculated using the relationship:

$$\text{mass of compound} = \text{number of moles of the compound} \times \text{mass of 1 mole of the compound}$$

Example

Calculate the mass of (a) 3 moles and (b) 0.2 moles of carbon dioxide gas, CO_2. (A_r: C=12; O=16)

    a   One mole of CO_2 contains 1 mole of carbon atoms and 2 moles of oxygen atoms. Therefore:

        mass of 1 mole of CO_2
        $= (1 \times 12) + (2 \times 16)$
        $= 44g$

        mass of 3 moles of CO_2
        $=$ number of moles $\times$ mass of 1 mole of CO_2
        $= 3 \times 44$
        $= 132g$

    b   mass of 0.2 moles of CO_2
        $=$ number of moles $\times$ mass of 1 mole of CO_2
        $= 0.2 \times 44$
        $= 8.8g$

If we know the mass of the compound then it is possible to calculate the number of moles of the compound using the relationship:

$$\text{number of moles of compound} = \frac{\text{mass of compound}}{\text{mass of 1 mole of the compound}}$$

Example

Calculate the number of moles of magnesium oxide, MgO, in (a) 80g and (b) 10g of the compound. (A_r: Mg=24; O=16)

    a   One mole of MgO contains 1 mole of magnesium atoms and 1 mole of oxygen atoms. Therefore:

        mass of 1 mole of MgO
        $= (1 \times 24) + (1 \times 16)$
        $= 40g$

        number of moles of MgO in 80g
        $= \dfrac{\text{mass of MgO}}{\text{mass of 1 mole of MgO}}$
        $= \dfrac{80}{40}$
        $= 2$

    b   number of moles of MgO in 10g
        $= \dfrac{\text{mass of MgO}}{\text{mass of 1 mole of MgO}}$
        $= \dfrac{10}{40}$
        $= 0.25$

Moles and gases

Many substances exist as gases. If we want to find the number of moles of a gas we can do this by measuring the volume rather than the mass.

Chemists have shown by experiment that:

One mole of any gas occupies a volume of approximately 24 dm³ (24 l) at room temperature and pressure (rtp).

Therefore, it is relatively easy to convert volumes of gases into moles and moles of gases into volumes using the following relationship:

$$\text{number of moles of a gas} = \frac{\text{volume of the gas (in dm}^3 \text{ at rtp)}}{24\,\text{dm}^3}$$

or

$$\text{volume of a gas (in dm}^3 \text{ at rtp)} = \text{number of moles of gas} \times 24\,\text{dm}^3$$

Examples

1   Calculate the number of moles of ammonia gas, NH_3, in a volume of 72 dm³ of the gas measured at rtp.

        number of moles of ammonia
        $= \dfrac{\text{volume of ammonia in dm}^3}{24\,\text{dm}^3}$
        $= \dfrac{72}{24}$
        $= 3$

2   Calculate the volume of carbon dioxide gas, CO_2, occupied by (a) 5 moles and (b) 0.5 moles of the gas measured at rtp.

a volume of CO_2
= number of moles of $CO_2 \times 24\,dm^3$
= 5×24
= $120\,dm^3$

b volume of CO_2
= number of moles of $CO_2 \times 24\,dm^3$
= 0.5×24
= $12\,dm^3$

The volume occupied by one mole of any gas must contain 6×10^{23} molecules. Therefore, it follows that equal volumes of all gases measured at the same temperature and pressure must contain the same number of molecules. This idea was first put forward by Avogadro and is called **Avogadro's Law**.

Moles and solutions

Chemists often need to know the concentration of a solution. Sometimes it is measured in grams per cubic decimetre ($g\,dm^{-3}$) but more often concentration is measured in moles per cubic decimetre ($mol\,dm^{-3}$). When 1 mole of a substance is dissolved in water and the solution is made up to $1\,dm^3$ ($1000\,cm^3$), a **1 molar (1M)** solution is produced. Chemists do not always need to make up such large volumes of solution. A simple method of calculating the concentration is by using the relationship:

$$\text{concentration} = \frac{\text{number of moles}}{\text{volume (in } dm^3)}$$

Example
Calculate the concentration (in $mol\,dm^{-3}$) of a solution of sodium hydroxide, NaOH, which was made by dissolving 10 g of solid sodium hydroxide in water and making up to $250\,cm^3$. (A_r: Na=23; O=16; H=1)

1 mole of NaOH contains 1 mole of sodium, 1 mole of oxygen and 1 mole of hydrogen. Therefore:

mass of 1 mole of NaOH = $(1 \times 23) + (1 \times 16) + (1 \times 1)$
= 40 g

number of moles of NaOH in 10 g
= $\dfrac{\text{mass of NaOH}}{\text{mass of 1 mole of NaOH}}$
= $\dfrac{10}{40}$
= 0.25

$(250\,cm^3 = \dfrac{250\,dm^3}{1000} = 0.25\,dm^3)$

concentration of the NaOH solution
= $\dfrac{\text{number of moles of NaOH}}{\text{volume of solution (} dm^3)}$
= $\dfrac{0.25}{0.25}$
= $1\,mol\,dm^{-3}$ (or 1M)

Sometimes chemists need to know the mass of a substance that has to be dissolved to prepare a known volume of solution at a given concentration. A simple method of calculating the number of moles and hence the mass of substance needed is by using the relationship:

$$\begin{array}{ccc} \text{number of} & = & \text{concentration} \times \text{volume of solution} \\ \text{moles} & & \text{(in } mol\,dm^{-3}) \quad \text{(in } dm^3) \end{array}$$

Example
Calculate the mass of potassium hydroxide, KOH, that needs to be used to prepare $500\,cm^3$ of a $2\,mol\,dm^{-3}$ (2M) solution in water. (A_r: K=39; O=16; H=1)

number of moles of KOH
= concentration of solution $\times$ volume of solution
  $(mol\,dm^{-3})$ $\qquad$ (dm^3)
= $2 \times \dfrac{500}{1000}$
= 1

1 mole of KOH contains 1 mole of potassium, 1 mole of oxygen and 1 mole of hydrogen. Therefore:

mass of 1 mole of KOH = $(1 \times 39) + (1 \times 16) + (1 \times 1)$
= 56 g

Therefore,

mass of KOH in 1 mole
= number of moles $\times$ mass of 1 mole
= 1×56
= 56 g

QUESTIONS

Use the values of A_r which follow to answer the questions below: C=12; Ne=20; Mg=24; O=16; S=32; Na=23; H=1; Fe=56; Cu=63.5; N=14; Zn=65; K=39.
One mole of any gas at rtp occupies $24\,dm^3$.

a Calculate the number of moles in:
   (i) 2 g of neon atoms
   (ii) 4 g of magnesium atoms
   (iii) 24 g of carbon atoms.
b Calculate the mass of:
   (i) 0.1 mole of oxygen molecules
   (ii) 5 moles of sulphur atoms
   (iii) 0.25 moles of sodium atoms.
c Calculate the number of moles in:
   (i) 9.8 g of sulphuric acid (H_2SO_4)
   (ii) 40 g of sodium hydroxide (NaOH)
   (iii) 720 g of iron(II) oxide (FeO).
d Calculate the mass of:
   (i) 2 moles of zinc oxide (ZnO)
   (ii) 0.25 mole of hydrogen sulphide (H_2S)
   (iii) 0.35 mole of copper(II) sulphate ($CuSO_4$).
e Calculate the number of moles at rtp in:
   (i) $2\,dm^3$ of carbon dioxide (CO_2)
   (ii) $240\,dm^3$ of sulphur dioxide (SO_2)
   (iii) $20\,cm^3$ of carbon monoxide (CO).

QUESTIONS (continued)

f Calculate the volume of:
  (i) 0.3 moles of hydrogen chloride (HCl)
  (ii) 4.4 g of carbon dioxide
  (iii) 34 g of ammonia (NH_3)

g Calculate the concentration of solutions containing:
  (i) 0.2 moles of sodium hydroxide dissolved in water and made up to 100 cm³
  (ii) 9.8 g of sulphuric acid dissolved in water and made up to 500 cm³.

h Calculate the mass of:
  (i) copper(II) sulphate ($CuSO_4$) which needs to be used to prepare 500 cm³ of a 0.1 M solution
  (ii) potassium nitrate (KNO_3) which needs to be used to prepare 200 cm³ of a 2 M solution.

Calculating formulae

If we have 1 mole of a compound, then the formula shows the number of moles of each element in that compound. For example, the formula for lead(II) bromide is $PbBr_2$. This means that 1 mole of lead(II) bromide contains 1 mole of lead ions and 2 moles of bromide ions. If we do not know the formula of a compound, we can find the masses of the elements present experimentally and these masses can be used to work out the formula of that compound.

Finding the formula

Magnesium oxide

When magnesium ribbon is heated strongly, it burns very brightly to form the white powder called magnesium oxide.

$$\text{magnesium} + \text{oxygen} \rightarrow \text{magnesium oxide}$$

The data shown in Table 5.1 were obtained from an experiment using the apparatus shown in Figure 5.4 to find the formula for this white powder, magnesium oxide. From these data we can calculate the number of moles of each of the reacting elements. (A_r: Mg = 24; O = 16).

	Mg	**O**
Masses reacting (g)	0.24	0.16
Number of moles	$\frac{0.24}{24}$	$\frac{0.16}{16}$
	= 0.01	= 0.01
Ratio of moles	1	1
Formula		MgO

This formula is called the **empirical formula** of the compound. It shows the simplest ratio of the atoms present.

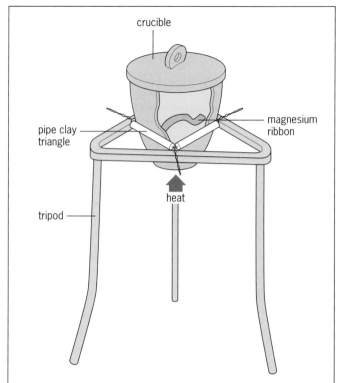

Figure 5.4 Apparatus used to determine magnesium oxide's formula

Table 5.1 Data from experiment shown in Figure 5.4

Mass of crucible	14.63 g
Mass of crucible and magnesium	14.87 g
Mass of crucible and magnesium oxide	15.03 g
Mass of magnesium used	0.24 g
Mass of oxygen which has reacted with the magnesium	0.16 g

Unknown compound 1

In another experiment an unknown organic compound was found to contain 0.12 g of carbon and 0.02 g of hydrogen. Calculate the empirical formula of the compound. (A_r: C = 12; H = 1)

	C	**H**
Masses (g)	0.12	0.02
Number of moles	$\frac{0.12}{12}$	$\frac{0.02}{1}$
	= 0.01	= 0.02
Ratio of moles	1	2
Empirical formula		CH_2

From our knowledge of bonding (Chapter 4, p. 49) we know that a molecule of this formula cannot exist. However, molecules with the following formulae do exist: C_2H_4, C_3H_6, C_4H_8, C_5H_{10}. All of these formulae show the same ratio of carbon atoms to hydrogen atoms, CH_2, as our unknown. To find out which of these formulae is the actual formula for the unknown organic compound, we need to know the mass of one mole of the compound.

Using a mass spectrometer, the relative molecular mass (M_r) of this organic compound was found to be 56. We need to find out the number of empirical formulae units present:

M_r of the empirical formula unit
$= (1 \times 12) + (2 \times 1)$
$= 14$

Number of empirical formula units present
$= \dfrac{M_r \text{ of compound}}{M_r \text{ of empirical formula unit}}$
$= \dfrac{56}{14}$
$= 4$

Therefore, the actual formula of the unknown organic compound is $4 \times CH_2 = C_4H_8$.

This substance is called butene. C_4H_8 is the **molecular formula** for this substance and shows the **actual** numbers of atoms of each element present in one molecule of the substance.

Sometimes the composition of a compound is given as a percentage by mass of the elements present. In cases such as this the procedure shown in the following example is followed.

Unknown compound 2
Calculate the empirical formula of an organic compound containing 92.3% carbon and 7.7% hydrogen by mass. The M_r of the organic compound is 78. What is its molecular formula? (A_r: C=12; H=1)

	C	H
% by mass	92.3	7.7
in 100g	92.3g	7.7g
moles	$\dfrac{92.3}{12}$	$\dfrac{7.7}{1}$
	$= 7.7$	$= 7.7$
Ratios of moles	1	1
Empirical formula	CH	

M_r of the empirical formula unit CH
$= 12 + 1$
$= 13$

Number of empirical formula units present
$= \dfrac{M_r \text{ of compound}}{M_r \text{ of empirical formula unit}}$
$= \dfrac{78}{13}$
$= 6$

The molecular formula of the organic compound is $6 \times CH = C_6H_6$. This is a substance called benzene.

QUESTIONS

Use the following values of A_r to answer the questions below: H=1; C=12; O=16; Ca=40.
 a Determine the empirical formula of an oxide of calcium formed when 0.4g of calcium reacts with 0.16g of oxygen.
 b Determine the empirical formula of an organic hydrocarbon compound which contains 80% by mass of carbon and 20% by mass of hydrogen. If the M_r of the compound is 30, what is its molecular formula?

Moles and chemical equations

When we write a balanced chemical equation we are indicating the numbers of moles of reactants and products involved in the chemical reaction. Consider the reaction between magnesium and oxygen.

$$\text{magnesium} + \text{oxygen} \rightarrow \text{magnesium oxide}$$
$$2Mg(s) + O_2(g) \rightarrow 2MgO(s)$$

This shows that 2 moles of magnesium react with 1 mole of oxygen to give 2 moles of magnesium oxide.

Using the ideas of moles and masses we can use this information to calculate the quantities of the different chemicals involved.

$2Mg(s)$	$+$	$O_2(g)$	$\rightarrow$	$2MgO(s)$
2 moles		1 mole		2 moles
2×24		$1 \times (16 \times 2)$		$2 \times (24+16)$
$= 48g$		$= 32g$		$= 80g$

You will notice that the total mass of reactants is equal to the total mass of product. This is true for any chemical reaction and it is known as the **Law of conservation of mass** for a chemical reaction. This law was understood by the Greeks but was first clearly formulated by Antoine Lavoisier in 1774. Chemists can use this idea to calculate masses of products formed and reactants used in chemical processes before they are carried out.

Example using a solid
Lime (calcium oxide, CaO) is used in the manufacture of mortar. It is manufactured in large quantities by Tilcon at the Swindon quarry in North Yorkshire (see Figure 5.1) by heating limestone (calcium carbonate, $CaCO_3$). The equation for the process is:

$CaCO_3(s)$	$\rightarrow$	$CaO(s)$	$+$	$CO_2(g)$
1 mole		1 mole		1 mole
$(40+12+(3\times16))$		$40+16$		$12+(2\times16)$
$= 100g$		$= 56g$		$= 44g$

Calculate the amount of lime produced when 10 tonnes of limestone are heated. (A_r: Ca = 40; C = 12; O = 16)

$$1 \text{ tonne (t)} = 1000\,kg$$
$$1\,kg = 1000\,g$$

From this relationship between grams and tonnes we can replace the masses in grams by masses in tonnes.

$$CaCO_3(s) \rightarrow CaO(s) + CO_2(g)$$

	100t	56t	44t
Hence 10t	5.6t	4.4t

The equation now shows that 100t of limestone will produce 56t of lime. Therefore, 10t of limestone will produce 5.6t of lime.

Example using a gas

Many chemical processes involve gases. The volume of a gas is measured more easily than its mass. This example shows how chemists work out the volumes of gaseous reactants and products needed using the idea of moles.

Some rockets use hydrogen gas as a fuel. When hydrogen burns in oxygen it forms steam. Calculate the volumes of (a) $O_2(g)$ used and (b) water, $H_2O(g)$, produced if 960 dm³ of hydrogen gas, $H_2(g)$, were burned in oxygen. (A_r: H = 1; O = 16) Assume 1 mole of any gas occupies a volume of 24 dm³.

$2H_2(g)$	+	$O_2(g)$	$\rightarrow$	$2H_2O(g)$
2 moles		1 mole		2 moles
2×24		1×24		2×24
= 48 dm³		= 24 dm³		= 48 dm³

Therefore

(×2)	96 dm³	48 dm³	96 dm³
(×10)	960 dm³	480 dm³	960 dm³

When 960 dm³ of hydrogen are burned in oxygen:

a 480 dm³ of oxygen are required and
b 960 dm³ of $H_2O(g)$ are produced.

Examples using a solution

Chemists usually carry out chemical reactions using solutions. If they know the concentration of the solution(s) they are using they can find out the quantities reacting.

1 Calculate the volume of 1 M H_2SO_4 required to react completely with 6 g of magnesium. (A_r: Mg = 24)

number of moles of magnesium

$$= \frac{\text{mass of magnesium}}{\text{mass of 1 mole of magnesium}}$$

$$= \frac{6}{24}$$

$$= 0.25$$

$Mg(s)$	+	$H_2SO_4(aq)$	$\rightarrow$	$MgSO_4(aq)$	+	$H_2(g)$
1 mole		1 mole		1 mole		1 mole
0.25 mol		0.25 mol		0.25 mol		0.25 mol

We can see that 0.25 mol of $H_2SO_4(aq)$ is required. Using:

volume of $H_2SO_4(aq)$ (dm³)

$$= \frac{\text{moles of } H_2SO_4}{\text{concentration of } H_2SO_4 \text{ (mol dm}^{-3})}$$

$$= \frac{0.25}{1}$$

$$= 0.25\,dm^3 \text{ or } 250\,cm^3$$

2 What is the concentration of sodium hydroxide solution used in the following neutralisation reaction: 40 cm³ of 0.2 M hydrochloric acid just neutralised 20 cm³ of sodium hydroxide solution.

number of moles of HCl used

$$= \text{concentration (mol dm}^{-3}) \times \text{volume (dm}^3)$$

$$= 0.2 \times 0.04$$

$$= 0.008$$

$HCl(aq)$	+	$NaOH(aq)$	$\rightarrow$	$NaCl(aq)$	+	$H_2O(l)$
1 mole		1 mole		1 mole ·		1 mole
0.08 mol		0.08 mol		0.08 mol		0.08 mol

You will see that 0.08 moles of NaOH would have been present. The concentration of the NaOH(aq) will be given by:

concentration of NaOH (mol dm⁻³)

$$= \frac{\text{number of moles of NaOH}}{\text{volume of NaOH (dm}^3)}$$

(volume of NaOH in dm³ $= \frac{20}{1000} = 0.02$)

$$= \frac{0.008}{0.02}$$

$$= 0.4\,mol\,dm^{-3} \text{ or } 0.4\,M$$

QUESTIONS

Use the following A_r values to answer the questions below: O = 16; S = 32; Cu = 63.5; Mg = 24; K = 39.

a Calculate the mass of sulphur dioxide produced by burning 16 g of sulphur in an excess of oxygen in the Contact process.

b Calculate the mass of sulphur which, when burned in excess oxygen, produces 640 g of sulphur dioxide in the Contact process.

c Calculate the mass of copper required to produce 159 g of copper(II) oxide when heated in excess oxygen.

d In the rocket mentioned previously in which hydrogen is used as a fuel, calculate the volume of hydrogen used to produce 24 dm³ of water ($H_2O(g)$).

e Calculate the volume of 2 M sulphuric acid required to react with 24 g of magnesium.

f What is the concentration of potassium hydroxide solution used in the following neutralisation reaction? 20 cm³ of 0.2 M hydrochloric acid just neutralised 15 cm³ of potassium hydroxide solution.

Checklist

After studying Chapter 5 you should know and understand the following terms.

Mole The amount of substance which contains 6×10^{23} atoms, ions or molecules. This number is called Avogadro's constant.

Atoms — 1 mole of atoms has a mass equal to the relative atomic mass (A_r) in grams.

Molecules — 1 mole of molecules has a mass equal to its relative molecular mass (M_r) in grams.

Calculating moles
Elements:

mass of element = number of × mass of 1 mole
(in grams)   moles   of the element

$$\text{number of moles} = \frac{\text{mass of the element}}{\text{mass of 1 mole of that element}}$$

Compounds:

mass of compound = number of × mass of 1 mole
(in grams)   moles   of compound

$$\text{number of moles} = \frac{\text{mass of compound}}{\text{mass of 1 mole of compound}}$$

Gases: 1 mole of any gas occupies $24\,dm^3$ (litres) at room temperature and pressure (rtp).

$$\text{number of moles of gas} = \frac{\text{volume of the gas (in } dm^3 \text{ at rtp)}}{24\,dm^3}$$

Solutions:

$$\text{concentration of a solution (in } mol\,dm^{-3}) = \frac{\text{number of moles of solute}}{\text{volume (in } dm^3)}$$

number of = concentration × volume of solution
moles   (in $mol\,dm^{-3}$)   (in dm^3)

Empirical formula A formula showing the simplest ratio of atoms present.

Molecular formula A formula showing the actual number of atoms of each element present in one molecule.

QUESTIONS

Use the data in the table below to answer the questions which follow:

Element	A_r
H	1
C	12
N	14
O	16
Na	23
Mg	24
Si	28
S	32
Cl	35.5
Fe	56

1 Calculate the mass of:
  a 1 mole of:
    (i) nitrogen molecules
    (ii) magnesium oxide.
  b 0.5 moles of:
    (i) sodium hydroxide
    (ii) water.

2 Calculate the volume occupied, at rtp, by the following gases. (One mole of any gas occupies a volume of $24\,dm^3$ at rtp.)
  a 12.5 moles of oxygen gas
  b 0.15 moles of carbon dioxide gas.

3 Calculate the number of moles of gas present in the following:
  a $24\,cm^3$ of sulphur dioxide
  b $120\,dm^3$ of hydrogen chloride.

4 Using the following experimental information, determine the empirical formula of an oxide of silicon.

Mass of crucible	15.20 g
Mass of crucible + silicon	15.48 g
Mass of crucible + oxide of silicon	15.80 g

5 a Calculate the empirical formula of an organic liquid containing 52.17% of carbon, 13% of hydrogen with the rest being oxygen.
  b The M_r of the liquid is 92. What is its molecular formula?

6 Iron is extracted from its ore, haematite, in the blast furnace. The main extraction reaction is:

$$Fe_2O_{3(s)} + 3CO_{(g)} \rightarrow 2Fe_{(s)} + 3CO_{2(g)}$$

  a Name the reducing agent in this process.
  b Name the oxide of iron shown in the equation.
  c Explain why this is a **redox** reaction.
  d Calculate the mass of iron which will be produced from 320 tonnes of haematite.

7 Consider the following information about the newly discovered element, stratium, whose symbol is St. 'Stratium is a solid at room temperature. It is easily cut by a penknife to reveal a shiny surface which tarnishes rapidly. It reacts violently with water, liberating a flammable gas and forms a solution with a pH of 12. When stratium reacts with chlorine, it forms a white crystalline solid containing 29.5% chlorine.'
(A_r: St = 85)
  a Calculate the empirical formula of stratium chloride.
  b To which group of the periodic table should stratium be assigned?
  c Write a word and balanced chemical equation for the reaction between stratium and chlorine.

QUESTIONS (continued)

d What other information in the description supports the assignment of group you have given to stratium?

e What type of bonding is present in stratium chloride?

f Write a word and balanced chemical equation for the reaction between stratium and water.

g Write the formulae for:
   (i) stratium oxide
   (ii) stratium nitrate
   (iii) stratium carbonate.

Look at the periodic table (p. 35) to find out the real name of stratium.

8 0.024 g of magnesium was reacted with excess dilute hydrochloric acid at room temperature and pressure. The hydrogen gas given off was collected.

a Write a word and balanced symbol equation for the reaction taking place.

b Draw a diagram of an apparatus which could be used to carry out this experiment.

c How many moles of magnesium were used in the experiment?

d Using the equation you have written in your answer to (a), calculate the number of moles of hydrogen and hence the volume of this gas produced in the reaction.

e Calculate the volume of 0.1 M hydrochloric acid which would be needed to react exactly with 0.024 g of magnesium.

6 ELECTROLYSIS

What do all the items in the photographs shown in Figure 6.1 have in common? They all involve electricity through a process known as **electrolysis**. Electrolysis is the process of splitting up (decomposing) substances by passing an electric current through them. The substance which is decomposed is called the **electrolyte**. An electrolyte is a substance that conducts electricity when in the molten state or in solution. The electricity is carried through the electrolyte by **ions**. The electric current enters and leaves the electrolyte through **electrodes** (Figure 6.2), which are usually made of unreactive metals such as platinum or of non-metal carbon. The names given to the two electrodes are **cathode**, the negative electrode which attracts **cations**, and **anode**, the positive electrode which attracts **anions**.

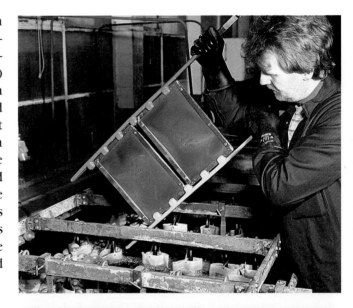

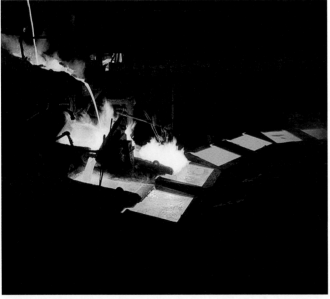

Figure 6.1 (**a**, top right) Rack of copper-plated sheets being removed from an electroplating bath. (**b**, bottom left) Molten aluminium produced from an electrolytic process. (**c**, bottom right) These teapots are being electroplated with silver

69

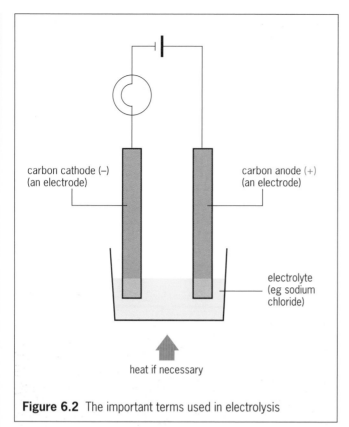

Figure 6.2 The important terms used in electrolysis

Electrolysis is very important in industry. To help you to understand what is happening in the process shown in the photographs above, we will first consider the electrolysis of aluminium oxide.

Electrolysis of aluminium oxide

Aluminium is the most abundant metallic element in the Earth's crust. It was discovered in 1825 by Hans Christian Oersted in Copenhagen, Denmark, but was first isolated by Friedrich Wöhler in 1827. It makes up

8% of the crust and is found in the minerals bauxite, cryolite and mica, as well as in clay (Figure 6.3).

In the nineteenth century Napoleon III owned a very precious dinner service. It was said to be made of a metal more precious than gold. The metal was aluminium. The reason it was precious was that it was very rarely found as the pure metal. Aluminium is a reactive metal and as such was very difficult to extract from its ore. Reactive metals hold on tightly to the element(s) they have combined with and many are extracted from their ores by electrolysis.

Today we use aluminium in very large quantities. The commercial extraction of aluminium has been made possible by two scientists, working independently of each other, who discovered a method using electrolysis. The two scientists were Charles Martin Hall (USA), who discovered the process in 1886, and the French chemist Paul Héroult, who discovered the process independently in the same year. The process they developed, often called the Hall–Héroult process, involves the electrolysis of aluminium oxide (alumina). The process involves the following.

- Bauxite, an impure form of aluminium oxide, is first treated with sodium hydroxide to obtain pure aluminium oxide, removing impurities such as iron(III) oxide and sand.
- The purified aluminium oxide is then dissolved in molten cryolite (Na_3AlF_6). Cryolite, a mineral found naturally in Greenland, is used to reduce the working temperature of the Hall–Héroult cell from 2017 °C (the melting point of pure aluminium oxide) to 800–1000 °C. Therefore, the cryolite provides a considerable saving in the energy requirements of the process.
- The molten mixture is then electrolysed in a cell similar to that shown in Figure 6.4.

Figure 6.3 Bauxite mining in Australia

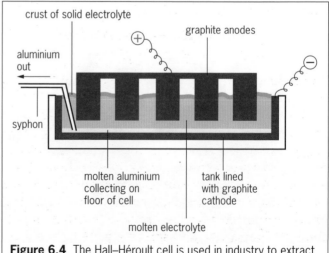

Figure 6.4 The Hall–Héroult cell is used in industry to extract aluminium by electrolysis

The anodes of this process are blocks of graphite which are lowered into the molten mixture from above. The cathode is the graphite lining of the steel vessel containing the cell.

Aluminium oxide is an ionic compound. When it is melted the ions become mobile, as the strong electrostatic forces of attraction between them are broken by the input of heat energy. During electrolysis the negatively charged oxide ions are attracted to the anode (the positive electrode), where they lose electrons to form oxygen gas.

$$\text{oxide ions} \rightarrow \text{oxygen molecules} + \text{electrons}$$
$$2O^{2-}(l) \rightarrow O_2(g) + 4e^-$$

The positive aluminium ions are attracted to the cathode (the negative electrode). They gain electrons to form molten aluminium metal.

$$\text{aluminium ions} + \text{electrons} \rightarrow \text{aluminium metal}$$
$$Al^{3+}(l) + 3e^- \rightarrow Al(l)$$

The gain of electrons which takes place at the cathode is called **reduction**. The loss of electrons which takes place at the anode is called **oxidation**. This is a further way of considering oxidation and reduction (Chapter 2, p. 15). A handy way of remembering it is **oil rig** (**o**xidation **i**s **l**oss, **r**eduction **i**s **g**ain of electrons).

The overall reaction which takes place in the cell is:

$$\text{aluminium oxide} \xrightarrow{\text{electrolysis}} \text{aluminium} + \text{oxygen}$$
$$2Al_2O_3(l) \longrightarrow 4Al(l) + 3O_2(g)$$

The molten aluminium collects at the bottom of the cell and it is syphoned out at regular intervals. No problems arise with other metals being deposited, since the cryolite is 'unaffected' by the flow of electricity. Problems do arise, however, with the graphite anodes. At the working temperature of the cell, the oxygen liberated reacts with the graphite anodes, producing carbon dioxide.

$$\text{carbon} + \text{oxygen} \rightarrow \text{carbon dioxide}$$
$$C(s) + O_2(g) \rightarrow CO_2(g)$$

The anodes burn away and have to be replaced on a regular basis.

This is a continuous process in which vast amounts of electricity are used. In order to make the process an economic one, a cheap form of electricity is required. Hydroelectric power (HEP) is usually used for this process. The plant shown in Figure 6.5 uses an HEP scheme to provide some of the electrical energy required for this process. Further details about HEP are given in Chapter 13, p. 163.

Using cheap electrical energy has allowed aluminium to be produced in such large quantities that it is

Figure 6.5 The aluminium smelting plant at Fort William, Scotland

Figure 6.6 Aluminium is used in the manufacture of aeroplane bodies

Figure 6.7 A red mud lake in Brazil

the second most widely used metal after iron. It is used in the manufacture of electrical cables, cars, bikes and cooking foil as well as in alloys (Chapter 9, p. 124) such as duralumin, which is used in the manufacture of aeroplane bodies (Figure 6.6).

Environmental problems associated with the location of aluminium plants are concerned with:

- the effects of the extracted impurities, which form a red mud (Figure 6.7)

- the fine cryolite dust, which is emitted through very tall chimneys so as not to affect the surrounding area (Figure 6.5)
- the claimed link between environmental aluminium and a degenerative brain disease called Alzheimer's disease — it is thought that aluminium is a major influence on the early onset of this disease.

QUESTION

a Produce a flow chart to summarise the processes involved in the extraction of aluminium metal.

Extraction of sodium

Sodium, a very reactive metal, was first discovered by Sir Humphry Davy in 1807. It is now extracted by electrolysing molten sodium chloride in the Down's cell. Sodium chloride may be obtained from the evaporation of sea water or mined as rock salt (Figure 6.8). The largest salt mines in England are those in Cheshire. The rock salt is mined from large underground caverns and brought to the surface by conveyor belts or trucks.

The first economic extraction of sodium was developed by Down in 1921 in North America using hydroelectric power generated by the Niagara Falls. This ensured the availability of cheap electricity.

The process involves the following.

- Sodium chloride, obtained from rock salt or sea water, is mixed with calcium chloride and melted. The calcium chloride is added to the sodium chloride electrolyte to reduce the working temperature of the cell from 801 °C (the melting point of sodium chloride) to 600 °C (the melting point of the mixture). This saves electrical energy and, therefore, makes the process more economical.
- The mixture is then electrolysed in a cell similar to the one in Figure 6.9.

The chloride ions are attracted to the anode, where they lose electrons and form chlorine gas.

$$\text{chloride ions} \rightarrow \text{chlorine molecules} + \text{electrons}$$
$$2Cl^-(l) \quad \rightarrow \quad Cl_2(g) \quad + \quad 2e^-$$

The positive sodium ions are attracted to the cathode. They gain electrons to form molten sodium metal.

$$\text{sodium ions} + \text{electrons} \rightarrow \text{sodium metal}$$
$$Na^+(l) \quad + \quad e^- \quad \rightarrow \quad Na(l)$$

The overall reaction which takes place in the cell is:

$$\text{sodium chloride} \xrightarrow{\text{electrolysis}} \text{sodium} + \text{chlorine}$$
$$2NaCl(l) \quad \longrightarrow \quad 2Na(l) + \quad Cl_2(g)$$

The cathode is a circle of steel around the graphite anode. At 600 °C sodium and chlorine would react violently together to re-form sodium chloride. To prevent this from happening, the Down's cell contains a steel gauze around the graphite anode to keep them apart. The molten sodium floats on the electrolyte and is run off for storage.

A problem arises, however, in that calcium ions are also attracted to the cathode, where they form calcium metal. Therefore, the sodium which is run off contains a significant proportion of calcium. Fortunately, the calcium crystallises out when the

Figure 6.8 Rock salt mining in Argentina

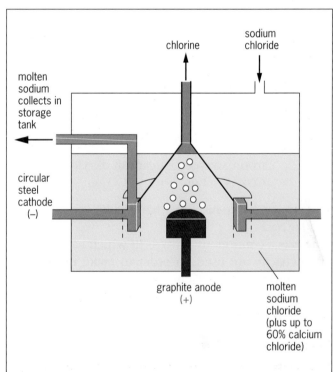

Figure 6.9 The Down's cell used for the electrolysis of sodium chloride

Figure 6.10 Molten sodium — the result of electrolysis of sodium chloride

mixture cools and relatively pure sodium metal remains (Figure 6.10).

Sodium is used as a liquid coolant in nuclear power stations as well as in street lighting and in the production of metals such as titanium. Chlorine gas (the co-product) is sold to make the process more economical (see p. 74 for the uses of chlorine).

(see p. 74 for the uses of chlorine)

QUESTIONS

a Find uses, other than those given in the text, for sodium.
b A student carries out, in a fume cupboard, the electrolysis of molten lead(II) bromide.
  (i) Draw a diagram to show a suitable apparatus the student could use to carry out this experiment.
  (ii) Write anode and cathode reactions to represent the processes taking place during the electrolysis.
  (iii) Why does this experiment need to be carried out in a fume cupboard?

Electrolysis of aqueous solutions

Other industrial processes involve the electrolysis of aqueous solutions. To help you to understand what is happening in these processes, we will first consider the electrolysis of water.

Electrolysis of water

Pure water is a very poor conductor of electricity because there are so few ions in it. However, it can be made to decompose if an electric current is

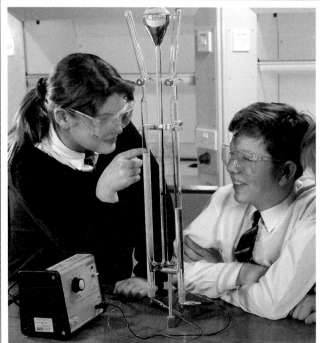

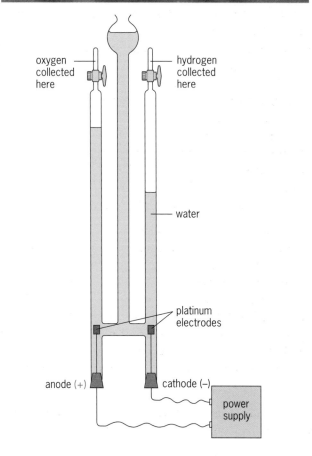

Figure 6.11 A Hofmann voltameter is used to electrolyse water

passed through it in a Hofmann voltameter, as shown in Figure 6.11.

To enable water to conduct electricity better, some dilute sulphuric acid (or sodium hydroxide solution) is added. When an electric current flows through this solution gases can be seen to be produced at the two electrodes and they are collected in the side arms of the apparatus. After about 20 minutes, roughly twice as much gas is produced at the cathode as at the anode.

The gas collected at the cathode burns with a squeaky pop, showing it to be hydrogen gas. For hydrogen to be collected in this way, the **positively** charged hydrogen ions must have been attracted to the cathode.

$$\text{hydrogen ions} + \text{electrons} \rightarrow \text{hydrogen molecules}$$
$$4H^+(aq) \quad + \quad 4e^- \quad \rightarrow \quad 2H_2(g)$$

If during this process the water molecules lose $H^+(aq)$, then the remaining portion must be hydroxide ions, $OH^-(aq)$. These ions are attracted to the anode. The gas collected at the anode relights a glowing splint, showing it to be oxygen.

This gas is produced in the following way.

$$\begin{array}{cccc}\text{hydroxide} \rightarrow & \text{water} & + \text{ oxygen } & + \text{electrons} \\ \text{ions} & \text{molecules} & \text{molecules} & \\ 4OH^-(aq) \rightarrow & 2H_2O(l) & + \quad O_2(g) & + \quad 4e^-\end{array}$$

This experiment was first carried out by Sir Humphry Davy, who confirmed by this experiment that the formula for water was H_2O.

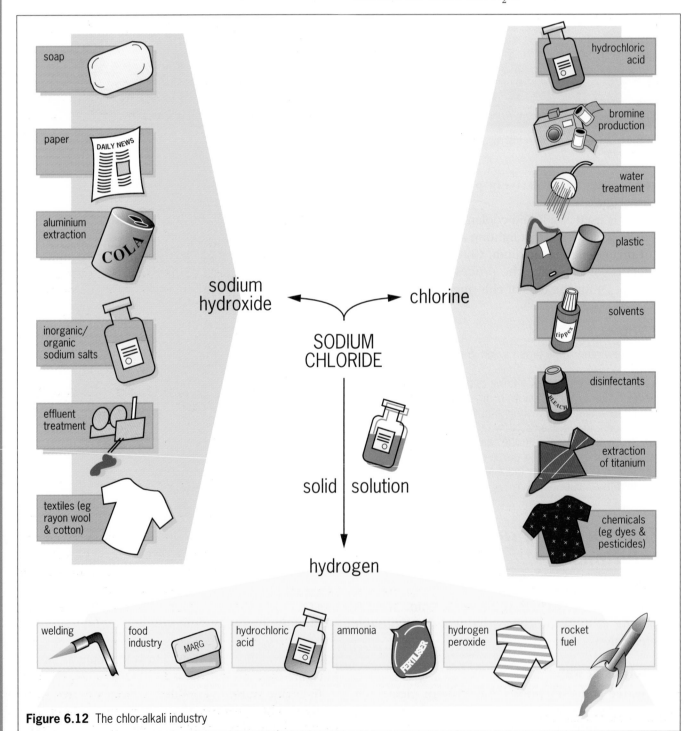

Figure 6.12 The chlor-alkali industry

Chlor-alkali industry

The electrolysis of saturated sodium chloride solution (brine) is the basis of a major industry. Three very important substances are produced — chlorine, sodium hydroxide and hydrogen. The electrolytic process is a very expensive one, requiring vast amounts of electricity. The process is economical only because all three products have a large number of uses (Figure 6.12).

There are two well-established methods for electrolysing brine, the **diaphragm cell** and the **mercury cell**. However, recent developments in electrolysis technology, by chemical engineers, have produced the **membrane cell** (Figure 6.13). This method is now preferred to the other two because it produces a purer product, it causes little pollution and it is cheaper to run.

The brine is first purified to remove calcium, strontium and magnesium compounds by a process of ion exchange.

The membrane cell is used continuously, with fresh brine flowing into the cell as the process breaks up the brine. The cell has been designed to ensure that the products do not mix. The ions in this concentrated sodium chloride solution are:

from the water: $H^+(aq)$ $OH^-(aq)$
from the sodium chloride: $Na^+(aq)$ $Cl^-(aq)$

When the current flows, the chloride ions, $Cl^-(aq)$, are attracted to the anode. Chlorine gas is produced by the electrode process.

chloride ions $\xrightarrow{\text{oxidation}}$ chlorine molecules + electrons
$2Cl^-(aq) \longrightarrow Cl_2(g) + 2e^-$

This leaves a high concentration of sodium ions, $Na^+(aq)$, around the anode.

The hydrogen ions, $H^+(aq)$, are attracted to the cathode and hydrogen gas is produced.

hydrogen ions + electrons $\xrightarrow{\text{reduction}}$ hydrogen molecules
$2H^+(aq) + 2e^- \longrightarrow H_2(g)$

This leaves a high concentration of hydroxide ions, $OH^-(aq)$, around the cathode. The sodium ions, $Na^+(aq)$, are drawn through the membrane, where they combine with the $OH^-(aq)$ to form sodium hydroxide, NaOH, solution.

QUESTIONS

a Suggest a reason for only 'roughly' twice as much hydrogen gas being produced at the cathode as oxygen gas at the anode in the electrolysis of water.
b Account for the following observations which were made when concentrated sodium chloride solution, to which a little universal indicator had been added, was electrolysed in the laboratory.
  (i) The universal indicator initially turns red in the region of the anode, but as the electrolysis proceeds it loses its colour.
  (ii) The universal indicator turns blue in the region of the cathode.
c Why is it important to remove compounds of calcium, strontium and magnesium before brine is electrolysed?

Purification of copper

Because copper is a very good conductor of electricity, it is used for electrical wiring and cables (Figure 6.14). However, even small amounts of impurities cut down this conductivity quite noticeably. The metal must be 99.99% pure to be used in this way. To ensure this level of purity, the newly extracted copper has to be purified by electrolysis (Figure 6.15).

The impure copper is used as the anode and is typically 1 m square, 35–50 mm thick and 330 kg in weight. The cathode is a 1 mm thick sheet and weighs about 5 kg; it is made from very pure copper.

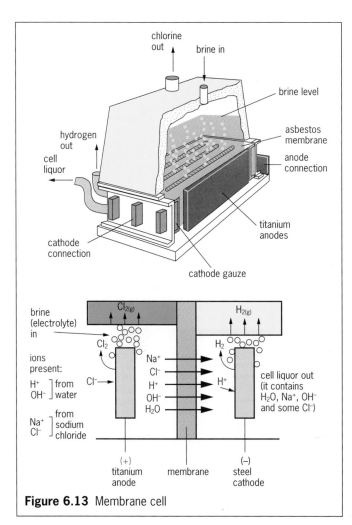

Figure 6.13 Membrane cell

The electrolyte is a solution of copper(II) sulphate acidified with sulphuric acid.

When the current flows, the copper moves from the impure anode to the pure cathode. Any impurities fall to the bottom of the cell and collect below the anode in the form of a slime. This slime is rich in precious metals and the recovery of these metals is an important aspect of the economics of the process. The electrolysis proceeds for about three weeks until the anodes are reduced to about 10% of their original size and the cathodes weigh between 100 and 120 kg.

The ions present in the solution are:

from the water: $H^+(aq)$ $OH^-(aq)$
from the copper(II) sulphate: $Cu^{2+}(aq)$ $SO_4^{2-}(aq)$

Figure 6.14 The copper used in electrical wiring has to be very pure

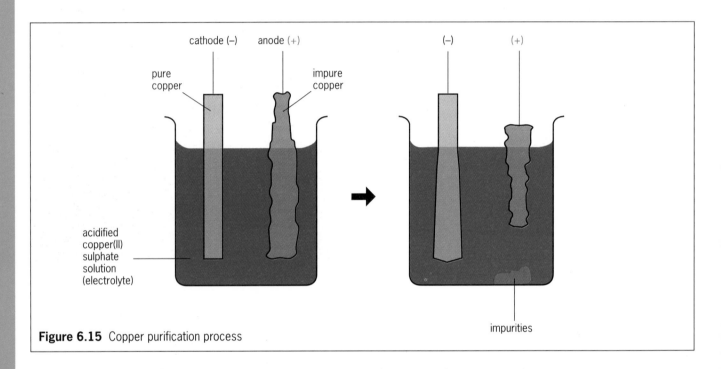

Figure 6.15 Copper purification process

During the process the impure anode loses mass because the copper atoms lose electrons and become copper ions, $Cu^{2+}(aq)$ (Figure 6.16).

$$\text{copper atoms} \rightarrow \text{copper ions} + \text{electrons}$$
$$Cu(s) \rightarrow Cu^{2+}(aq) + 2e^-$$

The electrons released at the anode travel around the external circuit to the cathode. There the electrons are passed on to the copper ions, $Cu^{2+}(aq)$, from the copper(II) sulphate solution and the copper is deposited on the cathode.

$$\text{copper ions} + \text{electrons} \rightarrow \text{copper atoms}$$
$$Cu^{2+}(aq) + 2e^- \rightarrow Cu(s)$$

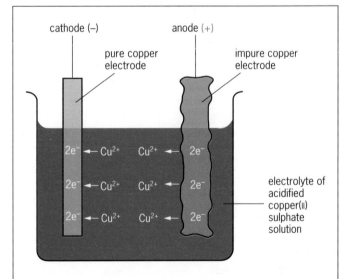

Figure 6.16 The movement of ions in the purification of copper by electrolysis

Electrolysis guidelines

The following points may help you work out the products of electrolysis in unfamiliar situations. They will also help you to remember what happens at each electrode.

- Non-metals are produced at the anode whereas metals and hydrogen gas are produced at the cathode.
- At the anode, chlorine, bromine and iodine (the halogens) are produced in preference to oxygen.
- At the cathode, hydrogen is produced in preference to metals unless unreactive metals such as copper and nickel are present.

QUESTIONS

a Using your knowledge of electrolysis, predict the likely products of the electrolysis of copper(II) chloride solution, using platinum electrodes. Write electrode equations for the formation of these products.

b Predict the products of electrolysis of a solution of copper(II) sulphate if carbon electrodes are used instead of those made from copper as referred to in the purification of copper section.

c Predict the products of the electrolysis of concentrated hydrochloric acid using platinum electrodes.

Figure 6.17 This tap has been chromium plated

Electroplating

Electroplating is the process involving electrolysis to plate, or coat, one metal with another. Often the purpose of electroplating is to give a protective coating to the metal beneath. For example, bath taps are chromium plated to prevent corrosion, and at the same time it gives them a more attractive finish (Figure 6.17).

The electroplating process is carried out in a cell such as the one shown in Figure 6.18. This is usually known as the 'plating bath' and it contains a suitable electrolyte.

For silver plating the electrolyte is a solution of a silver salt. The article to be plated is made the cathode in the cell so that the metal ions move to it when the current is switched on. The cathode reaction in this process is:

$$\text{silver ions} + \text{electrons} \rightarrow \text{silver atoms}$$
$$\text{Ag}^+(\text{aq}) \ + \ \text{e}^- \ \rightarrow \ \text{Ag(s)}$$

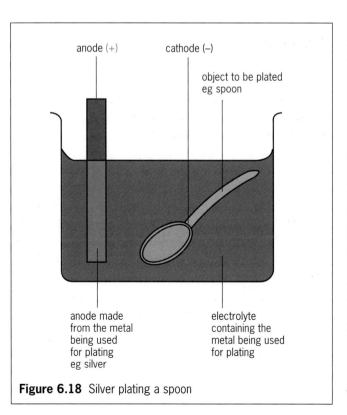

Figure 6.18 Silver plating a spoon

Figure 6.19 This plastic has been coated with copper by electrolysis

Figure 6.20 This leaf has been electroplated

Plating plastics

Nowadays it is not only metals that are electroplated. Plastics have been developed that are able to conduct electricity. For example, the plastic poly(pyrrole) can be electroplated in the same way as the metals we have discussed above (Figure 6.19).

QUESTIONS

a The leaf in Figure 6.20 has been copper plated. Suggest a suitable method for copper plating the leaf.

b Explain why copper(II) chloride solution would not be used as an electrolyte for copper plating.

Calculations in electrolysis

The quantity of electricity flowing through an electrolysis cell is measured in **coulombs** (C). If one **ampere** (A) is passed for one second, the quantity of electricity is said to be 1 coulomb.

quantity of electricity = current × time
    (coulombs)   (amperes) (seconds)

Therefore, if 2 amps flow for 1 second then the quantity of electricity passed is 2 coulombs.

In the purification of copper you saw that copper was deposited at the cathode. The electrode equation is:

$$Cu^{2+}(aq) + 2e^- \rightarrow Cu(s)$$
$$\text{1 mole} \quad \text{2 moles} \quad \text{1 mole}$$

This equation tells us that 1 mole of copper(II) ions combines with 2 moles of electrons to produce 1 mole of copper metal atoms (63.5 g).

Figure 6.21 In 1883 Michael Faraday was the first scientist to measure the masses of elements produced during electrolysis

A mole of electrons is called a **Faraday**. This unit is named after an English scientist, Michael Faraday (Figure 6.21), who carried out many significant experiments while investigating the nature of magnetism and electricity. So we can say that we need 2 Faradays of electricity to form 1 mole of copper atoms (63.5 g) at the cathode during this purification process.

From accurate electrolysis experiments it has been found that:

$$\text{1 Faraday} = 96\,500 \text{ coulombs}$$

Therefore, the quantity of electricity required to deposit 1 mole of copper atoms (63.5 g) is:

$$2 \times 96\,500 = 193\,000 \text{ coulombs (2 Faradays)}$$

Examples

1 Calculate the number of Faradays required to deposit 10 g of silver on the surface of a fork during an electroplating process. (A_r: Ag = 108)

Electrode equation: $Ag^+(aq) + e^- \longrightarrow Ag(s)$

1 mole of silver is deposited by 1 Faraday.
Therefore, 108 g of silver is deposited by 1 Faraday.
Hence 1 g of silver is deposited by $\frac{1}{108}$ Faradays.
Therefore, 10 g of silver are deposited by

$$\frac{1}{108} \times 10 = 0.093 \text{ Faradays}$$

2 Calculate the volume of oxygen gas, measured at room temperature and pressure (rtp), liberated at the anode in the electrolysis of acidified water by 2 Faradays. (1 mole of oxygen at rtp occupies a volume of 24 dm³.)

Electrode equation:
$$4OH^-(aq) \longrightarrow 2H_2O(l) + O_2(g) + 4e^-$$

1 mole of oxygen gas is liberated by 4 Faradays.
Therefore, 24 dm³ of oxygen is liberated by 4 Faradays.
Hence 12 dm³ of oxygen would be liberated by 2 Faradays.

3 The industrial production of aluminium uses a current of 25 000 amps. Calculate the time required to produce 10 kg of aluminium from the electrolysis of molten aluminium oxide.

Electrode equation: $Al^{3+}(l) + 3e^- \longrightarrow Al(l)$

1 mole of aluminium is produced by 3 Faradays.
27 g of aluminium is produced by 3 Faradays.
Hence 1 g of aluminium would be produced by $\frac{3}{27}$ Faradays.
Therefore, 10 000 g of aluminium would be produced by

$$\frac{3}{27} \times 10\,000 \text{ Faradays}$$
$$= 1111.1 \text{ Faradays}$$

So, the quantity of electricity
= number of Faradays × 96 500
= 1111.1 × 96 500
= 1.07×10^8 C
Time (seconds)
$= \dfrac{\text{coulombs}}{\text{amps}}$
$= \dfrac{1.07 \times 10^8}{25\,000}$
= 4280 seconds (71.3 minutes)

QUESTIONS

a Write equations which represent the discharge at the cathode of the following ions:
  (i) K⁺
  (ii) Pb²⁺
  (iii) Al³⁺
  and at the anode of:
  (i) Br⁻
  (ii) O²⁻
  (iii) F⁻.
b How many Faradays are required to discharge 1 mole of the following ions:
  (i) Mg²⁺?
  (ii) K⁺?
  (iii) Al³⁺?
  (iv) O²⁻?
c How many coulombs are required to discharge 1 mole of the following ions:
          (1 Faraday = 96 500 coulombs)
  (i) Cu²⁺?
  (ii) Na⁺?
  (iii) Pb²⁺?
d Calculate the number of Faradays required to deposit 6.35 g of copper on a metal surface. (A_r: Cu = 63.5)

Checklist

After studying Chapter 6 you should know and understand the following terms.

Electrolysis A process in which a chemical reaction is caused by the passage of an electric current.

Electrolyte A substance which will carry electric current only when it is molten or dissolved.

Electrode A point where the electric current enters and leaves the electrolytic cell.

Anode The positive electrode. It is positively charged because electrons are drawn away from it.

Cathode The negative electrode. It is negatively charged because an excess of electrons move towards it.

Reduction Takes place at the cathode and involves a positive ion gaining electrons.

Oxidation Takes place at the anode and involves a negative ion losing electrons.

Coulombs The unit used to measure the quantity of electricity:

$$1 \text{ coulomb} = 1 \text{ amp} \times 1 \text{ second}$$

Faraday One mole of electrons, which is equivalent to 96 500 coulombs.

Down's cell The electrolysis cell in which sodium is extracted from molten sodium chloride to which 60% calcium chloride has been added to reduce the working temperature to 600 °C. This cell has a graphite anode surrounded by a cylindrical steel cathode.

Hall–Héroult cell   The electrolysis cell in which aluminium is extracted from purified bauxite dissolved in molten cryolite at 900 °C. This cell has both a graphite anode and cathode.

Electroplating   The process of depositing metals from solution in the form of a layer on other surfaces such as metal or plastic.

Membrane cell   An electrolytic cell used for the production of sodium hydroxide, hydrogen and chlorine from brine in which the anode and cathode are separated by a membrane.

QUESTIONS

1 Copper is purified by electrolysis, as in the example shown below.

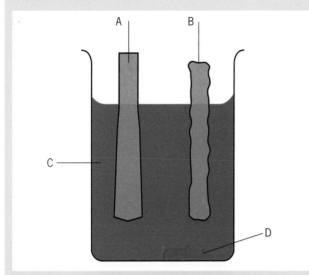

a Name the materials used for the electrodes **A** and **B**.
b Name the electrolyte **C** and substance **D**.
c Write equations for the reactions which take place at the cathode and anode.
d Draw a labelled diagram to show the cell after electrolysis has taken place.
e Why has electrolyte **C** to be acidified?
f Why does copper have to be 99.99% pure for use in electrical cables?

2 Copy and complete the table below, which shows the results of the electrolysis of four substances using inert electrodes.

Electrolyte	Product at anode (positive electrode)	Product at cathode (negative electrode)
Molten aluminium oxide		Aluminium
Concentrated sodium chloride solution	Chlorine	
Molten sodium chloride		
Silver nitrate solution		Silver

b State what you understand by the term 'inert electrodes'.

c Explain why the sodium chloride solution becomes progressively more alkaline during electrolysis.
d Explain why solid sodium chloride is a non-conductor of electricity, whereas molten sodium chloride and sodium chloride solution are good conductors of electricity.
e During the electrolysis of molten aluminium chloride the carbon anodes are burned away. Explain why this should happen and write balanced chemical equations for the reactions that take place.

3 Explain the meaning of each of the following terms. Use a suitable example, in each case, to help with your explanation.
a Anode.
b Cathode.
c Electrolysis.
d Electrolyte.
e Anion.
f Cation.
g Oxidation.
h Reduction.

4 The industrial production of sodium uses a current of 15 000 amps. Calculate the time required to produce 20 kg of sodium from the electrolysis of molten sodium chloride. (1 Faraday = 96 500 coulombs. A_r: Na = 23)

5 Calculate the volume of hydrogen gas, measured at room temperature and pressure, liberated at the cathode in the electrolysis of acidified water by 6 Faradays of electricity. (One mole of any gas at rtp occupies a volume of 24 dm³.)

6 Calculate the number of Faradays required to produce 10 g of gold deposited on the surface of some jewellery during an electroplating process. (A_r: Au = 197; note gold in solution is present as the Au^{3+} ion)

7 A pupil carried out an experiment in a fume cupboard to find out how electricity affected different substances. Some of the substances were in aqueous solution while others were electrolysed in the molten state. Carbon electrodes were used in each experiment and she wrote down her results in a table with the following headings.

Substance	What was formed at the cathode (–)	What was formed at the anode (+)

Make a table like the one shown and fill it in with what you think happened for each of the substances listed below.
a Molten lead bromide.
b Sugar solution.
c Silver nitrate solution.
d Copper(II) sulphate solution.
e Molten sodium iodide.
f Sodium chloride solution.

8 Sodium hydroxide is manufactured by the electrolysis of brine.
  a Draw and label a diagram of the cell used in this process. Make certain that you have labelled on the diagram:
    (i) the electrolyte
    (ii) the material of the electrodes
    (iii) the material of the membrane.
  b Write equations for the reactions which take place at the cathode and anode. State clearly whether a reaction is oxidation or reduction.
  c Give two large-scale uses of the products of this electrolysis process.
  d Comment on the following statement: 'This electrolysis process is a very expensive one'.

7 ACIDS, BASES AND SALTS

Figure 7.1 What do all these foods have in common?

Figure 7.2 Some common alkaline substances

Acids and alkalis

All the substances shown in Figure 7.1 contain an **acid** of one sort or another. Acids are certainly all around us. What properties do these substances have which make you think that they are acids or contain acids?

The word acid means 'sour' and all acids possess this property. They are also:

• soluble in water
• corrosive.

Alkalis are very different from acids. They are the chemical 'opposite' of acids.

• They will remove the sharp taste from an acid.
• They have a soapy feel.

Some common alkaline substances are shown in Figure 7.2.

It would be too dangerous to taste a liquid to find out if it was acidic. Chemists use substances called **indicators** which change colour when they are added to acids or alkalis. Many indicators are dyes which have been extracted from natural sources. For example, litmus is a purple dye which has been extracted from lichens. Litmus turns red when it is added to an acid and turns blue when added to an alkali. Some other indicators are shown in Table 7.1 along with the colours they turn in acids and alkalis.

Table 7.1 Indicators and their colours in acid and alkaline solution

Indicator	Colour in acid solution	Colour in alkaline solution
Phenolphthalein	Colourless	Pink
Methyl orange	Pink	Yellow
Red litmus	Red	Blue
Blue litmus	Red	Blue
Methyl red	Red	Yellow

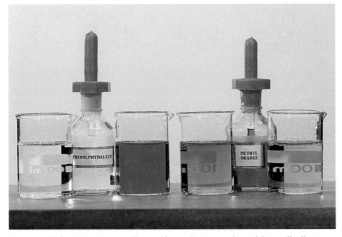

Figure 7.3 Indicators tell you if a substance is acid or alkaline

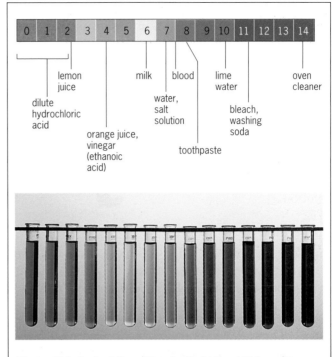

Figure 7.4 (**a**, top) The pH scale. (**b**, bottom) Universal indicator in solution, showing the colour range

These indicators tell chemists whether a substance is acid or alkaline (Figure 7.3). To obtain an idea of how acidic or alkaline a substance is, we use another indicator known as a **universal indicator**. This indicator is a mixture of many other indicators. The colour shown by this indicator can be matched against a **pH scale**. The pH scale was developed by a Scandinavian chemist called Sörenson, who was employed by the Carlsberg brewery. The pH scale runs from below 0 to 14. A substance with a pH of less than 7 is an acid. One with a pH of greater than 7 is alkaline. One with a pH of 7 is said to be neither acid nor alkaline, that is neutral. Water is the most common example of a neutral substance. Figure 7.4 shows the universal indicator colour range along with everyday substances which give the particular pH values.

Another way in which the pH of a substance can be measured is by using a pH meter (Figure 7.5). The pH electrode is placed into the solution whose pH you want to measure and a pH reading is given on the digital display.

Figure 7.5 A digital pH meter

Strong and weak acids

You will notice from Figure 7.4 that different acids have different pH values. A typical strong acid is hydrochloric acid. It is formed by dissolving hydrogen chloride gas in water. In hydrochloric acid the ions formed separate completely.

$$\text{hydrogen chloride gas} \xrightarrow{\text{water}} \text{hydrogen ions} + \text{chloride ions}$$

$$\text{HCl(g)} \xrightarrow{\text{water}} \text{H}^+\text{(aq)} + \text{Cl}^-\text{(aq)}$$

All the HCl molecules break up to form H⁺ ions and Cl⁻ ions. Any acid that behaves in this way is termed a **strong acid**. Both sulphuric acid and nitric acid behave in this way and are therefore termed strong acids. All these acids have a high concentration of hydrogen ions in solution (H⁺(aq)). Substances that are strong acids have a low pH.

A **weak acid** such as ethanoic acid, which is found in vinegar, produces only a few hydrogen ions when it dissolves in water.

$$\text{ethanoic acid} \xrightleftharpoons{\text{water}} \text{hydrogen ions} + \text{ethanoate ions}$$

$$\text{CH}_3\text{COOH(l)} \xrightleftharpoons{\text{water}} \text{H}^+\text{(aq)} + \text{CH}_3\text{COO}^-\text{(aq)}$$

The ⇌ sign means that the reaction is reversible. This means that if the ethanoic acid molecule breaks down to give hydrogen ions and ethanoate ions, they will then react together to re-form the ethanoic acid molecule. The fact that only a few ethanoic acid molecules break down and that the reaction is reversible means that only a few hydrogen ions are present in the solution. Other examples of weak acids are citric acid, (found in oranges and lemons), carbonic acid (found in soft drinks), sulphurous acid (acid rain) and ascorbic acid (vitamin C).

All acids produce hydrogen ions (H^+(aq)) when in aqueous solution. To say an acid is a strong acid does not mean it is concentrated. The strength of an acid tells you how easily it ionises to produce hydrogen ions. The concentration of an acid indicates the proportion of water and acid present in aqueous solution.

Strong and weak alkalis

An alkali is a substance that produces hydroxide ions (OH^-(aq)) when it is dissolved in water. Sodium hydroxide is a strong alkali because when it dissolves in water the ions separate completely.

sodium hydroxide $\xrightarrow{\text{water}}$ sodium ions+hydroxide ions

$$NaOH(aq) \xrightarrow{\text{water}} Na^+(aq) + OH^-(aq)$$

Substances which produce a high concentration of hydroxide ions in solution are **strong alkalis**, and they have a high pH. Both sodium hydroxide and potassium hydroxide are strong alkalis.

A **weak alkali**, such as ammonia, produces only a few hydroxide ions when it dissolves in water.

ammonia+water $\rightleftharpoons$ ammonium+hydroxide
ions ions

$$NH_3(g) + H_2O(l) \rightleftharpoons NH_4^+(aq) + OH^-(aq)$$

The ammonia molecules react with the water molecules to form ammonium ions and hydroxide ions. However, only a few ammonia molecules do this, so only a low concentration of hydroxide ions is produced.

Neutralising an acid

A common situation involving neutralisation of an acid is when you suffer from indigestion. This is caused by a build-up of acid in your stomach. Normally you treat it by taking an indigestion remedy containing a substance which will react with and neutralise the acid.

In the laboratory, if you wish to neutralise a common acid such as hydrochloric acid you can use an alkali such as sodium hydroxide. If the pH of the acid is measured as some sodium hydroxide solution is added to it, the pH increases. If equal volumes of the same concentration of hydrochloric acid and sodium hydroxide are added to one another, the resulting solution is found to have a pH of 7. The acid has been neutralised and a neutral solution has been formed.

hydrochloric+ sodium $\rightarrow$ sodium +water
acid hydroxide chloride
$$HCl(aq) + NaOH(aq) \rightarrow NaCl(aq) + H_2O(l)$$

As we have shown, when both hydrochloric acid and sodium hydroxide dissolve in water the ions separate completely. We may therefore write:

$$H^+(aq)+Cl^-(aq)+Na^+(aq)+OH^-(aq) \rightarrow Na^+(aq)+Cl^-(aq)+H_2O(l)$$

You will notice that certain ions are unchanged on either side of the equation. They are called **spectator ions** and are usually taken out of the equation. The equation now becomes:

$$H^+(aq)+OH^-(aq) \rightarrow H_2O(l)$$

This type of equation is known as an **ionic equation**. The reaction between any acid and alkali in aqueous solution can be summarised by this ionic equation. It shows the ion which causes acidity (H^+(aq)) reacting with the ion which causes alkalinity (OH^-(aq)) to produce neutral water (H_2O(l)).

QUESTIONS

a Alongside the names of various chemicals are shown their respective pH values in aqueous solution.
Potassium hydroxide pH 13
Hydrogen bromide pH 2
Calcium hydroxide pH 11
Sodium chloride pH 7
Hydrogen chloride pH 2
Magnesium hydroxide pH 10
Citric acid pH 4
Which of these substances is/are:
(i) a strong acid?
(ii) a weak acid?
(iii) a strong alkali?
(iv) a weak alkali?
(v) a neutral substance?
In each case write a chemical equation to show the molecules/ions present in solution.
b Write a chemical equation to represent the neutralisation of sulphuric acid by sodium hydroxide. Reduce this to the ionic equation.
Account for any difference you see between the ionic equation you have written and that shown above for the reaction between hydrochloric acid and sodium hydroxide.

Formation of salts

In the example above, sodium chloride was produced as part of the neutralisation reaction. Compounds formed in this way are known as **normal salts**. A normal salt is a compound that has been formed when all the hydrogen ions of an acid have been replaced by metal ions or by the ammonium ion (NH_4^+).

Normal salts can be classified as those which are soluble in water or those which are insoluble in water. The following salts are soluble in cold water:

- all nitrates
- all common sodium, potassium and ammonium salts
- all chlorides except lead, silver and mercury
- all sulphates except lead, barium and calcium.

Salts are very useful substances, as you can see from Table 7.2 and Figure 7.6.

If the acid being neutralised is hydrochloric acid, salts called **chlorides** are formed. Other types of salts can be formed with other acids. A summary of the different types of salt along with the acid they have been formed from is shown in Table 7.3.

Table 7.2 Useful salts

Salt	Use
Silver bromide	In photography
Calcium carbonate	Extraction of iron, making cement, glass making
Sodium carbonate	Glass making, softening water, making modern washing powders
Ammonium chloride	In torch batteries
Calcium chloride	In the extraction of sodium, drying agent (anhydrous)
Sodium chloride	Making hydrochloric acid, for food flavouring, hospital saline, in the solvay process for the manufacture of sodium carbonate
Tin(II) fluoride	Additive to toothpaste
Potassium nitrate	In fertiliser and gunpowder manufacture
Sodium stearate	In soap manufacture
Ammonium sulphate	In fertilisers
Calcium sulphate	For making plaster boards, plaster casts for injured limbs
Iron(II) sulphate	In 'iron' tablets
Magnesium sulphate	In medicines

Table 7.3 Types of salt and the acids they are formed from

Acid	Type of salt	Example
Carbonic acid	Carbonates	Sodium carbonate (Na_2CO_3)
Ethanoic acid	Ethanoates	Sodium ethanoate (CH_3COONa)
Hydrochloric acid	Chlorides	Potassium chloride (KCl)
Nitric acid	Nitrates	Potassium nitrate (KNO_3)
Sulphuric acid	Sulphates	Sodium sulphate (Na_2SO_4)

Methods of preparing soluble salts

There are four general methods of preparing soluble salts.

Acid + metal

This method can only be used with the less reactive metals. It would be very dangerous to use a reactive metal such as sodium in this type of reaction. The metals usually used in this method of salt preparation are the **MAZIT** metals, that is, **m**agnesium, **a**luminium, **z**inc, **i**ron and **t**in. A typical experimental method is given below.

Excess magnesium ribbon is added to dilute nitric acid. During this addition an effervescence is observed due to the production of hydrogen gas. In

Figure 7.6 Some uses of salts — sodium chloride is a flavour enhancer (top) and calcium sulphate is used in plaster casts

this reaction the hydrogen ions from the nitric acid gain electrons from the metal atoms as the reaction proceeds.

$$\text{hydrogen ions} + \text{electrons} \rightarrow \text{hydrogen gas}$$
$$\text{(from metal)}$$
$$2H^+ + 2e^- \rightarrow H_2(g)$$

How would you test the gas to show that it was hydrogen? What would be the name and formula of the compound produced during the test you suggested?

$$\text{magnesium} + \text{nitric} \rightarrow \text{magnesium} + \text{hydrogen}$$
$$\text{acid} \qquad \text{nitrate}$$
$$Mg(s) + 2HNO_3(aq) \rightarrow Mg(NO_3)_2(aq) + H_2(g)$$

The excess magnesium is removed by filtration (Figure 7.7). The magnesium nitrate solution is evaporated slowly to form a saturated solution of the salt (Figure 7.8).

The hot concentrated magnesium nitrate solution produced is tested by dipping a cold glass rod into it. If salt crystals form at the end of the rod the solution is ready to crystallise and is left to cool. Any crystals produced on cooling are filtered and dried between clean tissues.

Acid + carbonate

This method can be used with any metal carbonate and any acid, providing the salt produced is soluble. The typical experimental procedure is similar to that carried out for an acid and a metal. For example, copper(II) carbonate would be added in excess to dilute nitric acid. Effervescence would be observed due to the production of carbon dioxide.

How would you test the gas to show it was carbon dioxide? Write an equation to help you explain what is happening during the test you have chosen.

$$\text{copper(II)} + \text{nitric} \rightarrow \text{copper(II)} + \text{carbon} + \text{water}$$
$$\text{carbonate} \quad \text{acid} \qquad \text{nitrate} \quad \text{dioxide}$$
$$CuCO_3(s) + 2HNO_3(aq) \rightarrow Cu(NO_3)_2(aq) + CO_2(g) + H_2O(l)$$

Metal carbonates contain carbonate ions, CO_3^{2-}. In this reaction the carbonate ions react with the hydrogen ions in the acid.

$$\text{carbonate} + \text{hydrogen} \rightarrow \text{carbon} + \text{water}$$
$$\text{ions} \qquad \text{ions} \qquad \text{dioxide}$$
$$CO_3^{2-}(aq) + 2H^+(aq) \rightarrow CO_2(g) + H_2O(l)$$

Acid + alkali (soluble base)

This method is generally used for preparing the salts of very reactive metals, such as potassium or sodium. It would certainly be too dangerous to add the metal directly to the acid. In this case, we solve the problem indirectly and use an alkali which contains the particular reactive metal whose salt you wish to prepare.

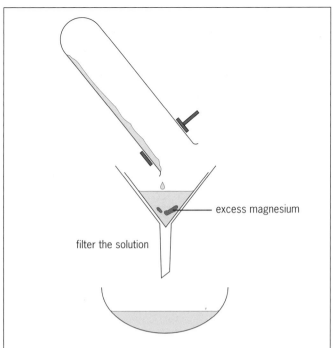

Figure 7.7 The excess magnesium is filtered in this way

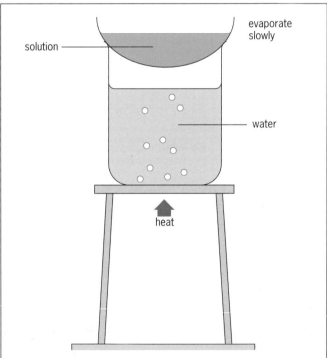

Figure 7.8 The solution of magnesium nitrate is concentrated by slow evaporation

A **base** is a substance which neutralises an acid, producing a salt and water as the only products. If the base is soluble the term alkali can be used, but there are several bases which are insoluble. In general, most metal oxides and hydroxides (as well as ammonia solution) are bases. Some examples of soluble and insoluble bases are shown in Table 7.4. Salts can only be formed by this method if the base is soluble.

Table 7.4 Examples of soluble and insoluble bases

Soluble bases (alkalis)	Insoluble bases
Sodium hydroxide (NaOH)	Iron(III) oxide (Fe_2O_3)
Potassium hydroxide (KOH)	Copper(II) oxide (CuO)
Calcium hydroxide ($Ca(OH)_2$)	Lead(II) oxide (PbO)
Ammonia solution ($NH_{3(aq)}$)	Magnesium oxide (MgO)

Because in this neutralisation reaction both reactants are in solution, a special technique called **titration** is required. Acid is slowly and carefully added to a measured volume of alkali using a burette (Figure 7.9) until the indicator, usually phenolphthalein, changes colour. An indicator is used to show when the alkali has been neutralised completely by the acid. This is called the **end-point**. Now that you know where the end-point is, you can add the same volume of acid to the measured volume of alkali but this time without the indicator. The solution which is produced can then be evaporated slowly to obtain the salt (Figure 7.10). For example,

$$\begin{array}{cccc} \text{hydrochloric} + & \text{sodium} & \rightarrow & \text{sodium} + \text{water} \\ \text{acid} & \text{hydroxide} & & \text{chloride} \end{array}$$

$$HCl_{(aq)} + NaOH_{(aq)} \rightarrow NaCl_{(aq)} + H_2O_{(l)}$$

As previously discussed on p. 84, this reaction can best be described by the ionic equation:

$$H^+_{(aq)} + OH^-_{(aq)} \rightarrow H_2O_{(l)}$$

Acid + insoluble base

This method can be used to prepare a salt of an unreactive metal, such as lead or copper. In these cases it is not possible to use a direct reaction of the metal with an acid so the acid is neutralised using the particular metal oxide (Figure 7.11). The method is generally the same as that for a metal carbonate and an acid, though some warming of the reactants may be necessary. An example of such a reaction is the neutralisation of sulphuric acid by copper(II) oxide.

$$\begin{array}{cccc} \text{sulphuric} + & \text{copper(II)} & \rightarrow & \text{copper(II)} + \text{water} \\ \text{acid} & \text{oxide} & & \text{sulphate} \end{array}$$

$$H_2SO_{4(aq)} + CuO_{(s)} \rightarrow CuSO_{4(aq)} + H_2O_{(l)}$$

Metal oxides contain the oxide ion, O^{2-}. The ionic equation for this reaction is therefore:

$$2H^+_{(aq)} + O^{2-}_{(s)} \rightarrow H_2O_{(l)}$$

Figure 7.9 The acid is added to the alkali until the indicator just changes colour

Figure 7.10 After slow evaporation the salt is left behind in the evaporating basin

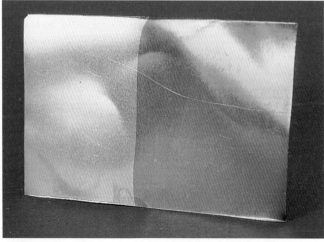

Figure 7.11 The citric acid in the lemon juice reacts with the oxide on the surface of the copper and the tarnish dissolves away

Methods of preparing insoluble salts

An insoluble salt, such as barium sulphate, can be made by **precipitation**. In this case, solutions of the two chosen soluble salts are mixed (Figure 7.12). To produce lead(II) iodide, lead nitrate and sodium iodide can be used. The lead(II) iodide precipitate can be filtered off, washed with distilled water and dried. The reaction that has occurred is:

$$\text{lead(II)} + \text{sodium} \rightarrow \text{lead(II)} + \text{sodium}$$
$$\text{nitrate} \quad \text{iodide} \quad \text{iodide} \quad \text{nitrate}$$
$$Pb(NO_3)_2(aq) + 2NaI(aq) \rightarrow PbI_2(s) + 2NaNO_3(aq)$$

The ionic equation for this reaction is:

$$Pb^{2+}(aq) + 2I^-(aq) \rightarrow PbI_2(s)$$

This method is sometimes known as **double decomposition** and may be summarised as follows

$$\text{soluble salt} + \text{soluble salt} \rightarrow \text{insoluble salt} + \text{soluble salt}$$
$$\text{(AX)} \qquad \text{(BY)} \qquad \text{(BX)} \qquad \text{(AY)}$$

It should be noted that even salts like barium sulphate dissolve to a very small extent. For example, 1 litre of water will dissolve 2.2×10^{-3} g of barium sulphate at 25 °C. This substance and substances like this are said to be **sparingly soluble**.

More about salts

You have already seen on p. 85 in Table 7.2 that salts are useful substances. Some of the salts shown in that table occur naturally and are mined, for example calcium sulphate (gypsum) and calcium carbonate (limestone). Many of the others must be made by the chemical industry, for example ammonium nitrate, iron(II) sulphate and silver bromide.

With acids such as sulphuric acid, which has two replaceable hydrogen ions per molecule, it is possible to replace only **one** of these with a metal ion. The salt produced is called an **acid salt**. An acid salt is one in which not all of the replaceable hydrogen ions of the acid have been replaced by metal ions or the ammonium ion. Some examples of acid salts are shown in Table 7.5.

Table 7.5 Examples of acid salts

Acid	Type of acid salt	Example
Carbonic acid (H_2CO_3)	Hydrogencarbonate	Sodium hydrogencarbonate ($NaHCO_3$)
Sulphuric acid (H_2SO_4)	Hydrogensulphate	Potassium hydrogensulphate ($KHSO_4$)

Sodium hydrogencarbonate is the salt used as the raising agent in the baking of cakes and bread.

Figure 7.12 When barium chloride solution is added to sodium sulphate a white precipitate of barium sulphate is produced

Testing for different salts

Sometimes we want to analyse a salt and find out what is in it. There are simple chemical tests which allow us to identify the anion part of the salt. These are often called **spot tests**.

Testing for a sulphate

You have seen that barium sulphate is a useful insoluble salt (p. 85). Therefore, if you take a solution of a suspected soluble sulphate and add it to a soluble barium salt (such as barium chloride) then a white precipitate of barium sulphate will be produced.

$$\text{barium ion} + \text{sulphate ion} \rightarrow \text{barium sulphate}$$
$$Ba^{2+}(aq) + SO_4^{2-}(aq) \rightarrow BaSO_4(s)$$

A few drops of dilute hydrochloric acid are also added to this mixture. If the precipitate does not dissolve, then it is barium sulphate and the unknown salt was in fact a sulphate. If the precipitate does dissolve, then the unknown salt may have been a sulphite (containing the SO_3^{2-} ion).

Testing for a chloride

Earlier in this chapter you saw that silver chloride is an insoluble salt (p. 85). Therefore, if you take a solution of a suspected soluble chloride and add to it a small volume of a soluble silver salt (such as silver nitrate) a white precipitate of silver chloride will be produced.

$$\text{chloride ion} + \text{silver ion} \rightarrow \text{silver chloride}$$
$$Cl^-(aq) + Ag^+(aq) \rightarrow AgCl(s)$$

If left to stand, the precipitate goes grey (Figure 7.13).

Testing for a carbonate

If a small amount of an acid is added to some of the suspected carbonate (either solid or in solution) then effervescence occurs. If it is a carbonate then carbon dioxide gas is produced, which will turn limewater 'milky' (a cloudy white precipitate of calcium carbonate forms).

$$\begin{array}{ccc} \text{carbonate} + \text{hydrogen} & \rightarrow & \text{carbon} + \text{water} \\ \text{ions} & \text{ions} & \text{dioxide} \end{array}$$
$$CO_3^{2-}(aq) + 2H^+(aq) \rightarrow CO_2(g) + H_2O(l)$$

Testing for an iodide

If a small amount of acidified lead(II) nitrate is added to some of the suspected soluble iodide, a yellow precipitate of lead(II) iodide is produced.

$$\text{lead ions} + \text{iodide ions} \rightarrow \text{lead(II) iodide}$$
$$Pb^{2+}(aq) + 2I^-(aq) \rightarrow PbI_2(s)$$

Testing for a nitrate

By using Devarda's alloy (45% Al, 5% Zn, 50% Cu) in alkaline solution, nitrates are converted to ammonia. The ammonia can be identified using damp indicator paper, which turns blue.

Figure 7.13 If left to stand the white precipitate of silver chloride goes grey. This photochemical change plays an essential part in black and white photography

Crystal hydrates

Some salts, such as sodium chloride, copper carbonate and sodium nitrate, crystallise in the anhydrous form (without water). However, many salts produce **hydrates** when they crystallise from solution. A hydrate is a salt which incorporates water into its crystal structure. This water is referred to as **water of crystallisation**. The shape of the crystal hydrate is very much dependent on the presence of water of crystallisation. Some examples of crystal hydrates are given in Table 7.6 and Figure 7.14.

Table 7.6 Examples of crystal hydrates

Salt hydrate	Formula
Cobalt(II) chloride hexahydrate	$CoCl_2.6H_2O$
Copper(II) sulphate pentahydrate	$CuSO_4.5H_2O$
Iron(II) sulphate heptahydrate	$FeSO_4.7H_2O$
Magnesium sulphate heptahydrate	$MgSO_4.7H_2O$
Sodium carbonate decahydrate	$Na_2CO_3.10H_2O$
Sodium hydrogensulphate monohydrate	$NaHSO_4.H_2O$
Sodium sulphate decahydrate	$Na_2SO_4.10H_2O$

QUESTIONS

a Complete the word equations and write balanced chemical equations for the following soluble salt preparations:
magnesium + sulphuric acid →
calcium carbonate + hydrochloric acid →
zinc oxide + hydrochloric acid →
potassium hydroxide + nitric acid →
Also write ionic equations for each of the reactions.

b Lead carbonate and lead iodide are insoluble. Which two soluble salts could you use in the preparation of each substance? Write
(i) a word equation
(ii) a symbol equation
(iii) an ionic equation
to represent the reactions taking place.

c An analytical chemist working for an environmental health organisation has been given a sample of water which is thought to have been contaminated by a sulphate and a carbonate.
(i) Describe how she could confirm the presence of these two types of salt by simple chemical tests.
(ii) Write ionic equations to help you explain what is happening during the testing process.

Figure 7.14 Hydrate crystals (left to right): cobalt nitrate, calcium nitrate and nickel sulphate (top) and manganese sulphate, copper sulphate and chromium potassium sulphate (bottom)

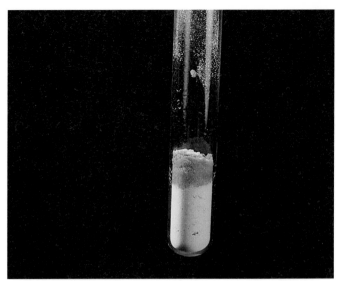

Figure 7.15 Anhydrous copper(II) sulphate is a white powder which turns blue when water is added to it

When many hydrates are heated the water of crystallisation is driven away. For example, if crystals of copper(II) sulphate hydrate (blue) are heated strongly, they lose their water of crystallisation. Anhydrous copper(II) sulphate remains as a white powder:

copper(II) sulphate → anhydrous copper(II)+ water
   pentahydrate   sulphate
$$CuSO_4.5H_2O(s) \rightarrow CuSO_4(s) + 5H_2O(g)$$

When water is added to anhydrous copper(II) sulphate it turns blue and the pentahydrate is produced (Figure 7.15). This is an extremely exothermic process.

$$CuSO_4(s) + 5H_2O(l) \rightarrow CuSO_4.5H_2O(s)$$

Because the colour change only takes place in the presence of water, the reaction is used to test for the presence of water.

Some crystal hydrates **effloresce**, that is they lose some or all of their water of crystallisation to the atmosphere. For example, when colourless sodium carbonate decahydrate crystals are left out in the air they become coated with a white powder, which is the monohydrate (Figure 7.16). The process is called **efflorescence**.

$$Na_2CO_3.10H_2O(s) \rightarrow Na_2CO_3.H_2O(s) + 9H_2O(g)$$

With some substances, not necessarily salt hydrates, the reverse of efflorescence occurs. For example, if anhydrous calcium chloride is left in the air, it absorbs water vapour and eventually forms a very concentrated solution. This process is called **deliquescence**, and substances which behave like this are said to be **deliquescent**. Solid sodium hydroxide will also deliquesce.

Figure 7.16 A white powder forms on the surface of sodium carbonate decahydrate when it is left in air

There are some substances which, if left out in the atmosphere, absorb moisture but do not change their state. For example, concentrated sulphuric acid, a colourless, viscous liquid, absorbs water vapour from the air and becomes a solution. Substances which do this are said to be **hygroscopic**.

Calculation of water of crystallisation

Sometimes it is necessary to work out the percentage, by mass, of water of crystallisation in a hydrated salt. The method is the same as that used in Chapter 5, p. 65, but this time the 'H$_2$O' is treated as an element in the calculation.

Example

Calculate the percentage by mass of water in the salt hydrate $MgSO_4.7H_2O$. (A_r: H=1, O=16, Mg=24, S=32)

$$M_r \text{ for } MgSO_4.7H_2O = 24+32+(4\times16)+(7\times18)$$
$$= 246$$

The mass of water as a fraction of the total mass of hydrate

$$= \frac{126}{246}$$

The percentage of water present

$$= \frac{126}{246} \times 100 = 51.2\%$$

Solubility of salts in water

Water is a very good solvent and will dissolve a whole range of solutes, including sodium chloride and copper(II) sulphate, as well as other substances such as sugar. You can dissolve more sugar than sodium chloride in $100\,cm^3$ of water at the same temperature. The sugar is more **soluble** than the sodium chloride at the same temperature. We say that the sugar has a greater **solubility** than the sodium chloride. The solubility of a solute in water at a given temperature is the number of grams of that solute which can be dissolved in $100\,g$ of water to produce a saturated solution at that temperature.

Calculating solubility

Example

21.5 g of sodium chloride dissolve in 60 g of water at 25 °C. Calculate the solubility of sodium chloride in water at that temperature.

If 60 g of water dissolves 21.5 g of sodium chloride, then 1 g of water will dissolve:

$$\frac{21.5}{60} \text{ g of sodium chloride}$$

Therefore, 100 g of water will dissolve:

$$\frac{21.5}{60} \times 100 = 35.8\,g$$

The solubility of sodium chloride at 25 °C = 35.8 g per 100 g of water.

Solubility curves

It is a well-known fact that usually the amount of solute that a solvent will dissolve increases with temperature. We say that the solubility increases with increasing temperature. Figure 7.17 shows how the solubilities of copper(II) sulphate and potassium nitrate vary with temperature. These graphs of solubility against temperature are known as **solubility curves**. By using curves of this type you can find the solubility of the solute at any temperature.

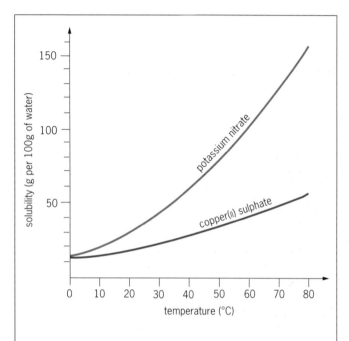

Figure 7.17 Solubility curves for copper(II) sulphate and potassium nitrate

From a solubility curve such as Figure 7.17, it can be seen that the solubility of copper(II) sulphate at 30 °C is 22 g per 100 g of water whereas that of potassium nitrate is 44 g per 100 g of water.

When a saturated solution is cooled, some of the solute crystallises out of solution. Thus, by using solubility curves it is possible to determine the amount of solute that will crystallise out of solution at different temperatures. For example, if potassium nitrate is cooled from 70 °C to 40 °C then the difference between the two solubilities at these temperatures tells you the amount that crystallises out.

Solubility at 70 °C = 135 g
Solubility at 40 °C = 62 g
Difference = 73 g

Therefore the amount of potassium nitrate which crystallises out during cooling from 70 °C to 40 °C is 73 g.

QUESTIONS

a Calculate the solubility, at the temperature given, of the following salts in water:
   (i) 94.5 g potassium nitrate dissolved in 70 g of water at 68 °C
   (ii) 9.2 g of ammonium chloride dissolved in 20 g of water at 40 °C
   (iii) 12.5 g of copper(II) sulphate dissolved in 25 g of water at 60 °C.
b Use the data given below to plot a solubility curve of ammonium chloride.

Temperature (°C)	0	10	20	30	40	50	60	70	80
Solubility (g per 100 g of water)	29.2	33.0	37.1	41.8	45.8	50.8	55.2	60.2	65.6

c Using your solubility curve, find the solubility of ammonium chloride at:
   (i) 15 °C
   (ii) 35 °C
   (iii) 65 °C.
d Using your answers to (c), calculate the amount of ammonium chloride that will crystallise out of solution when cooled from:
   (i) 65 °C to 35 °C
   (ii) 35 °C to 15 °C
   (iii) 65 °C to 15 °C.
e Calculate the amount of ammonium chloride that would crystallise out of solution if 25 g of the saturated solution were cooled from 52 °C to 23 °C.

Titration

On p. 87 you saw that it was possible to prepare a soluble salt by reacting an acid with a soluble base (alkali). The method used was that of **titration**. Titration can also be used to find the concentration of the alkali used. In the laboratory, the titration of hydrochloric acid with sodium hydroxide is carried out in the following way.

1 25 cm³ of sodium hydroxide solution is pipetted into a conical flask to which a few drops of phenolphthalein indicator have been added (Figure 7.18a). Phenolphthalein is pink in alkali and colourless in acid.
2 0.100 M hydrochloric acid is placed in the burette using a filter funnel until it is filled up exactly to the zero mark (Figure 7.18b).
3 The filter funnel is now removed.
4 The hydrochloric acid is added to the sodium hydroxide solution in small quantities — usually no more than 0.5 cm³ at a time (Figure 7.18c). The contents of the flask must be swirled after each addition of acid.

5 The acid is added until the alkali has been neutralised completely. This is shown by the pink colour of the indicator just disappearing.
6 The final reading on the burette at the end-point is recorded and further titrations carried out until consistent results are obtained (within 0.1 cm³ of each other). Below some sample data are shown.

Volume of sodium hydroxide solution = 25 cm³
Average volume of 0.10 M hydrochloric acid added = 22 cm³

The neutralisation reaction which has taken place is:

hydrochloric + sodium → sodium + water
acid   hydroxide   chloride
$HCl(aq) + NaOH(aq) \rightarrow NaCl(aq) + H_2O(l)$

From this equation it can be seen that 1 mole of hydrochloric acid neutralises 1 mole of sodium hydroxide.

Now you can work out the number of moles of the acid using the formula given in Chapter 5, p. 63.

$$moles = \frac{volume}{1000} \times concentration$$
$$= \frac{22 \times 0.100}{1000}$$
$$= 2.2 \times 10^{-3}$$

number of moles = number of moles
of hydrochloric acid   of sodium hydroxide

Therefore, the number of moles of sodium hydroxide = 2.2×10^{-3}

2.2×10^{-3} moles of sodium hydroxide is present in 25 cm³ of solution.

Therefore, in 1 cm³ of sodium hydroxide solution we have

$$\frac{2.2 \times 10^{-3}}{25} moles$$

Therefore, in 1 litre of sodium hydroxide solution we have

$$\frac{2.2 \times 10^{-3}}{25} \times 1000 = 0.088 \, moles$$

The concentration of sodium hydroxide solution is 0.088 mol dm⁻³. Therefore, it is a 0.088 M solution of sodium hydroxide.

You can simplify the calculation by substituting in the following mathematical equation:

$$\frac{M_1 V_1}{M_{acid}} = \frac{M_2 V_2}{M_{alkali}}$$

where:

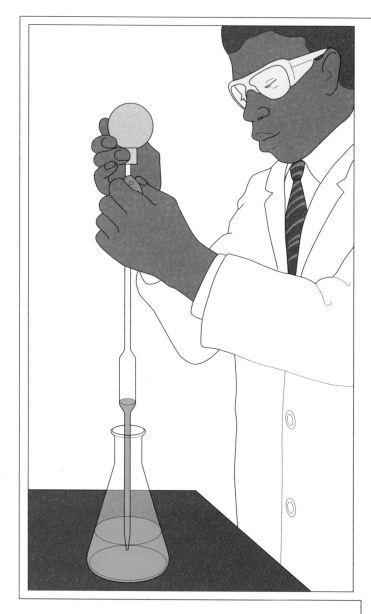

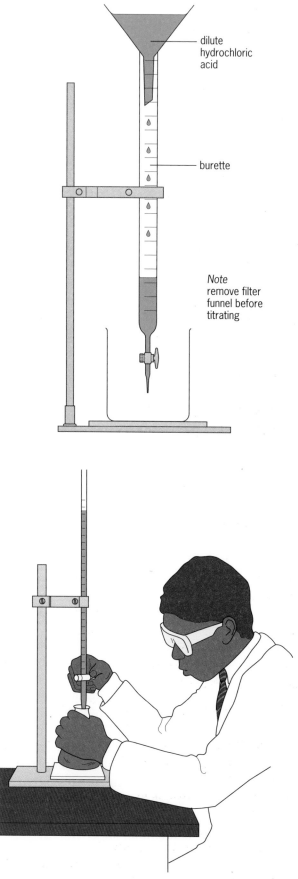

dilute hydrochloric acid

burette

Note
remove filter funnel before titrating

M_1 = molarity (concentration) of the acid used
V_1 = volume of acid used (cm³)
M_{acid} = number of moles of acid shown in the chemical equation
M_2 = molarity (concentration) of the alkali used
V_2 = volume of the alkali used (cm³)
M_{alkali} = number of moles of alkali shown in the chemical equation

In the example:

$$M_1 = 0.100\,\text{M}$$
$$V_1 = 22\,\text{cm}^3$$
$$M_{acid} = 1\,\text{mole}$$
$$M_2 = \text{unknown}$$
$$V_2 = 25\,\text{cm}^3$$
$$M_{alkali} = 1\,\text{mole}$$

Substituting in the equation:

$$\frac{0.100 \times 22}{1} = \frac{M_2 \times 25}{1}$$

Figure 7.18 (**a**, left) Exactly 25 cm³ of sodium hydroxide solution is pipetted into a conical flask. (**b**, top) The burette is filled up to the zero mark with 0.1 M hydrochloric acid. (**c**, bottom) The titration is carried out accurately

Rearranging:

$$M_2 = \frac{0.100 \times 22 \times 1}{1 \times 25}$$
$$M_2 = 0.088$$

The molarity of the sodium hydroxide solution
= 0.088 M.

QUESTIONS

a 28 cm³ of 0.200 M hydrochloric acid just neutralised 25 cm³ of a potassium hydroxide solution. What is the concentration of this potassium hydroxide solution?
b 20 cm³ of 0.100 M sulphuric acid just neutralised 25 cm³ of a sodium hydroxide solution. What is the concentration of this sodium hydroxide solution?

Checklist

After studying Chapter 7 you should know and understand the following terms.

Acid A substance which dissolves in water, producing $H^+(aq)$ ions as the only positive ions.

Base A substance which neutralises an acid, producing a salt and water as the only products.

Alkali A soluble base, which produces $OH^-(aq)$ ions in water.

Indicator A substance used to show whether a substance is acidic or alkaline (basic), for example phenolphthalein.

pH scale A scale running from below 0 to 14, used for expressing the acidity or alkalinity of a solution.

Strong acid One which produces a high concentration of $H^+(aq)$ ions in water solution, for example hydrochloric acid.

Strong alkali One which produces a high concentration of $OH^-(aq)$ ions in water solution, for example sodium hydroxide.

Weak acid One which produces a low concentration of $H^+(aq)$ in water solution, for example ethanoic acid.

Weak alkali One which produces a low concentration of $OH^-(aq)$ in water solution, for example ammonia solution.

Neutralisation The process in which the acidity or alkalinity of a substance is destroyed. Destroying acidity means removing $H^+(aq)$ by reaction with a base, carbonate or metal. Destroying alkalinity means removing the $OH^-(aq)$ by reaction with an acid.

$$H^+(aq) + OH^-(aq) \rightarrow H_2O(l)$$

Normal salt A substance formed when all the replaceable hydrogen of an acid is completely replaced by metal ions or the ammonium ion (NH_4^+).

Acid salt A substance formed when only some of the replaceable hydrogen of an acid is replaced by metal ions or the ammonium ion (NH_4^+).

Ionic equation The simplified equation of a reaction which we can write if the chemicals involved are ionic substances.

Double decomposition The process by which an insoluble salt is prepared from solutions of two suitable soluble salts.

Testing for a sulphate If a white precipitate is produced when dilute hydrochloric acid and barium chloride solution are added to the suspected sulphate, the solution contains a sulphate.

Testing for a chloride If a white precipitate is produced when dilute nitric acid and silver nitrate solution are added to the suspected chloride, the solution contains a chloride.

Testing for a carbonate If effervescence occurs when an acid is added to the suspected carbonate and the gas produced tests positively for carbon dioxide, the substance is a carbonate.

Salt hydrates Salts containing water of crystallisation.

Water of crystallisation Water incorporated into the structure of substances as they crystallise, for example copper(II) sulphate pentahydrate ($CuSO_4.5H_2O$).

Efflorescence The process during which a substance loses water of crystallisation to the atmosphere.

Deliquescence The process during which a substance absorbs water vapour from the atmosphere and eventually forms a very concentrated solution.

Hygroscopic The ability to absorb water vapour from the atmosphere without forming solutions or changing state, for example concentrated sulphuric acid.

Solubility The solubility of a solute in a solvent at a given temperature is the number of grams of that solute which can dissolve in 100 g of solvent to produce a saturated solution at that temperature.

Solubility curve This is a graph of solubility against temperature.

Titration A method of volumetric analysis in which a volume of one reagent (for example an acid) is added to a known volume of another reagent (for example an alkali) slowly from a burette until an end-point is reached. If an acid and alkali are used, then an indicator is used to show that the end-point has been reached.

QUESTIONS

1 Explain with the aid of an example what you understand by the terms:
a strong acid
b weak acid
c strong alkali
d weak alkali
e concentrated acid
f neutralisation.

2 a Copy out and complete the table, which covers the different methods of preparing salts.

Method of preparation	Name of salt prepared	Two substances used in the preparation
Acid + alkali	Sodium sulphate	. . . . and . . . .
Acid + metal	. . . .	. . . . and dilute hydrochloric acid
Acid + insoluble base	Copper(II) sulphate	. . . . and . . . .
Acid + carbonate	Magnesium	. . . . and . . . .
Precipitation	Barium sulphate	. . . . and . . . .

b Write word and balanced chemical equations for each reaction shown in your table. Also write ionic equations where appropriate.

3 Study the following scheme.

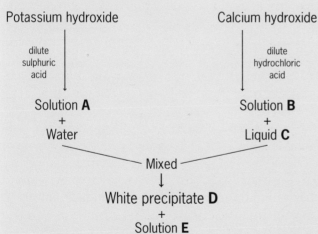

Potassium hydroxide → dilute sulphuric acid → Solution **A** + Water

Calcium hydroxide → dilute hydrochloric acid → Solution **B** + Liquid **C**

Mixed →
White precipitate **D** + Solution **E**

a Give the names and formulae of substances **A** to **E**.
b Describe a test which could be used to identify the presence of water.
c Which indicator is suitable for the initial reaction between the hydroxides and the dilute acids shown?
d Write balanced chemical equations for the reactions taking place in the scheme.
e Write an ionic equation for the production of the white precipitate **D**.

4 In a titration involving $20\,cm^3$ sodium hydroxide solution against a $1\,M$ solution of sulphuric acid, $30\,cm^3$ of the acid was found to just neutralise the alkali completely.
a Write a word and balanced chemical equation for the reaction.
b Name a suitable indicator for the titration and state the colour change you would observe.
c Calculate the molarity of the alkali.
d Describe a chemical test which you could use to identify the type of salt produced during the reaction.

5 Explain the following with the aid of balanced chemical equations and ionic equations, where appropriate.
a Vinegar can be used to descale kettles (Chapter 8).
b Indigestion remedies contain calcium carbonate.
c 'Epsom Salts' contain $MgSO_4$ as the heptahydrate.
d Rainwater is acidic.
e Mineral waters are fizzy as well as acidic.
f Alkalis are bases which are soluble in water.

6 Potassium nitrate is used in fertilisers. The table below shows the solubility of this substance in water at various temperatures.

Temperature (°C)	0	10	20	30	40	50	60	70
Solubility (g per 100 g H_2O)	13		32	46	64	86	110	138

a Plot these results as a solubility curve for potassium nitrate.
b Use your graph to answer the following questions.
(i) What is the solubility of potassium nitrate at $10\,°C$, $25\,°C$, $44\,°C$ and $63\,°C$?
(ii) Using your answers to (i), calculate the amount of potassium nitrate that will crystallise out of solution when it is cooled from $63\,°C$ to $25\,°C$.
(iii) Calculate the amount of potassium nitrate which would crystallise out of solution if 20 g of saturated solution was cooled from $68\,°C$ to $36\,°C$.
c Describe the method you would use in the laboratory to prepare a pure sample of potassium nitrate. Write word and balanced chemical equations for the method you have chosen.

7 Read the following passage and then answer the questions which follow.
Sodium carbonate decahydrate *effloresces* quite readily. With some substances, such as solid sodium hydroxide, the reverse of efflorescence occurs—they *deliquesce*. There are some substances, such as concentrated sulphuric acid, which when left open to the atmosphere are diluted—they are *hygroscopic*.
a What are the meanings of the terms in italics?
b What precautions should be taken to ensure that these substances are not involved in the processes described above?
c Which of the other salts shown in Table 7.6 on p. 89 are likely to effloresce? Give a reason for your answer.

QUESTIONS (continued)

8 Magnesium sulphate crystals exist as the *heptahydrate*, $MgSO_4.7H_2O$. It is a salt *hydrate*. If it is heated quite strongly, the *water of crystallisation* is driven off and the *anhydrous* salt remains.

  a Explain the meaning of the terms shown in italics.
  b Describe the experiment you would carry out to collect a sample of the water given off when the salt hydrate was heated strongly. Your description should include a diagram of the apparatus used and a chemical equation to represent the process taking place.
  c Describe a chemical test you could carry out to show that the colourless liquid given off was water.
  d Describe one other test you could carry out to show that the colourless liquid obtained in this experiment was **pure** water.
  e Sometimes it is necessary to work out the percentage by mass of water of crystallisation as well as the number of moles of water present in a hydrated crystal.

  (i) Use the information given below to calculate the percentage, by mass, of water of crystallisation in a sample of hydrated copper(II) sulphate.

Mass of crucible	= 13.10 g
Mass of crucible + hydrated $CuSO_4$	= 15.60 g
Mass after heating	= 14.70 g

  (ii) Calculate the number of moles of water of crystallisation driven off during the experiment as well as the number of moles of anhydrous salt remaining. (A_r: H = 1; O = 16; S = 32; Cu = 63.5)
  (iii) Using the information you have obtained in (ii), write down, in the form $CuSO_4.xH_2O$, the formula of hydrated copper(II) sulphate.

8 INORGANIC CARBON CHEMISTRY

Limestone

Environmental Problems in the Yorkshire Dales

Local residents have been protesting against proposals to site a new limestone quarry in the beautiful Yorkshire Dales. However, limestone is a useful and sought-after mineral, so demand has encouraged mining in National Park areas.

As the newspaper article above says, limestone is found in the Yorkshire Dales, an example being the cliffs at Malham Cove (Figure 8.1). In England, it is also found in the Derbyshire Peaks, Sussex Downs and Mendip Hills. As well as giving beautiful and varied countryside, limestone is a very useful raw material.

Limestone is one form of calcium carbonate ($CaCO_3$); however, calcium carbonate is also found in nature as chalk, calcite and marble. It is the second most abundant mineral in the Earth's crust after the different types of silicates (which include clay, granite and sandstone).

Deposits of chalk were formed from the shells of dead marine creatures that lived many millions of years ago. In several places in Great Britain, the chalk was covered with other types of rock and, therefore, was put under great pressure. This caused the relatively soft chalk to change into the harder material limestone.

Marble, one of the other major forms of calcium carbonate, formed in places where the chalk was subjected not only to a high pressure but also to a high temperature.

Figure 8.1 These cliffs at Malham Cove are made from limestone

In a typical year, about 80 million tonnes of limestone are quarried in Great Britain. Although limestone is cheap to quarry, as it is found near the surface, there are some environmental costs in its extraction. List the environmental issues which could arise through the quarrying of limestone.

97

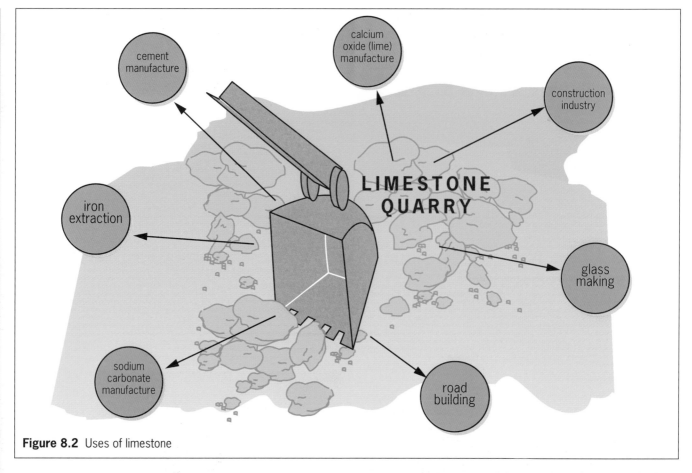

Figure 8.2 Uses of limestone

Direct uses of limestone

Limestone has a variety of uses in, for example, making of cement, road building, glass making and the extraction of iron (Figure 8.2).

Neutralising acid soil

Powdered limestone is most often used to neutralise acid soil (Figure 8.3) because it is cheaper than any form of lime (calcium oxide), which has to be produced by heating limestone (see p. 99). The reaction of limestone with acidic soil can be shown by the following ionic equation.

$$\begin{array}{ccc} \text{carbonate ion} + \text{hydrogen ion} & \rightarrow & \text{carbon} + \text{water} \\ \text{\small (from CaCO}_3\text{)} \quad \text{\small (from acid soils)} & & \text{dioxide} \\ CO_3^{2-}(s) \quad + \quad 2H^+(aq) & \rightarrow & CO_2(g) + H_2O(l) \end{array}$$

Figure 8.3 Spreading limestone onto soil

Manufacture of iron and steel

In the blast furnace, limestone is used to remove earthy and sandy materials found in the iron ore. A liquid slag is formed, which is mainly calcium silicate. More details of the extraction of iron and its conversion into steel are given in Chapter 9.

Manufacture of cement and concrete

Limestone (or chalk) is mixed with clay (or shale) in a heated rotary kiln, using coal or oil as the fuel (Figure 8.4). The material produced is called cement. It contains a mixture of calcium aluminate $(Ca(AlO_2)_2)$ and calcium silicate $(CaSiO_3)$. The dry

Figure 8.4 A modern rotary kiln

product is ground to a powder and then a little calcium sulphate ($CaSO_4$) is added to slow down the setting rate of the cement. When water is added to the mixture complex chemical changes occur, resulting in a hard interlocking mass of crystals of hydrated calcium aluminate and silicate.

Concrete is a mixture of cement with stone chippings and gravel which give it body. After the ingredients have been mixed with water they are poured into wooden moulds and allowed to set hard. Reinforced concrete is made by allowing concrete to set around steel rods or mesh to give greater tensile strength, which is required for the construction of large bridges (Figure 8.5) and tall buildings.

Manufacture of sodium carbonate — the Solvay process

Sodium carbonate (Na_2CO_3) is one of the world's most important industrial chemicals. It is used in the manufacture of soaps, detergents, dyes, drugs and other chemicals. The world production of sodium carbonate now exceeds 30 million tonnes per year.

Sodium carbonate is manufactured ingeniously using calcium carbonate, sodium chloride, carbon dioxide and ammonia in a continuous process. This ensures that sodium carbonate is made as economically as possible. Both the carbon dioxide and the ammonia are recycled continuously.

Indirect uses of limestone

Lime manufacture

When calcium carbonate is heated strongly it thermally dissociates (breaks up reversibly) to form calcium oxide (lime) and carbon dioxide.

calcium carbonate $\rightleftharpoons$ calcium oxide + carbon dioxide

$$CaCO_3(s) \rightleftharpoons CaO(s) + CO_2(g)$$

This reaction can go in either direction, depending on the temperature and pressure used. This reaction is an important industrial process and takes place in a lime kiln (Figure 8.6).

Calcium oxide (CaO) is a base and is still used by some farmers to spread on fields to neutralise soil

Figure 8.5 The Humber suspension bridge

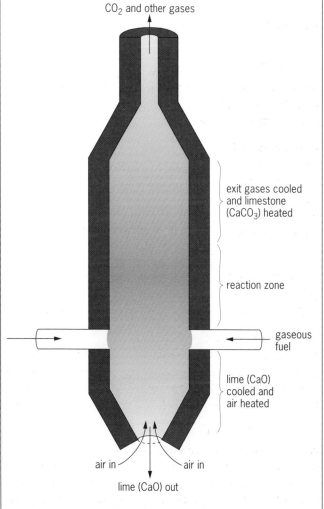

Figure 8.6 The calcium oxide produced from this process is known as quicklime or lime

acidity and to improve drainage of water through soils that contain large amounts of clay. It also has a use as a drying agent in industry. Soda glass is made by heating sand with soda (sodium carbonate, Na_2CO_3) and lime (Figure 8.7). For further discussion of glasses see Chapter 4, p. 56.

Large amounts of calcium oxide are also converted into calcium hydroxide ($Ca(OH)_2$) which is called **slaked lime**.

Manufacture of calcium hydroxide — slaked lime

Calcium hydroxide is a cheap industrial alkali (Figure 8.8). It is used in large quantities to make bleaching powder, by some farmers to reduce soil acidity, in the manufacture of whitewash, in glass manufacture and in water purification. Calcium hydroxide, a white powder, is produced by adding an equal amount of water to calcium oxide.

This process can be shown on the small scale in the laboratory by heating a lump of limestone very strongly to convert it to a calcium oxide. Water can then be carefully added dropwise to the calcium oxide. An exothermic reaction takes place as the water and calcium oxide react together (slaking) to form calcium hydroxide.

$$\text{calcium oxide} + \text{water} \rightarrow \text{calcium hydroxide}$$
$$CaO(s) \quad + H_2O(l) \rightarrow \quad Ca(OH)_2(s)$$

A weak solution of calcium hydroxide in water is called **limewater**. It is used to test for carbon dioxide gas, as a white solid of calcium carbonate is formed if carbon dioxide gas is mixed with it (Figure 8.9).

$$\begin{array}{ccccc} \text{calcium} & + \text{carbon} & \rightarrow & \text{calcium} & + \text{water} \\ \text{hydroxide} & \text{dioxide} & & \text{carbonate} & \end{array}$$
$$Ca(OH)_2(aq) + CO_2(g) \rightarrow CaCO_3(s) + H_2O(l)$$

Calcium hydroxide (slaked lime) is mixed with sand to give mortar. When it is mixed with water and

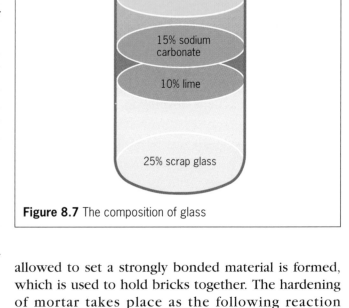

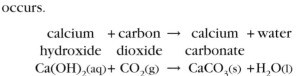

Figure 8.7 The composition of glass

allowed to set a strongly bonded material is formed, which is used to hold bricks together. The hardening of mortar takes place as the following reaction occurs.

$$\begin{array}{ccccc} \text{calcium} & + \text{carbon} & \rightarrow & \text{calcium} & + \text{water} \\ \text{hydroxide} & \text{dioxide} & & \text{carbonate} & \end{array}$$
$$Ca(OH)_2(aq) + CO_2(g) \rightarrow CaCO_3(s) + H_2O(l)$$

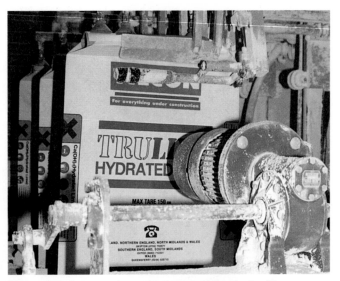

Figure 8.8 Large-scale production of calcium hydroxide

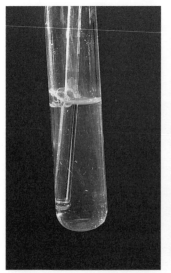

Figure 8.9 Testing for carbon dioxide

a Some suggested building alternatives to limestone are sandstone and granite. What would be the benefits of using either of these two materials instead of limestone?

b Devise an experiment that you could carry out in the laboratory to establish whether or not powdered limestone is better at curing soil acidity than calcium oxide.

c Why is calcium oxide described as a 'base' in this reaction?

Figure 8.10 Three naturally occurring carbonates — dolomite (left) with magnesite in front and malachite (right)

Carbonates and hydrogencarbonates

Carbonates form an important range of compounds. They are all salts of carbonic acid (H_2CO_3) and contain the carbonate ion (CO_3^{2-}). Many of them occur naturally in rock formations. For example, in addition to limestone, chalk and marble, malachite is copper(II) carbonate and dolomite is magnesium carbonate (Figure 8.10).

Properties of carbonates

- Most metal carbonates thermally decompose when heated to form the metal oxide and carbon dioxide gas. For example,

$$\text{copper(II)} \rightarrow \text{copper(II)} + \text{carbon}$$
$$\text{carbonate} \quad \text{oxide} \quad \text{dioxide}$$
$$CuCO_3(s) \rightarrow CuO(s) + CO_2(g)$$

Group 1 metal carbonates, except for lithium carbonate, do not dissociate on heating. It is generally found that the carbonates of the more reactive metals are much more difficult to dissociate than, for example, copper(II) carbonate.

- Carbonates are generally insoluble in water except for those of sodium, potassium and ammonium.
- Carbonates react with acids to form salts, carbon dioxide and water (Chapter 7, p. 86). For example, calcium carbonate reacts with dilute hydrochloric acid to form calcium chloride, carbon dioxide and water.

$$\text{calcium} + \text{hydrochloric} \rightarrow \text{calcium} + \text{carbon} + \text{water}$$
$$\text{carbonate} \quad \text{acid} \quad \text{chloride} \quad \text{dioxide}$$
$$CaCO_3(s) + 2HCl(aq) \rightarrow CaCl_2(aq) + CO_2(g) + H_2O(l)$$

- This reaction is used in the laboratory preparation of carbon dioxide. It is also used as a test for a carbonate because the reaction produces carbon dioxide which causes effervescence and if bubbled through limewater turns it chalky white.

Hydrogencarbonates

Almost all the hydrogencarbonates are known only in solution because they are generally too unstable to exist as solids. They contain the hydrogen carbonate ion (HCO_3^-).

Sodium hydrogencarbonate ($NaHCO_3$) is the most common solid hydrogencarbonate. It is found in indigestion remedies because it reacts with the excess stomach acid (HCl) which causes the indigestion.

$$\text{sodium} + \text{hydrochloric} \rightarrow \text{sodium} + \text{water} + \text{carbon}$$
$$\text{hydrogen-} \quad \text{acid} \quad \text{chloride} \quad \text{dioxide}$$
$$\text{carbonate}$$
$$NaHCO_3(s) + HCl(aq) \rightarrow NaCl(aq) + H_2O(l) + CO_2(g)$$

Sodium hydrogencarbonate is the substance which is put into 'plain' flour to make it into 'self-raising' flour. When heated, the sodium hydrogencarbonate in the flour decomposes to give carbon dioxide gas, which makes the bread or cakes rise.

$$\text{sodium} \rightarrow \text{sodium} + \text{water} + \text{carbon}$$
$$\text{hydrogencarbonate} \quad \text{carbonate} \quad \text{dioxide}$$
$$2NaHCO_3(s) \rightarrow Na_2CO_3(s) + H_2O(l) + CO_2(g)$$

Calcium hydrogencarbonate and magnesium hydrogencarbonate are two of the hydrogencarbonates found only in solution. They are the substances mainly responsible for the hardness in water.

Hardness in water

Rainwater dissolves carbon dioxide as it falls through the atmosphere. A small fraction of this dissolved carbon dioxide reacts with the water to produce carbonic acid, which is a weak acid.

$$\text{water} + \text{carbon dioxide} \rightleftharpoons \text{carbonic acid}$$
$$H_2O(l) + CO_2(g) \rightleftharpoons H_2CO_3(aq)$$

As this solution passes over and through rocks containing limestone and dolomite, the weak acid in the rain attacks these rocks and very slowly dissolves them. The dissolved substances are called calcium and magnesium hydrogencarbonates.

$$\text{calcium} + \text{carbonic} \rightarrow \text{calcium}$$
$$\text{carbonate} \quad \text{acid} \quad \text{hydrogencarbonate}$$
$$CaCO_3(s) + H_2CO_3(aq) \rightarrow \quad Ca(HCO_3)_2(aq)$$

Some of the rock strata may contain gypsum (calcium sulphate, $CaSO_4.2H_2O$), anhydrite ($CaSO_4$) or kieserite ($MgSO_4.H_2O$), which are very sparingly soluble in water. The presence of these dissolved substances causes the water to become **hard**. Hardness in water can be divided into two types — **temporary** and **permanent**. Temporary hardness is so called because it can be removed easily by boiling. Permanent hardness is so called because it is much more difficult to remove and certainly cannot be removed by boiling.

Temporary hardness is caused by the presence of dissolved calcium or magnesium hydrogencarbonates. Permanent hardness is caused by the presence of dissolved calcium or magnesium sulphates. When water containing these substances is evaporated, a white solid deposit is left behind (Figure 8.11). These white deposits are calcium or magnesium sulphates and/or calcium carbonate. Calcium carbonate causes the 'furring' in kettles that occurs in hard water areas. Also, blockages in hot water pipes are caused by this process (Figure 8.12).

Stalactites and stalagmites are found in underground caverns in limestone areas. They are formed from the

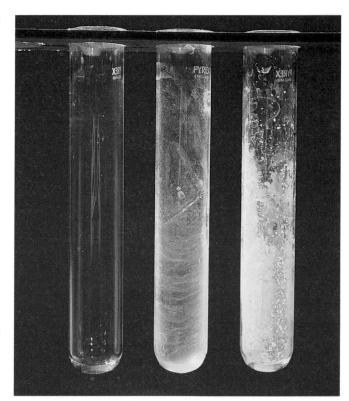

Figure 8.11 Evaporating pure water leaves no deposit, while temporary or permanent hard water leaves white deposits behind

slow decomposition of water containing calcium or magnesium hydrogencarbonates (Figure 8.13).

$$\text{calcium} \rightarrow \text{calcium} + \text{carbon} + \text{water}$$
$$\text{hydrogencarbonate} \quad \text{carbonate} \quad \text{dioxide}$$
$$Ca(HCO_3)_2(aq) \rightarrow CaCO_3(s) + CO_2(g) + H_2O(l)$$

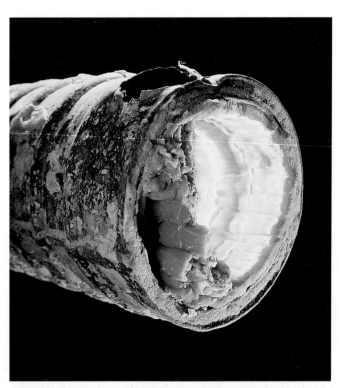

Figure 8.12 The thick deposits of limescale in this pipe have resulted from heating hot water

Figure 8.13 Stalactites and stalagmites have formed over hundreds of thousands of years

Effect of hard water on soap

If you live in one of the hard water areas shown in Figure 8.14 then you may have noticed that it is difficult to make the water lather. Instead, the water becomes cloudy. This cloudiness is caused by the presence of a solid material (a precipitate) formed by the reaction of the dissolved substances in the water with soap (sodium stearate). This white precipitate is known as **scum** (Figure 8.15).

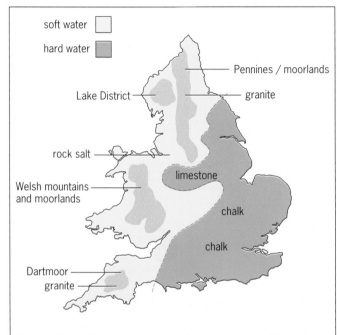

Figure 8.14 The hard and the soft water areas in England and Wales

Figure 8.15 Soap and hard water form scum

sodium + calcium → calcium + sodium
stearate hydrogen- stearate hydrogen-
(soap) carbonate (scum) carbonate

$$2NaSt(aq) + Ca(HCO_3)_2(aq) \rightarrow Ca(St)_2(s) + 2NaHCO_3(aq)$$

St = stearate   NaSt = $C_{17}H_{35}COO^-Na^+$

The amount of soap required to just produce a lather can be used to estimate the hardness in water.

To overcome the problem of scum formation, soap-less detergents have been produced which do not produce a scum because they do not react with the substances in hard water. For further discussion of soapless detergents see Chapter 14, p. 183.

Removal of hardness

Temporary hardness is easily removed from water by boiling. When heated the calcium hydrogencarbonate decomposes, producing insoluble calcium carbonate.

calcium $\xrightarrow{\text{heat}}$ calcium + water + carbon
hydrogencarbonate   carbonate   dioxide

$$Ca(HCO_3)_2(aq) \rightarrow CaCO_3(s) + H_2O(l) + CO_2(g)$$

The substances in permanently hard water are not decomposed when heated and therefore cannot be removed by boiling. Both types of hardness can be removed by any of the following.

- **Addition of washing soda ($Na_2CO_3.10H_2O$) crystals.** In each case the calcium or magnesium ion, which actually causes the hardness, is removed as a precipitate and can, therefore, no longer cause hardness.

calcium ions + carbonate ions → calcium carbonate
(from hard water)   (from washing soda)

$$Ca^{2+}(aq) + CO_3^{2-}(aq) \rightarrow CaCO_3(s)$$

- **Ion exchange**. The water is passed through a container filled with a suitable resin containing sodium ions. The calcium or magnesium ion causing the hardness is exchanged for the sodium ions in the resin (Figure 8.16).

calcium + sodium-resin → calcium-resin + sodium
ion   ion

$$Ca^{2+}(aq) + Na_2-R(s) \rightarrow Ca^{2+}-R(s) + 2Na^+(aq)$$

When all the sodium ions have been removed from the resin, it can be regenerated by pouring a solution of a suitable sodium salt through it.

- **Distillation**. The water is distilled away from the dissolved substances. This method is far too expensive to be used on a large scale.

Advantages and disadvantages of hard water

As we mentioned earlier, there are some problems associated with hard water; however, there are advantages as well. These are detailed in Table 8.1.

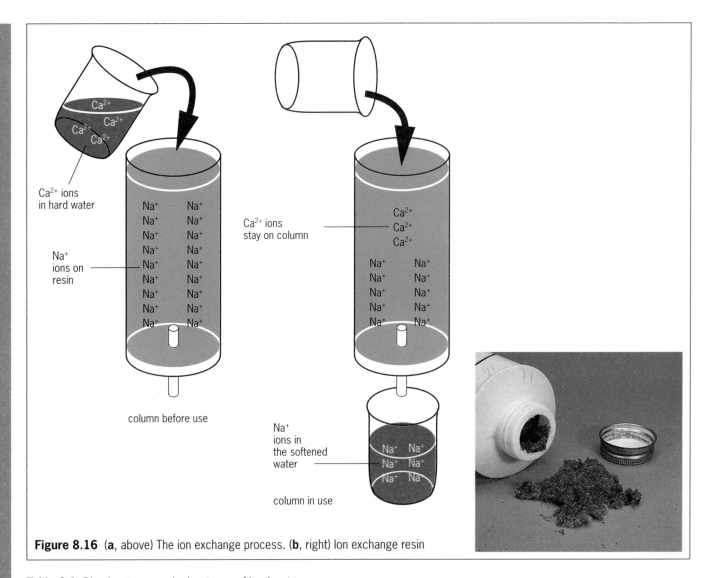

Figure 8.16 (**a**, above) The ion exchange process. (**b**, right) Ion exchange resin

Labels in figure:

Ca²⁺ ions in hard water

Na⁺ ions on resin

column before use

Ca²⁺ ions stay on column

Na⁺ ions in the softened water

column in use

Table 8.1 Disadvantages and advantages of hard water

Disadvantages	Advantages
Wastes soap	Has a nice taste
Causes kettles to fur	Calcium ions in hard water are required by the body for bones and teeth
Can cause hot water pipes to block	Coats lead pipes with a thin layer of lead(II) sulphate or lead(II) carbonate and cuts down the possibility of lead poisoning
Can spoil the finish of some fabrics	Is good for brewing beer

QUESTIONS

a Write word and balanced chemical equations to show the effect of heat on zinc carbonate ($ZnCO_3$) and nickel carbonate ($NiCO_3$).

b One of the substances found in some temporary hard waters is magnesium hydrogencarbonate. Write a word and a balanced chemical equation to show the effect of heat on this substance in aqueous solution.

c Turn to Chapter 14, p. 183, to explain why soapless detergents do not form a scum with hard water.

Carbon dioxide

Carbon forms two oxides — carbon monoxide (CO) and carbon dioxide (CO_2). Carbon dioxide is the more important of the two, and in industry large amounts of carbon dioxide are obtained from the liquefaction of air. Air contains approximately 0.03% by volume of carbon dioxide. This value has remained almost constant for a long period of time and is maintained via the **carbon cycle** (Figure 8.17). However, scientists have recently detected a slight increase in the amount of carbon dioxide in the atmosphere.

Carbon dioxide is produced by burning fossil fuels. It is also produced by all living organisms through **aerobic respiration**. Animals take in oxygen and breathe out carbon dioxide.

glucose + oxygen → carbon dioxide + water + energy

$$C_6H_{12}O_6(aq) + 6O_2(g) \rightarrow 6CO_2(g) + 6H_2O(l)$$

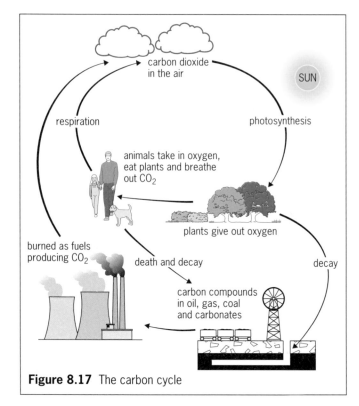

Figure 8.17 The carbon cycle

This carbon dioxide is taken in by plants through their leaves and used together with water, taken in through their roots, to synthesise sugars. This is the process of **photosynthesis**, and it takes place only in sunlight and only in green leaves, as they contain **chlorophyll** (the green pigment) which catalyses the process.

$$\text{carbon dioxide} + \text{water} \xrightarrow[\text{chlorophyll}]{\text{sunlight}} \text{glucose} + \text{oxygen}$$
$$6CO_2(g) + 6H_2O(l) \longrightarrow C_6H_{12}O_6(aq) + 6O_2(g)$$

This cycle has continued in this manner for millions of years. However, scientists have detected an imbal-ance in the carbon cycle due to the increase in the amount of carbon dioxide produced through burning fossil fuels and the deforestation of large areas of tropical rain forest. The Earth's climate is affected by the levels of carbon dioxide (and water vapour) in the atmosphere. If the amount of carbon dioxide, in particular, builds up in the air, it is thought that the average temperature of the Earth will rise. This effect is known as the **greenhouse effect** (Figure 8.18).

Some energy from the Sun is absorbed by the Earth and its atmosphere. The remainder is reflected back into space. The energy that is absorbed helps to heat up the Earth. The Earth radiates some heat energy back into space but the 'greenhouse gases', including carbon dioxide, prevent it from escaping. This effect is similar to that observed in a greenhouse where sunlight (visible/ultraviolet radiation) enters through the glass panes but heat (infrared radiation) has difficulty escaping through the glass.

The long-term effect of the higher temperatures will be the gradual melting of ice caps and consequent flooding in low-lying areas of the Earth. There will also be changes in the weather patterns which would affect agriculture worldwide.

Uses of carbon dioxide
Carbon dioxide has some important uses.

- **Carbonated drinks**. Large quantities are used to make soda and mineral waters as well as beer. The carbon dioxide gas is bubbled into the liquid under pressure, which increases its solubility.
- **Fire extinguishers**. It is used in extinguishers for use on electrical fires. Carbon dioxide is denser

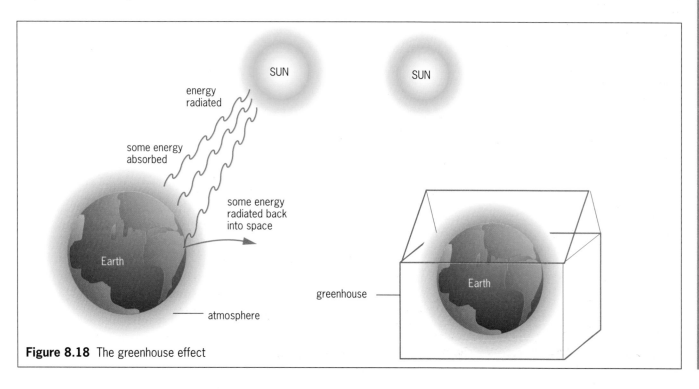

Figure 8.18 The greenhouse effect

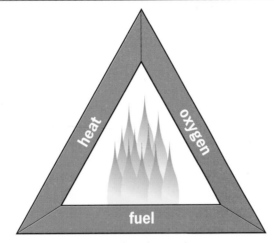

Figure 8.19 (**a**, top) Any fire needs fuel, oxygen and heat. If any of these is removed, the fire triangle is destroyed and the fire will be extinguished. (**b**, bottom) Carbon dioxide fire extinguishers starve the fire of oxygen. They are used mainly for electrical fires

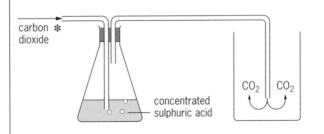

Figure 8.20 Preparation and collection of carbon dioxide gas

than air and forms a layer around the burning material. It covers the fire and starves it of oxygen. Carbon dioxide does not burn and so the fire is put out (Figure 8.19).

- **Refrigerants**. Solid carbon dioxide (dry ice) is used for refrigerating ice cream, meat and soft fruits. It is used for this purpose because it is colder than ice and it sublimes (p. 4), and so it does not pass through a potentially damaging liquid stage.
- **Special effects**. Carbon dioxide is used to create the 'smoke' effect you may see at pop concerts and on television. Dry ice is placed in boiling water and it forms thick clouds of white 'smoke'. It stays close to the floor due to the fact that carbon dioxide is denser than air.
- **Heat transfer agents**. Carbon dioxide gas is used for transferring heat in some nuclear power stations.

Laboratory preparation of carbon dioxide gas

In the laboratory the gas is made by pouring dilute hydrochloric acid onto marble chips ($CaCO_3$).

calcium + hydrochloric → calcium + water + carbon
carbonate   acid   chloride   dioxide
$$CaCO_3(s) + 2HCl(aq) → CaCl_2(aq) + H_2O(l) + CO_2(g)$$

A suitable apparatus for preparing carbon dioxide is shown in Figure 8.20.

Properties of carbon dioxide gas

Physical properties
Carbon dioxide is:

- a colourless gas
- sparingly soluble in water
- denser than air.

Chemical properties
- When bubbled into water it dissolves slightly and some of the carbon dioxide reacts, forming a solution of the weak acid carbonic acid which shows a pH of 4 or 5.

water + carbon dioxide ⇌ carbonic acid
$$H_2O(l) + CO_2(g) ⇌ H_2CO_3(aq)$$

- It will only support the combustion of strongly burning substances such as magnesium. This burning reactive metal decomposes the carbon dioxide to provide oxygen for its continued burning in the gas. This reaction is accompanied by much crackling (Figure 8.21).

magnesium + carbon → magnesium + carbon
                  dioxide   oxide
$$2Mg(s) + CO_2(g) → 2MgO(s) + C(s)$$

Figure 8.21 When magnesium burns in carbon dioxide gas, magnesium oxide (white) and carbon (black) are produced

This reaction is a good example of a **redox** process (Chapter 2, p. 15). Which of these substances has been oxidised and which has been reduced? Which of these substances is the oxidising agent and which is the reducing agent?

• When carbon dioxide is bubbled through limewater (calcium hydroxide solution), a white precipitate is formed. This white solid is calcium carbonate ($CaCO_3$). This is used as a test to show that a gas is carbon dioxide.

carbon + calcium → calcium + water
dioxide hydroxide carbonate
$$CO_2(g) + Ca(OH)_2(aq) \rightarrow CaCO_3(s) + H_2O(l)$$

If carbon dioxide is bubbled through this solution continuously then it will eventually become clear. This is because of the formation of soluble calcium hydrogencarbonate solution.

calcium + water + carbon → calcium
carbonate dioxide hydrogencarbonate
$$CaCO_3(s) + H_2O(l) + CO_2(g) \rightarrow Ca(HCO_3)_2(aq)$$

QUESTIONS

a Describe an experiment that you could carry out to show that carbonic acid is a weak acid.
b Why is it not possible to use dilute sulphuric acid to make carbon dioxide from limestone in the laboratory?
c When carbon dioxide is 'poured' from a gas jar onto a burning candle the candle goes out. What properties of carbon dioxide does this experiment show?
d In the test for carbon dioxide a precipitate of calcium carbonate is formed when the gas is bubbled through limewater. Write an ionic equation to show the formation of this precipitate.

Checklist

After studying Chapter 8 you should know and understand the following terms.

Limestone This is a form of calcium carbonate ($CaCO_3$). Other forms include chalk, calcite and marble. It is used directly to neutralise soil acidity and in the manufacture of iron and steel, glass, cement, concrete, sodium carbonate and lime.

Thermal decomposition The breakdown of a substance under the influence of heat.

Lime A white solid known chemically as calcium oxide (CaO). It is produced by heating limestone. It is used to counteract soil acidity and to manufacture calcium hydroxide (slaked lime). It is also used as a drying agent in industry.

Carbonate A salt of carbonic acid containing the carbonate ion, CO_3^{2-}, for example $CuCO_3$.

Hardness of water This is caused by the presence of calcium (or magnesium) ions in water, which form a 'scum' with soap and prevent the formation of a lather. There are two types of hardness:
• temporary hardness—caused by the presence of dissolved calcium (or magnesium) hydrogencarbonate
• permanent hardness—this results mainly from dissolved calcium (or magnesium) sulphate.

Removal of hardness Temporary hardness is removed by boiling. Both temporary and permanent hardness are removed by:
• addition of washing soda (sodium carbonate)
• ion exchange
• distillation.

Carbon dioxide A colourless, odourless gas, soluble in water, producing a weak acid called carbonic acid. It makes up 0.03% of the atmosphere. It is produced by respiration in animals and by the burning of fossil fuels. It is consumed by plants in photosynthesis.

Photosynthesis The chemical process by which green plants synthesise their carbon compounds from atmospheric carbon dioxide using light as the energy source and chlorophyll as the catalyst.

Greenhouse effect The absorption of reflected infrared radiation from the Earth by gases in the atmosphere such as carbon dioxide (a greenhouse gas) leading to atmospheric warming.

QUESTIONS

1 This question is about the limestone cycle.

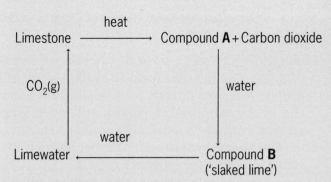

a Name and give the formula of:
  (i) compound **A**
  (ii) compound **B**.
b Write balanced chemical equations for the formation of both compounds **A** and **B**.
c Name and give the symbol/formula for the ions present in limewater.
d Describe with the aid of a balanced chemical equation what happens if carbon dioxide is bubbled through limewater.

2 The diagram shown below is a simplified version of the carbon cycle.

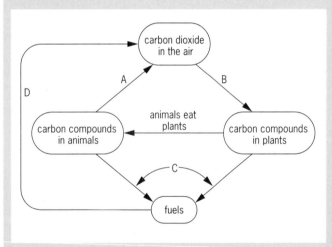

a Name the processes **A**, **B**, **C** and **D**.
b Write balanced chemical equations to represent the processes taking place in **A** and **B**.
c There are a number of fuels which could be placed in the 'fuels' box. Name three such fuels.
d 'The air contains approximately 0.03% of carbon dioxide and this amount is almost constant.' Explain why this is a true statement.
e (i) What is the 'greenhouse effect'?
  (ii) Discuss the possible consequences of the greenhouse effect if nothing is done to counteract it.

3 Limestone is an important raw material used in many different industries.
a One of the properties of limestone is that it reacts with acids.
  (i) Why do farmers spread powdered limestone on their fields?
  (ii) How can buildings made of limestone be affected by 'acid rain'?
  (iii) Write an ionic equation which would represent the reactions taking place in both (i) and (ii).
b Limestone is used in the manufacture of iron (Chapter 9, p. 117).
  (i) Why is it added to the blast furnace along with coke and haematite?
  (ii) Write chemical equations for the reactions it is involved in, in this process.
c (i) Name a building material made by heating a mixture of limestone and clay in a rotary kiln.
  (ii) What substance is the dry product produced in the rotary kiln added to?
    Explain why this substance is added.

4

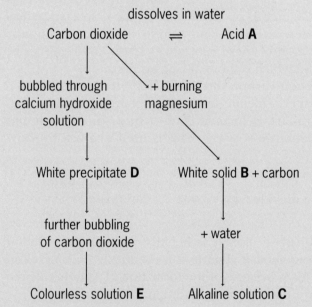

a Name and give the formulae of substances **A** to **E**.
b Write balanced chemical equations for the reactions in which compounds **B**, **C** and **E** were formed.
c Where would you expect to find acid **A**?
d Universal indicator solution was added to solution **C**. What colour did it go?
e Upon addition of dilute hydrochloric acid to solution **C**, a neutralisation reaction took place.
  (i) Write a balanced chemical equation for the reaction taking place.
  (ii) Name the salt produced in this reaction.

5 The following question is about carbon dioxide.

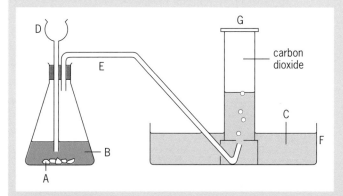

a Name and give the formula of each of the substances **A**, **B** and **C**.

b Identify by name the different pieces of apparatus **D**, **E**, **F** and **G**.

c Draw and label the apparatus that should be used if a dry sample of carbon dioxide gas was required.

d When a gas jar containing carbon dioxide is held over a burning wooden splint and the cover removed, the flame goes out. State two properties of carbon dioxide illustrated by this observation.

e Carbon dioxide is also produced when copper(II) carbonate is heated strongly.
  (i) Write a balanced chemical equation for the reaction taking place.
  (ii) Name the process which is taking place as the copper(II) carbonate is heated.
  (iii) Calculate the volume of carbon dioxide that would be produced (measured at room temperature and pressure (rtp)) if 12.35 g of copper(II) carbonate were heated strongly. (One mole of any gas occupies 24 dm^3 at rtp. A_r: Cu = 63.5; C = 12; O = 16)

6 Explain the following.

a Industry normally requires water which has been softened.

b Hard water is good for the promotion of healthy bones and teeth.

c Hard water causes kettles to fur.

d Hard water wastes soap.

e Hard water can coat lead pipes and reduce the possibility of lead poisoning.

7 The results of testing five samples of water from different parts of the country are shown in the table below. The soap solution was gradually added to 25 cm^3 of each sample of water with shaking until a permanent lather (one which lasts for 20 seconds) was obtained.

Water sample (25 cm^3)	Volume of soap solution added (cm^3)	
	Before boiling	After boiling
A	12	1
B	13	6
C	11	11
D	14	3
E	16	16

a (i) Which samples are permanently hard?
  (ii) Which samples are temporarily hard?
  (iii) Which sample contains both temporary and permanent hardness?

b Name a compound which could be present in sample **D** but not in sample **E**.

c Name a compound which could be present in sample **E** but not in sample **D**.

d Explain how the compound you have named in (c) gets into the water.

e Sample **E** was distilled. The water collected was tested with soap solution. What volume of soap solution might you expect to be added to produce a permanent lather? Comment on your answer.

8 Lime (calcium oxide) is produced in very large quantities in a lime kiln. The equation for the reaction taking place is:

$$CaCO_3(s) \rightarrow CaO(s) + CO_2(g)$$

a How much lime would be produced by heating 110 tonnes of limestone? (A_r: C = 12; Ca = 40; O = 16)

b Why is the carbon dioxide swept out of the lime kiln?

c Give three uses of lime.

d (i) What problems are associated with the large-scale quarrying of limestone?
  (ii) What steps have been taken to overcome or reduce the problems you have outlined in (i)?

9 METAL REACTIVITY

You have already seen in Chapter 2, p. 13, that metals usually have similar physical properties. However, they differ in other ways. Look closely at the three photographs below in Figure 9.1 (a–c).

Potassium is soft and reacts violently with both air and water. Iron also reacts with air and water but much more slowly, forming rust. Gold, however, remains totally unchanged after many hundreds of years. Potassium is said to be more reactive than iron and, in turn, iron is said to be more reactive than gold.

Figure 9.1 (**a**, right) Potassium burning in air/oxygen. (**b**, bottom left) Iron rusts when left unprotected. (**c**, bottom right) Gold is used in leaf form on this giant Buddha as it is unreactive

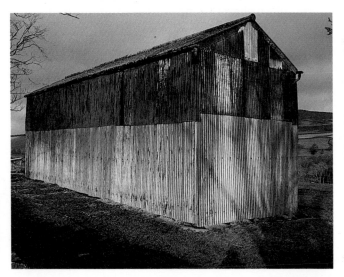

Metal reactions

By carrying out reactions in the laboratory with other metals and air, water and dilute acid, it is possible to produce an order of reactivity of the metals.

With acid

Look closely at the photograph in Figure 9.2 showing magnesium metal reacting with dilute hydrochloric acid. You will notice effervescence, which is caused

Figure 9.2 Effervescence occurs when magnesium is put into acid

by bubbles of hydrogen gas being formed as the reaction between the two substances proceeds. The other product of this reaction is the salt, magnesium chloride.

magnesium + hydrochloric → magnesium + hydrogen
acid   chloride

$$Mg(s) + 2HCl(aq) \rightarrow MgCl_2(aq) + H_2(g)$$

If a metal reacts with dilute hydrochloric acid then hydrogen and the metal chloride are produced.

If similar reactions are carried out using other metals with acid, an order of reactivity can be produced. This is known as a **reactivity series**. An order of reactivity, giving the most reactive metal first, using results from experiments with dilute acid, is shown in Table 9.1. The table also shows how the metals react with air/oxygen and water/steam, and, in addition, the ease of extraction of the metal.

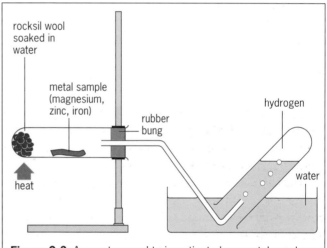

Figure 9.3 Apparatus used to investigate how metals such as magnesium react with steam

Table 9.1 Order of reactivity of metals

Reactivity series	Reaction with dilute acid	Reaction with air/oxygen	Reaction with water	Ease of extraction
Potassium (K)	Produce H₂ with decreasing vigour	Burn very brightly and vigorously	Produce H₂ with decreasing vigour with cold water	Difficult to extract
Sodium (Na)				
Calcium (Ca)		Burn to form an oxide with decreasing vigour		Easier to extract
Magnesium (Mg)			React with steam with decreasing vigour	
(Aluminium (Al)*)				
Zinc (Zn)				
Iron (Fe)				
Lead (Pb)		React slowly to form the oxide	Do not react with cold water or steam	
Copper (Cu)	Do not react with dilute acids			Found as the element (native)
Silver (Ag)		Do not react		
Gold (Au)				
Platinum (Pt)				

Increasing Reactivity

*Because aluminium reacts so readily with the oxygen in the air, a protective oxide layer is formed on its surface. This often prevents any further reaction and disguises aluminium's true reactivity. This gives us the use of a light and strong metal (Figure 9.5)

With air/oxygen

Many metals react directly with oxygen to form oxides. For example, magnesium burns brightly in oxygen to form the white powder magnesium oxide.

$$\text{magnesium} + \text{oxygen} \rightarrow \text{magnesium oxide}$$
$$2Mg(s) + O_2(g) \rightarrow 2MgO(s)$$

With water/steam

Reactive metals such as potassium, sodium and calcium react with cold water to produce the metal hydroxide and hydrogen gas. For example, the reaction of sodium with water produces sodium hydroxide and hydrogen.

$$\text{sodium} + \text{water} \rightarrow \text{sodium hydroxide} + \text{hydrogen}$$
$$2Na(s) + 2H_2O(l) \rightarrow 2NaOH(aq) + H_2(g)$$

The moderately reactive metals, magnesium, zinc and iron, react slowly with water. They will, however, react more rapidly with steam (Figure 9.3). In their reaction with steam, the metal oxide and hydrogen are formed. For example, magnesium produces magnesium oxide and hydrogen gas.

$$\text{magnesium} + \text{steam} \rightarrow \text{magnesium oxide} + \text{hydrogen}$$
$$Mg(s) + H_2O(g) \rightarrow MgO(s) + H_2(g)$$

Generally, it is the unreactive metals that we find the most uses for; for example, the metals iron and copper can be found in many everyday objects (Figure 9.4). However, magnesium is one of the metals used in the construction of Concorde (Figure 9.5).

Both sodium and potassium are so reactive that they have to be stored under oil to prevent them from coming into contact with water or air. However, because they have low melting points and are good conductors of heat, they are used as coolants for nuclear reactors.

QUESTIONS

a Write balanced chemical equations for the reactions between:
  (i) iron and dilute hydrochloric acid
  (ii) calcium and oxygen
  (iii) potassium and water.
b Make a list of six things you have in your house made from copper or iron.
  Give a use for each of the other unreactive metals shown in the reactivity series.

Figure 9.4 (Above)
(**a**, left) This stove is made of iron. (**b**, right) Copper pots

Figure 9.5 (Right) Concorde is made of an alloy which contains magnesium

Using the reactivity series

What predictions can be made using the reactivity series? It is useful in predicting how metals react.

Competition reactions in the solid state

If a more reactive metal is heated with the oxide of a less reactive metal, then it will remove the oxygen from it (as the oxide anion). You can see from the reactivity series that iron is less reactive than aluminium. If iron(III) oxide is mixed with aluminium and the mixture is heated using a magnesium fuse (Figure 9.6), a very violent reaction occurs as the competition between the aluminium and the iron for the oxygen takes place.

The aluminium, being the more reactive metal, takes the oxygen from the less reactive iron. It is a very exothermic reaction. When the reaction is over, a solid lump of iron is left along with a lot of white aluminium oxide powder.

$$\text{iron(III) oxide} + \text{aluminium} \xrightarrow{\text{heat}} \text{aluminium oxide} + \text{iron}$$
$$Fe_2O_3(s) + 2Al(s) \longrightarrow Al_2O_3(s) + 2Fe(s)$$

This is a redox reaction (see Chapter 2, p. 15 for a discussion of this type of reaction).

This particular reaction is known as the **Thermit reaction** (Figure 9.7). Since large amounts of heat are given out and the iron is formed in a molten state, this reaction is used to weld together damaged railway

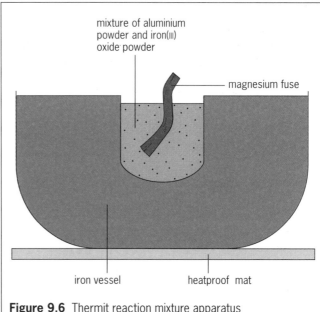

Figure 9.6 Thermit reaction mixture apparatus

lines. It is also used for incendiary bombs (Figure 9.8).

Some metals, such as chromium and titanium, are prepared from their oxides using this type of competition reaction.

Competition reactions in aqueous solutions

In another reaction, metals compete with each other for other anions. This type of reaction is known as a **displacement reaction**. As in the previous type of competitive reaction, the reactivity series can be used to predict which of the metals will 'win'.

Figure 9.7 The Thermit reaction is used to weld damaged railway lines

Figure 9.8 Incendiary bombs were used during World War II

113

In a displacement reaction, a more reactive metal will displace a less reactive metal from a solution of its salt. Zinc is above copper in the reactivity series. Figure 9.9 shows what happens when a piece of zinc metal is left to stand in a solution of copper(II) nitrate. The copper(II) nitrate slowly loses its blue colour as the zinc continues to displace the copper from the solution. The solution eventually becomes colourless zinc nitrate.

zinc + copper(II) nitrate $\rightarrow$ zinc nitrate + copper

$Zn(s) + Cu(NO_3)_2(aq) \rightarrow Zn(NO_3)_2(aq) + Cu(s)$

The ionic equation for this reaction is:

zinc + copper ions $\rightarrow$ zinc ions + copper

$Zn(s) + Cu^{2+}(aq) \rightarrow Zn^{2+}(aq) + Cu(s)$

This is also a redox reaction involving the transfer of two electrons from the zinc metal to the copper ions. The zinc is oxidised to zinc ions in aqueous solution, while the copper ions are reduced. (See Chapter 6, p. 71, for a discussion of oxidation and reduction in terms of electron transfer.)

It is possible to confirm the reactivity series for metals using competition reactions of the types discussed in this section.

Figure 9.9 Zinc displaces copper

QUESTIONS

a Predict whether or not the following reactions will take place:
  (i) magnesium + copper(II) oxide
  (ii) iron + aluminium oxide
  (iii) calcium + magnesium oxide.
  Complete the word equations, and write balanced chemical and ionic equations for those reactions which do take place.

b Predict whether or not the following reactions will take place:
  (i) magnesium + calcium nitrate solution
  (ii) iron + copper(II) nitrate solution
  (iii) copper + silver nitrate solution.
  Complete the word equations, and write balanced chemical and ionic equations for those reactions which do take place.

For example,

iron(III) + sodium $\rightarrow$ iron(III) + sodium
chloride hydroxide   hydroxide   chloride

$FeCl_3(aq) + 3NaOH(aq) \rightarrow Fe(OH)_3(s) + 3NaCl(aq)$

The ionic equation for this reaction is:

iron(III) ions + hydroxide ions $\rightarrow$ iron(III) hydroxide

$Fe^{3+}(aq) + 3OH^-(aq) \rightarrow Fe(OH)_3(s)$

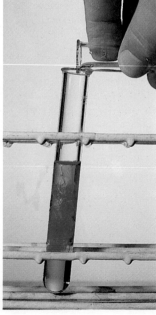

Figure 9.10 (a, left) Iron(III) hydroxide is precipitated. **(b**, right) Copper(II) hydroxide is precipitated

Identifying metal ions

When an alkali dissolves in water, it produces hydroxide ions. It is known that most metal hydroxides are insoluble. So if hydroxide ions from a solution of an alkali are added to a solution of a metal salt, an insoluble, often coloured, metal hydroxide is precipitated from solution (Figure 9.10).

Table 9.2 shows some of the colours of insoluble metal hydroxides. The colours of these insoluble metal hydroxides can be used to identify the metal cations present in solution.

Amphoteric hydroxides

The hydroxides of metals are basic and they react with acids to form salts (Chapter 7, p. 87). The hydroxides of some metals, however, will also react with strong bases, such as sodium hydroxide, to form soluble salts. Hydroxides of this type are said to be **amphoteric**. For example,

zinc + hydrochloric → zinc + water
hydroxide acid chloride
$$Zn(OH)_2(aq) + 2HCl(aq) \rightarrow ZnCl_2(aq) + 2H_2O(l)$$

and

zinc hydroxide + sodium hydroxide → sodium zincate
$$Zn(OH)_2(aq) + 2NaOH(aq) \rightarrow Na_2Zn(OH)_4(aq)$$

Other amphoteric hydroxides are lead hydroxide $(Pb(OH)_2)$ and aluminium hydroxide $(Al(OH)_3)$. This sort of behaviour can be used to help identify these metal cations, as their hydroxides are soluble in strong bases.

It should be noted that the oxides of the metals used as examples above are also amphoteric.

QUESTIONS

a Write ionic equations for the reactions which take place to produce the metal hydroxides shown in Table 9.2.
b Describe what you would see when sodium hydroxide is added slowly to a solution containing lead(II) nitrate.

Discovery of metals

Metals have been used since prehistoric times. Primitive iron tools have been excavated. These were probably made from small amounts of native iron found in lumps of rock from meteorites. It was not until about 2500 BC that iron became more widely used. This date marks the dawn of the iron age, when people learned how to get iron from its **ores** in larger quantities by reduction using charcoal. An ore is a naturally occurring mineral from which a metal can be extracted.

Over the centuries other metals, which like iron are also relatively low in the reactivity series, were isolated in a similar manner. These included copper, lead, tin and zinc. However, due to the relatively low abundance of the ores containing these metals, they were not extracted and used in large amounts.

Table 9.2 Some of the colours of insoluble metal hydroxides

Name	Formula	Colour of hydroxide
Aluminium hydroxide	$Al(OH)_3$	White
Copper(II) hydroxide	$Cu(OH)_2$	Blue
Iron(II) hydroxide	$Fe(OH)_2$	Green
Calcium hydroxide	$Ca(OH)_2$	White
Iron(III) hydroxide	$Fe(OH)_3$	Brown (rust)
Zinc hydroxide	$Zn(OH)_2$	White

Metals high in the reactivity series have proved very difficult to isolate. It was not until more recent times, through Sir Humphry Davy's work on electrolysis, that potassium (1807), sodium (1807), calcium (1808) and magnesium (1808) were isolated. Aluminium, the most plentiful reactive metal in the Earth's crust, was not extracted from its ore until 1827, by Friedrich Wöhler (p. 70), and the extremely reactive metal rubidium was not isolated until 1861 by Robert Bunsen and Gustav Kirchhoff.

Extraction of metals from their ores

The majority of metals are too reactive to exist on their own in the Earth's crust, and they occur naturally in rocks as compounds in ores (Figure 9.11). These ores are usually carbonates, oxides or sulphides of the metal, mixed with impurities.

Some metals, such as gold and silver, occur in a native form (Figure 9.12) as the free metal. They are

Figure 9.11 Metal ores — chalcopyrite (left) and galena

Figure 9.12 Gold crystals

Table 9.3 Some common ores

Metal	Name of ore	Chemical name of compound in ore	Formula	Usual method of extraction
Aluminium	Bauxite	Aluminium oxide	$Al_2O_3.2H_2O$	Electrolysis of oxide dissolved in molten cryolite
Copper	Copper pyrites	Copper iron sulphide	$CuFeS_2$	The sulphide ore is roasted in air
Iron	Haematite	Iron(III) oxide	Fe_2O_3	Heat oxide with carbon
Sodium	Rock salt	Sodium chloride	NaCl	Electrolysis of molten sodium chloride
Zinc	Zinc blende	Zinc sulphide	ZnS	Sulphide is roasted in air and the oxide produced is heated with carbon

very unreactive and have withstood the action of water and the atmosphere for many thousands of years without reacting to become compounds.

Some of the common ores are shown in Table 9.3.

Large lumps of the ore are first crushed and ground up by very heavy machinery. Some ores are already fairly concentrated when mined. For example, in some parts of the world, haematite contains over 80% Fe_2O_3. However, other ores, such as copper pyrites, are often found to be less concentrated, with only 1% or less of the copper compound, and so they have to be concentrated before the metal can be extracted. The method used to extract the metal from its ore depends on the position of the metal in the reactivity series.

Figure 9.13 A blast furnace

Extraction of reactive metals

Because reactive metals, such as sodium, hold on to the element(s) they have combined with, they are usually difficult to extract. For example, sodium chloride (as rock salt) is an ionic compound with the Na^+ and Cl^- ions strongly bonded to one another. Consequently, the separation of these ions and the subsequent isolation of the sodium metal is difficult.

Electrolysis of the molten, purified ore is the method used in these cases. During this process, the metal is produced at the cathode while a non-metal is produced at the anode. As you might expect, extraction of metal by electrolysis is expensive. In order to keep costs low, many metal smelters using electrolysis are situated in regions where there is hydro-electric power. Hydroelectric power is discussed further in Chapter 13, p. 163.

For further discussion of the extraction of the reactive metals sodium and aluminium, see Chapter 6, p. 70.

Extraction of fairly reactive metals

Metals towards the middle of the reactivity series, such as iron and zinc, may be extracted by reducing the metal oxide with the non-metal carbon.

Iron

Iron is extracted mainly from its oxides, haematite (Fe_2O_3) and magnetite (Fe_3O_4), in a blast furnace

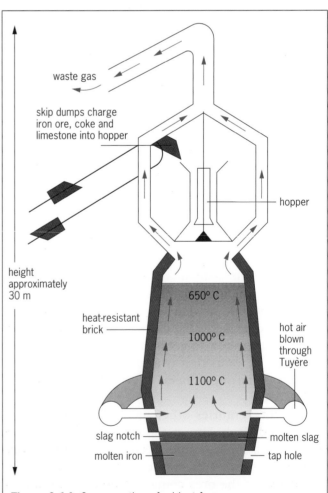

Figure 9.14 Cross-section of a blast furnace

(Figures 9.13 and 9.14). The blast furnace is a steel tower approximately 30 m high lined with heat-resistant bricks. It is loaded with the 'charge' of iron ore (usually haematite), coke (made by heating coal) and limestone (calcium carbonate). A blast of hot air is sent in near the bottom of the furnace through holes (tuyères) which makes the 'charge' glow, as the coke burns in the preheated air.

$$\text{carbon} + \text{oxygen} \rightarrow \text{carbon dioxide}$$
$$C(s) \quad + \quad O_2(g) \quad \rightarrow \qquad CO_2(g)$$

A number of chemical reactions then follow.

- The limestone begins to decompose:

$$\begin{array}{c}\text{calcium} \rightarrow \text{calcium} + \text{carbon} \\ \text{carbonate} \quad \text{oxide} \quad \text{dioxide} \end{array}$$
$$CaCO_3(s) \rightarrow CaO(s) + CO_2(g)$$

- The carbon dioxide gas produced reacts with more hot coke higher up in the furnace, producing carbon monoxide.

$$\text{carbon dioxide} + \text{coke} \rightarrow \text{carbon monoxide}$$
$$CO_2(g) \qquad + C(s) \rightarrow \qquad 2CO(g)$$

- Carbon monoxide is a reducing agent. It rises up the furnace and reduces the iron(III) oxide ore. This takes place at a temperature of around 700 °C:

$$\begin{array}{c}\text{iron(III)} + \text{carbon} \rightarrow \text{iron} + \text{carbon} \\ \text{oxide} \quad \text{monoxide} \qquad \text{dioxide} \end{array}$$
$$Fe_2O_3(s) + 3CO(g) \rightarrow 2Fe(l) + 3CO_2(g)$$

The molten iron produced trickles to the bottom of the furnace.

- The calcium oxide formed from the limestone reacts with acidic impurities, for example silicon(IV) oxide (sand) (SiO$_2$), in the iron ore to form a liquid **slag**, which is mainly calcium silicate.

$$\begin{array}{c}\text{calcium} + \text{silicon(IV)} \rightarrow \text{calcium} \\ \text{oxide} \quad \text{oxide} \quad \text{silicate} \end{array}$$
$$CaO(s) + SiO_2(s) \rightarrow CaSiO_3(l)$$

This material also trickles to the bottom of the furnace, but because it is less dense than the molten iron, it floats on top of it. The molten iron, as well as the molten slag, may be **tapped off** (run off) at intervals.

The waste gases, mainly nitrogen and oxides of carbon, escape from the top of the furnace. They are used in a heat exchange process to heat incoming air and so help to reduce the energy costs of the process. Slag is the other waste material. It is used by builders and road makers (Figure 9.15) for foundations.

The extraction of iron is a continuous process and is much cheaper to run than an electrolytic method.

Figure 9.15 Slag is used in road foundations

The iron obtained by this process is known as 'pig' or cast iron and contains about 4% carbon (as well as some other impurities). The name pig iron arises from the fact that if it is not subsequently converted into steel it is poured into moulds called pigs. Because of its brittle and hard nature, the iron produced by this process has limited use. Gas cylinders are sometimes made of cast iron, since they are unlikely to get deformed during their use.

The majority of the iron produced in the blast furnace is converted into different steel alloys (p. 123) such as manganese and tungsten steels as well as stainless steel (p. 122).

Zinc

The principal source of zinc is zinc sulphide or zinc blende. This ore occurs mainly in Australia, Canada and the US. The zinc ore is first concentrated by a process called **froth flotation**. The crushed ore is fed into tanks of water containing a chemical collector and frothing agent. The collector sticks to the surface of the zinc sulphide particles giving them a water-repellant coating. Air is blown through so that the whole mixture froths up. The zinc sulphide particles are forced up the tank by the air bubbles and are skimmed off and dried.

This ore now contains 55-75% of zinc sulphide. The zinc sulphide is then heated very strongly in a current of air in a furnace (Figure 9.16) to convert it to the oxide:

zinc sulphide + oxygen → zinc oxide + sulphur dioxide
$$2ZnS(s) + 3O_2(s) → 2ZnO(s) + 2SO_2(g)$$

The sulphur dioxide is a useful co-product and is used in the manufacture of sulphuric acid. The zinc oxide is mixed with powdered coke in a furnace and heated very strongly to a temperature of approximately 1400 °C. The zinc oxide is reduced by the coke to zinc.

zinc oxide + coke (carbon) → zinc + carbon monoxide
$$ZnO(s) + C(s) → Zn(g) + CO(g)$$

The mixture of zinc vapour and carbon monoxide passes through an outlet near the top of the furnace and the zinc metal cools and condenses. The heating costs of the furnace are reduced by burning the carbon monoxide which is produced.

Zinc is used in alloys such as brass. It is also used to galvanise steel and for electrodes in batteries.

Extraction of unreactive metals

Copper

Copper is quite a long way down the reactivity series. Copper can be found as the free metal element or 'native' in the US. It is principally extracted, however, from copper pyrites, $CuFeS_2$. The crushed ore is concentrated by froth flotation. A chemical known as a collector is added to an ore/water mixture. The collector sticks to the surface of the copper pyrites particles, giving them a water-repellant coating. Detergent is added and air is blown into the mixture to make it froth. The copper pyrites particles are concentrated in the froth and can be removed easily. They are then roasted in a limited supply of air to ensure conversion of copper pyrites to copper(I) sulphide:

copper + oxygen → copper(I) + sulphur + iron(II)
pyrites   sulphide   dioxide   oxide
$$2CuFeS_2(s) + 4O_2(g) → Cu_2S(s) + 3SO_2(g) + 2FeO(s)$$

Silicon is then added and the mixture is heated in the absence of air. The iron(II) oxide is converted into iron(II) silicate ($FeSiO_3$), which is run off. The remaining copper(I) sulphide is then reduced to copper by heating in a controlled amount of air.

copper(I) + oxygen → copper + sulphur
sulphide   dioxide
$$Cu_2S(s) + O_2(g) → 2Cu(s) + SO_2(g)$$

Copper is then refined by electrolysis (Chapter 6, p. 75) to give a product which is at least 99.92%

Figure 9.16 Zinc is extracted in a furnace like this one

Figure 9.17 Copper is alloyed to produce brass

pure. The purified copper is easily drawn into wires (it is highly ductile), which makes it useful for electrical wiring. It is also used in alloys, such as bronze and brass (Figure 9.17). Copper is also used to make water and central heating pipes, as well as steam boilers (it is a good conductor of heat).

Silver and gold

These are very unreactive metals. Silver exists mainly as silver sulphide, Ag_2S (silver glance). The extraction involves treatment of the pulverised ore with sodium cyanide. Zinc is then added to displace the silver from solution. The pure metal is obtained by electrolysis. Silver also exists to a small extent native in the Earth's crust. Gold is nearly always found native (Figure 9.18). It is also obtained in significant amounts during the electrolytic refining of copper as well as during the extraction of lead.

Silver and gold, because of their resistance to corrosion, are used to make jewellery. Both of these metals are used in the electronics industry because of their high electrical conductivity.

Figure 9.18 This man is panning for gold

QUESTIONS

a How does the method used for extracting a metal from its ore depend on the metal's position in the reactivity series?

b 'It is true to say that almost all the reactions by which a metal is extracted from its ore are reduction reactions.' Discuss this statement with respect to the extraction of iron, aluminium and zinc.

c Suggest a method which could be used to extract magnesium from magnesium chloride.

d With reference to the properties of metals and alloys, suggest reasons why:
  (i) gold is used in the electronics industry
  (ii) copper is used to make water and central-heating pipes.

Recycling metals

Recycling 'banks' have become commonplace in recent years (Figure 9.19). Why should we really want to recycle metals? Certainly, if we extract less metals from the Earth then the existing reserves will last that much longer. Also, recycling metals prevents the creation of a huge environmental problem (Figure 9.20). However, one of the main considerations is that it saves money.

The main metals which are recycled include aluminium and iron. Aluminium is saved by many households as drinks cans and milk bottle tops, to be melted down and recast. Iron is collected at local authority tips in the form of discarded household goods and it also forms a large part of the materials collected by scrap metal dealers. Iron is recycled to steel. Many steel-making furnaces run mainly on scrap iron.

Figure 9.19 Aluminium can recycling

Figure 9.20 If we did not recycle metals, then this sight would be commonplace

Rusting of iron

After a period of time, objects made of iron or steel will become coated with rust. The rusting of iron is a serious problem and wastes enormous amounts of money in the UK each year. It is estimated that upwards of £500 million a year is spent on replacing iron and steel structures.

Rust is an orange–red powder consisting mainly of hydrated iron(III) oxide ($Fe_2O_3.xH_2O$). Both water and oxygen are essential for iron to rust, and if one of these two substances is not present then rusting will not take place. The rusting of iron is encouraged by salt. Figure 9.21 shows an experiment to show that oxygen (from the air) and water are needed for iron to rust.

Rust prevention

To prevent iron rusting, it is necessary to stop oxygen (from the air) and water coming into contact with it. There are several ways of doing this.

Painting

Ships, lorries, cars, bridges and many other iron and steel structures are painted to prevent rusting (Figure 9.22). However, if the paint is scratched, the iron beneath it will start to rust (Figure 9.23) and corrosion can then spread under the paintwork which is still sound. This is why it is essential that the paint is kept in good condition and checked regularly.

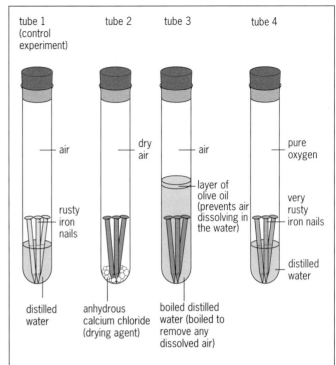

Figure 9.21 Rusting experiment with nails

Figure 9.22 Painting keeps the air and water away from the steel body panels of this car

Figure 9.23 A brand new car is protected against corrosion (top). However, if the paintwork is damaged, then rusting will result

Oiling/greasing

The iron and steel in the moving parts of machinery are coated with oil to prevent them from coming into contact with air or moisture (Figure 9.24). This is the most common way of protecting moving parts of machinery, but the protective film must be renewed.

Coating with plastic

The exteriors of refrigerators, freezers and many other items are coated with plastic, such as PVC, to prevent the steel structure rusting.

Plating

Cans for food can be made from steel coated with tin. The tin is deposited onto the steel used to make food cans by dipping the steel into molten tin. Some car bumpers, as well as bicycle handlebars, are electroplated with chromium to prevent rusting. The chromium gives a decorative finish as well as protecting the steel beneath.

Galvanising

Some steel girders, used in the construction of bridges and buildings, are **galvanised**. Coal bunkers and steel dustbins are also galvanised. This involves dipping the object into molten zinc (Figure 9.25). The thin layer of the more reactive zinc metal coating the steel object slowly corrodes and loses electrons to the iron, thereby protecting it. This process continues even when much of the layer of zinc has been scratched away, so the iron continues to be protected.

Sacrificial protection

Bars of zinc are attached to the hulls of ships and to oil rigs (as shown in Figure 9.26a). Zinc is above iron in the reactivity series and will react in preference to it and so is corroded. As long as some of the zinc bars remain in contact with the iron structure, the structure will be protected from rusting. When the zinc runs out, it must be renewed. Gas and water pipes made of iron and steel are connected by a wire to blocks of magnesium to obtain the same result. In both cases, as the more reactive metal corrodes it loses electrons to the iron and so protects it (Figure 9.26b).

Figure 9.24 Oil and grease lubricates and protects machinery from rusting

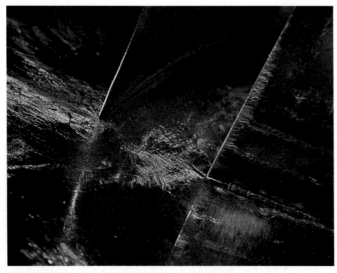

Figure 9.25 A metal object being galvanised

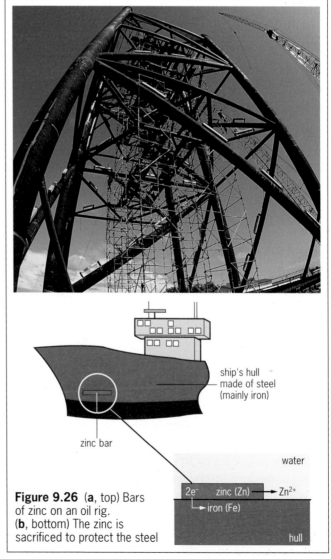

ship's hull made of steel (mainly iron)

zinc bar

water

$2e^-$ zinc (Zn) → Zn^{2+}
→ iron (Fe)

hull

Figure 9.26 (**a**, top) Bars of zinc on an oil rig. (**b**, bottom) The zinc is sacrificed to protect the steel

121

Corrosion

Rusting is the most common form of **corrosion**. Corrosion is the name given to the process which takes place when metals and alloys are chemically attacked by oxygen, water or any other substances found in their immediate environment. The metals in the reactivity series will corrode to a greater or lesser extent. Generally, the higher the metal is in the reactivity series, the more rapidly it will corrode. If sodium and potassium were not stored under oil they would corrode very rapidly indeed. Magnesium, calcium and aluminium are usually covered by a thin coating of oxide after initial reaction with oxygen in the air. Freshly produced copper is pink in colour. However, it soon turns brown due to the formation of copper(II) oxide on the surface of the metal (see Figure 3.23).

In more exposed environments, copper roofs and pipes quickly become covered in verdigris. Verdigris is green in colour (Figure 9.27) and is composed of copper salts formed on copper. The composition of verdigris varies depending on the atmospheric conditions, but includes mixed copper(II) carbonate and copper(II) hydroxide ($CuCO_3.Cu(OH)_2$).

Gold and platinum are unreactive and do not corrode, even after hundreds of years.

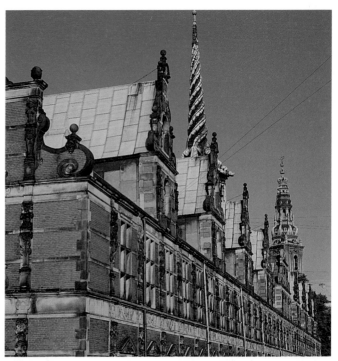

Figure 9.27 Verdigris soon covers copper roofs in exposed environments

Figure 9.28 Stainless steel cutlery and saucepan

QUESTIONS

a If the experiment shown in Figure 9.21 was left to run for 2 weeks, what changes would you expect to see on each nail after this time had elapsed?

b Rusting is another redox reaction. Explain the process of rusting in terms of oxidation and reduction (Chapter 6, p. 71).

c Design an experiment to help you decide whether iron rusts faster in sea water than in deionised (or distilled) water.

d Why is the corrosion of aluminium quite useful to us?

Alloys

The majority of the metallic substances used today are **alloys**. Alloys are mixtures of two or more metals and are formed by mixing molten metals thoroughly. It is generally found that alloying produces a metallic substance that has more useful properties than the original pure metals it was made from. Steel, which is a mixture of the metal iron and the non-metal carbon, is also considered to be an alloy.

Of all the alloys we use, steel is perhaps the most important. Many steels have been produced; they contain not only iron but also carbon and other metals. For example, nickel and chromium are the added metals when stainless steel is produced (Figure 9.28). The chromium prevents the steel from rusting while the nickel makes it harder.

Production of steel

The 'pig iron' obtained from the blast furnace contains between 3% and 5% of carbon and other impurities, such as sulphur, silicon and phosphorus. These impurities make the iron hard and brittle. In order to improve the quality of the metal, most of the impurities must be removed and in doing this, steel is produced.

The impurities are removed in the **basic oxygen process**. In this process, molten pig iron from the blast furnace is poured into the basic oxygen furnace

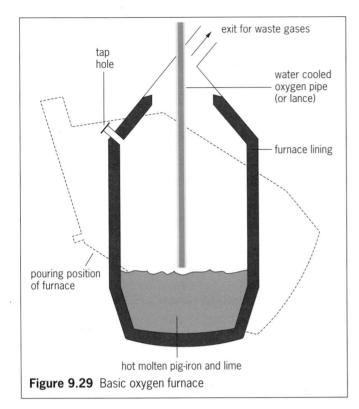

Figure 9.29 Basic oxygen furnace

(labels on figure): tap hole; exit for waste gases; water cooled oxygen pipe (or lance); furnace lining; pouring position of furnace; hot molten pig-iron and lime

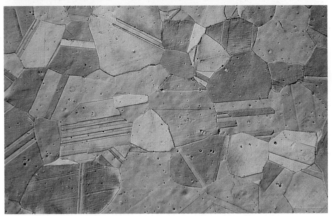

Figure 9.30 (**a**, top) Bronze is often used in sculptures. (**b**, bottom) A polarised light micrograph of brass showing the distinct grain structure of this alloy

(Figure 9.29). A water-cooled 'lance' is introduced into the furnace and oxygen at 5–15 atm pressure is blown onto the surface of the molten metal. Carbon is oxidised to carbon monoxide and carbon dioxide, while sulphur is oxidised to sulphur dioxide. Silicon and phosphorus are oxidised to silicon(IV) oxide and phosphorus pentoxide, which are solid oxides. Some calcium oxide (lime) is added to remove these solid oxides as slag. The slag may be skimmed or poured off the surface. Samples are continuously taken and checked for carbon content. When the required amount of carbon has been reached, the blast of oxygen is turned off.

The basic oxygen furnace can convert up to 300 tonnes of pig iron to steel per hour.

There are various types of steel that differ only in their carbon content. The differing amounts of carbon present confer different properties on the steel and they are used for different purposes (Table 9.4). If other types of steel are required then up to 30% scrap steel, along with other metals (such as tungsten), are added and the carbon is burned off.

Alloys to order

Just as the properties of iron can be changed by alloying, so the same can be done with other useful metals. Metallurgists have designed alloys to suit a wide variety of different uses. Many thousands of alloys are now made, with the majority being 'tailor-made' to do a particular job (Figure 9.30).

Table 9.5 shows some of the more common alloys, together with some of their uses.

Table 9.4 Different types of steel

Steel	Typical composition	Properties	Uses
Mild steel	99.5% iron, 0.5% carbon	Easily worked Lost most of brittleness	Car bodies, large structures
Hard steel	99% iron, 1% carbon	Tough and brittle	Cutting tools, chisels, razor blades
Manganese steel	87% iron, 13% manganese	Tough, springy	Drill bits, springs
Stainless steel	74% iron, 18% chromium, 8% nickel	Tough, does not corrode	Cutlery, kitchen sinks, surgical instruments
Tungsten steel	95% iron, 5% tungsten	Tough, hard, even at high temperatures	Edges of high-speed cutting tools

Table 9.5 Uses of common alloys

Alloy	Composition	Use
Brass	65% copper, 35% zinc	Jewellery, machine bearings, electrical connections, door furniture
Bronze	90% copper, 10% tin	Castings, machine parts
Cupro-nickel	30% copper, 70% nickel	Turbine blades
	75% copper, 25% nickel	Coinage metal
Duralumin	95% aluminium, 4% copper, 1% magnesium, manganese and iron	Aircraft construction, bicycle parts
Magnalium	70% aluminium, 30% magnesium	Aircraft construction
Pewter	30% lead, 70% tin, a small amount of antimony	Plates, ornaments and drinking mugs
Solder	70% lead, 30% tin	Connecting electrical wiring

QUESTIONS

a Calcium oxide is a base. It combines with solid, acidic oxides in the basic oxygen furnace.
Write a chemical equation for one of these oxides reacting with the added lime.
b 'Many metals are more useful to us when mixed with some other elements.' Discuss this statement with respect to stainless steel.

Checklist

After studying Chapter 9 you should know and understand the following terms.

Reactivity series of metals An order of reactivity, giving the most reactive metal first, based on results from experiments with oxygen, water and dilute hydrochloric acid.

Competition reactions Reactions in which metals compete for oxygen or anions. The more reactive metal:

- takes the oxygen from the oxide of a less reactive metal
- displaces the less reactive metal from a solution of that metal salt — this type of competition reaction is known as a displacement reaction.

Metal ion precipitation These are reactions in which certain metal cations form insoluble hydroxides. The colours of these insoluble hydroxides can be used to identify the metal cations which are present; for example, copper(II) hydroxide is a blue precipitate.

Amphoteric hydroxide A hydroxide which can behave as an acid (react with an alkali) or a base (react with an acid), for example zinc hydroxide.

Ore A naturally occurring mineral from which a metal can be extracted.

Metal extraction The method used to extract a metal from its ore depends on the position of the metal in the reactivity series.

- Reactive metals are usually difficult to extract. The preferred method is by electrolysis of the molten ore (electrolytic reduction); for example, sodium from molten sodium chloride.
- Moderately reactive metals (those near the middle of the reactivity series) are extracted using a chemical reducing agent (eg carbon) in a furnace; for example, iron from haematite in the blast furnace.
- Unreactive metals, eg gold and silver, occur in an uncombined (native) state as the free metal.

Blast furnace A furnace for smelting iron ores such as haematite (Fe_2O_3) and magnetite (Fe_3O_4) to produce pig (or cast) iron. In a modified form it can be used to extract metals such as zinc.

Recycling metals Metal drink cans such as those made of aluminium are collected in large 'banks' for the sole purpose of recycling them. Reusing the metal in this way saves money.

Rust A loose, orange–brown, flaky layer of hydrated iron(III) oxide found on the surface of iron or steel. The conditions necessary for rusting to take place are the presence of oxygen and water. The rusting process is encouraged by other substances such as salt. It is an oxidation process.

Rust prevention To prevent iron rusting it is necessary to stop oxygen and water coming into contact with it. The methods employed include painting, oiling/greasing, coating with plastic, plating, galvanising and sacrificial protection.

Corrosion The name given to the process that takes place when metals and alloys are chemically attacked by oxygen, water or any other substances found in their immediate environment.

Alloy Generally, a mixture of two or more metals (for example, brass is an alloy of zinc and copper) or a metal and a non-metal (for example, steel is an alloy of iron and carbon, sometimes with other metals included). They are formed by mixing the molten substances thoroughly. Generally, it is found that alloying produces a metallic substance which has more useful properties than the original pure metals it was made from.

QUESTIONS

1 Use the following list of metals to answer the questions (a) to (i): iron, calcium, potassium, gold, aluminium, magnesium, sodium, zinc, platinum.
  a Which of the metals is found native?
  b Which of the metals is found in nature as the ore:
  (i) haematite?
  (ii) bauxite?
  c Which metal has a carbonate found in nature called marble?
  d Which of the metals will not react with oxygen to form an oxide?
  e Which of the metals will react violently with cold water?
  f Choose one of the metals in your answer to (e) and write a balanced chemical equation for the reaction which takes place.
  g Which of the metals has a protective oxide coating on its surface?
  h Which of the metals reacts very slowly with cold water but extremely vigorously with steam?
  i Which of the metals is used to galvanise iron?

2

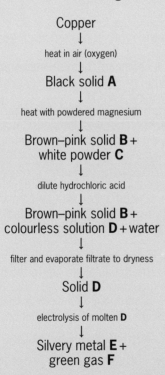

Copper
↓
heat in air (oxygen)
↓
Black solid **A**
↓
heat with powdered magnesium
↓
Brown–pink solid **B** + white powder **C**
↓
dilute hydrochloric acid
↓
Brown–pink solid **B** + colourless solution **D** + water
↓
filter and evaporate filtrate to dryness
↓
Solid **D**
↓
electrolysis of molten **D**
↓
Silvery metal **E** + green gas **F**

  a Name and give the formulae of the substances **A** to **F**.
  b Write balanced chemical equations for the reactions in which:
  (i) black solid **A** was formed
  (ii) white powder **C** and brown–pink solid **B** were formed
  (iii) colourless solution **D** was formed.
  c The reaction between black solid **A** and magnesium is a redox reaction. With reference to this reaction, explain what you understand by this statement.
  d Write anode and cathode reactions for the processes which take place during the electrolysis of molten **D**.
  e Suggest a use for:
  (i) brown–pink solid **B**
  (ii) silvery metal **E**
  (iii) green gas **F**.

3 Explain the following.
  a Recycling metals can save money.
  b Metals such as gold and silver occur native in the Earth's crust.
  c The parts of shipwrecks made of iron rust more slowly in deep sea water.
  d Zinc bars are attached to the hulls of ships to prevent the hulls from rusting.
  e Copper roofs quickly become covered with a green coating when exposed to the atmosphere.

4 Iron is extracted from its ores haematite and magnetite. Usually it is extracted from haematite (iron(III) oxide). The ore is mixed with limestone and coke and reduced to the metal in a blast furnace. The following is a brief outline of the reactions involved.

$$\text{coke} + \text{oxygen} \rightarrow \text{gas } \mathbf{X}$$
$$\text{gas } \mathbf{X} + \text{coke} \rightarrow \text{gas } \mathbf{Y}$$
$$\text{iron(III) oxide} + \text{gas } \mathbf{Y} \rightarrow \text{iron} + \text{gas } \mathbf{X}$$

  a Name the gases **X** and **Y**.
  b Write balanced chemical equations for the reactions shown above.
  c The added limestone is involved in the following reactions:

$$\text{limestone} \rightarrow \text{calcium oxide} + \text{gas } \mathbf{Y}$$
$$\text{calcium oxide} + \text{silicon(IV) oxide} \rightarrow \text{slag}$$

  (i) Give the chemical names for limestone and slag.
  (ii) Write balanced chemical equations for the reactions shown above.
  (iii) Describe what happens to the liquid iron and slag when they reach the bottom of the furnace.
  d Explain why blast furnaces are usually found near to or on coal fields.

5 The iron obtained from the blast furnace is known as pig or cast iron. Because of the presence of impurities, such as carbon, it has a hard and brittle nature. Most of this type of iron is therefore converted into steel in the basic oxygen process. During this process either all or some of the carbon is removed. Calculated quantities of other elements are then added to produce the required type of steel.
  a Explain the meaning of the term alloy as applied to steel.
  b Name two impurities, other than carbon, which are present in cast iron and which are removed completely during the steel manufacture.
  c Describe the method of steel manufacture used which removes the impurities referred to in (b).
  d Name two metallic elements which may be added to the basic oxygen furnace to produce different varieties of steel.
  e Give two uses of stainless steel.
  f Give two advantages of stainless steel compared to cast iron.

QUESTIONS (continued)

6 Aluminium is extracted in the Hall–Héroult cell. A diagram of this cell is shown below. To help you answer this question you should refer to Chapter 6, p. 70.

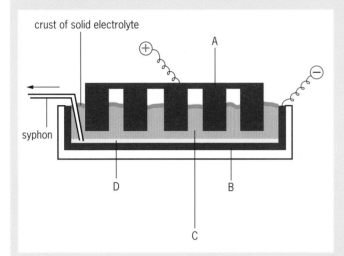

crust of solid electrolyte

syphon

a **Name:**
   (i) the substance formed at **A**
   (ii) electrolyte **C**
   (iii) substance **D**
   (iv) the material that electrodes **A** and **B** are made of.
b Identify electrodes **A** and **B**.
c What is the approximate working temperature of the cell?
d Explain why this working temperature is below the normal melting point of the purified ore in electrolyte **C**.
e Write an electrode equation for the formation of aluminium at electrode **B**.
f Why do the electrodes at **A** have to be replaced periodically? Write a balanced chemical equation to support your answer.
g Give a reason for the location of an aluminium extraction plant at Holyhead, Anglesey.

7 Magnesium can be reacted with steam using the apparatus shown below. When gas **A** is collected, mixed with air and ignited it gives a small pop. A white solid **B** remains in the test tube when the reaction has stopped.

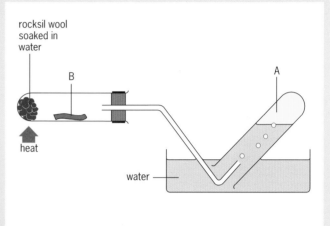

rocksil wool soaked in water

heat

water

a Name and give the formula of gas **A**.
b **(i)** Name the product formed when gas **A** burns in air.
   (ii) Write a balanced chemical equation for this reaction.
c **(i)** Name white solid **B**.
   (ii) Write a balanced chemical equation to represent the reaction between magnesium and steam.
d Name two other metals which could be safely used to replace magnesium and produce another sample of gas **A**.
e When magnesium reacts with dilute hydrochloric acid, gas **A** is produced again. Write a balanced chemical equation to represent this reaction and name the other product of this reaction.

8 Zinc is extracted by roasting zinc blende in a current of air. This converts the sulphide ore to the oxide. Zinc oxide is then mixed with powdered coke in a furnace and heated very strongly. The oxide is converted to the metal according to the following equation.

$$ZnO(s) + C(s) \longrightarrow Zn(g) + CO(g)$$

Approximately 500 tonnes of zinc are produced per day from this process.
a **(i)** What is the name of the chemical process by which zinc oxide is converted to zinc?
   (ii) At what temperature does this process take place?
b It is a very expensive business to maintain the high temperature needed to efficiently convert zinc oxide to zinc. How are the heating costs offset?
c Calculate the quantity of zinc oxide required to produce 500 tonnes of zinc.
   (A_r: Zn = 65; O = 16; C = 12)

10 ATMOSPHERE AND OCEANS

The developing atmosphere

About 4500 million years ago the planets in the solar system were formed. Each planet had a thick layer of gases, mainly hydrogen and helium, surrounding its core. This layer of gas is known as the **primary atmosphere**. Over a period of time intense solar activity caused these lighter gases to be removed from this primary atmosphere of planets nearest to the Sun. During this time the Earth was cooling to become a molten mass upon which a thin crust formed.

Figure 10.1 The Earth's atmosphere formed over many millions of years and involved various atmospheric phenomena

Figure 10.2 (**a**, left) Volcanic activity expelled gases through the crust to form a secondary atmosphere. (**b**, right) Derwent Water in Cumbria. The Earth's surface has evolved as a result of lakes and other large expanses of water

Volcanic activity through the crust pushed out huge quantities of gases, such as ammonia, nitrogen, methane, carbon monoxide, carbon dioxide and a small amount of sulphur dioxide, which formed a **secondary atmosphere** around the Earth (Figure 10.2a). About 3800 million years ago, when the Earth had cooled below 100 °C, the water vapour in this secondary atmosphere condensed and fell as rain. This caused the formation of the first oceans, lakes and seas on the now rapidly cooling Earth. The structure of the surface of the Earth, as we know it today, has evolved as a result of the presence of these large expanses of water (Figure 10.2b). Eventually, early forms of life developed in these oceans, lakes and seas at depths which prevented potentially harmful ultraviolet light from the Sun affecting them. About 3000 million years ago the first forms of bacteria appeared. These were the predecessors of algae-like organisms which used the light from the Sun, via photosynthesis, to produce their own food, and oxygen was released into the atmosphere as a waste product. The process of photosynthesis can be described by the following equation.

$$\text{carbon dioxide} + \text{water} \xrightarrow[\text{chlorophyll}]{\text{sunlight}} \text{glucose} + \text{oxygen}$$

$$6CO_2(g) + 6H_2O(l) \longrightarrow C_6H_{12}O_6(aq) + 6O_2(g)$$

The ultraviolet radiation broke down some of the oxygen molecules, which had been released, into oxygen atoms.

$$\text{oxygen molecules} \xrightarrow{\text{UV light}} \text{oxygen atoms}$$

$$O_2(g) \longrightarrow 2O(g)$$

Some of the highly reactive oxygen atoms reacted with molecules of oxygen to form ozone molecules, $O_3(g)$.

$$\begin{array}{ccc} \text{oxygen} + \text{oxygen} & \rightarrow & \text{ozone} \\ \text{atom} \quad \text{molecule} & & \\ O(g) + O_2(g) & \rightarrow & O_3(g) \end{array}$$

Ozone is an unstable molecule which readily decomposes, under the action of ultraviolet radiation, into single oxygen atoms and oxygen molecules.

$$\begin{array}{ccc} \text{ozone} \xrightarrow{\text{UV light}} & \text{oxygen} + \text{oxygen} \\ & \text{atom} \quad \text{molecule} \\ O_3(g) \longrightarrow & O(g) + O_2(g) \end{array}$$

However, the single oxygen atoms would react quickly, reforming ozone molecules.

Ozone is an important gas in the atmosphere (stratosphere). It prevents harmful ultraviolet radiation from reaching the Earth. Over many millions of years the amount of ultraviolet radiation was reduced significantly. About 400 million years ago the first land plants appeared on the Earth, and so the amount of oxygen and hence ozone increased.

Oxygen is a reactive gas, and over millions of years organisms adapted to make use of it. The oxygen from the atmosphere was used, along with the carbon they obtained from their food, to produce energy in a process known as **respiration**. The process of respiration can be represented as:

$$\begin{array}{ccc} \text{glucose} + \text{oxygen} \rightarrow & \text{carbon} + \text{water} + \text{energy} \\ & \text{dioxide} \\ C_6H_{12}O_6(aq) + 6O_2(g) \rightarrow & 6CO_2(g) + 6H_2O(l) + \text{energy} \end{array}$$

a Recently 'holes' have been observed in the ozone layer. Why is this a potential problem?

Table 10.1 Composition of the atmosphere

Component	%
Nitrogen	78.08
Oxygen	20.95
Argon	0.93
Carbon dioxide	0.03
Neon	0.002
Helium	0.0005
Krypton	0.0001
Xenon plus minute amounts of other gases.	0.00001

The structure of the atmosphere

The gases in the atmosphere are held in an envelope around the Earth by its gravity. The atmosphere is 80 km thick and it is divided into four layers (Figure 10.3).

About 75% of the mass of the atmosphere is found in the layer nearest the Earth called the **troposphere**. Beyond the troposphere the atmosphere reaches into space but becomes extremely thin beyond the mesosphere.

The composition of the atmosphere

If a sample of dry, unpolluted air was taken from anywhere in the troposphere and analysed, the composition by volume of the sample would be similar to that shown in Table 10.1.

Measuring the percentage of oxygen in the air

When $100 \, cm^3$ of air is passed backwards and forwards over heated copper turnings it is found that the amount of gas decreases (Figure 10.4). This is because the reactive part of the air, the oxygen gas, is reacting with the copper to form black copper(II) oxide (Figure 10.5). In such an experiment the volume of gas in the syringe decreases from $100 \, cm^3$ to about $79 \, cm^3$, showing that the air contained $21 \, cm^3$ of oxygen gas. The percentage of oxygen gas in the air is:

$$\frac{21}{100} \times 100 = 21\%$$

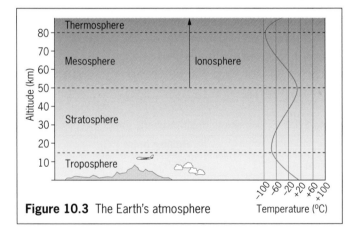

Figure 10.3 The Earth's atmosphere

Figure 10.5 Copper turnings before and after reaction

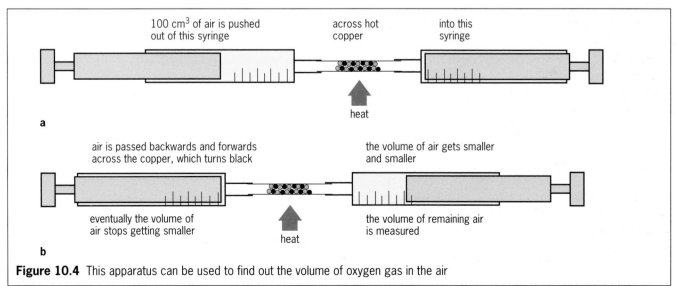

Figure 10.4 This apparatus can be used to find out the volume of oxygen gas in the air

The composition of the atmosphere is affected by the following factors:

- respiration
- photosynthesis
- volcanic activity
- radioactive decay, in which helium is formed
- human activity, involving burning of fossil fuels, in which carbon dioxide and water vapour are produced as well as other gases (p. 161).

Compare the components of our atmosphere with those of the other planets in the solar system (Table 10.2).

Table 10.2 Atmospheres of the other planets in the solar system

Planet	Atmosphere
Mercury	No atmosphere — the gases were burned off by the heat of the Sun
Venus	Carbon dioxide and sulphur dioxide
Mars	Mainly carbon dioxide
Jupiter	Ammonia, helium, hydrogen, methane
Saturn	Ammonia, helium, hydrogen, methane
Uranus	Ammonia, helium, hydrogen, methane
Neptune	Helium, hydrogen, methane
Pluto	No atmosphere; it is frozen

QUESTIONS

a Draw a pie chart to show the data given in Table 10.1
b Is air a compound or a mixture? Explain your answer.
c Design an experiment to find out how much oxygen there is in exhaled air.

Fractional distillation of liquid air

Air is the major source of oxygen, nitrogen and the noble gases. The gases are obtained by fractional distillation of liquid air but it is a difficult process, involving several different steps (Figure 10.6).

- The air is passed through fine filters to remove any dust.

- The air is cooled to about –80 °C to remove water vapour and carbon dioxide as solids. If these are not removed, then serious blockages of pipes can result.
- Next, the cold air is compressed to about 100 atm of pressure. This warms up the air, so it is passed into a heat exchanger to cool it down again.
- The cold, compressed air is allowed to expand rapidly and that cools it still further.
- The process of compression followed by expansion is repeated until the air reaches a temperature below –200 °C. At this temperature the majority of the air liquefies (Table 10.3).
- The liquid air is passed into a fractionating column and it is fractionally distilled.
- The gases are then stored separately in large tanks and cylinders.

Table 10.3 Boiling points of atmospheric gases

Gas	Boiling point (°C)
Helium	–269
Neon	–246
Nitrogen	–196
Argon	–186
Oxygen	–183
Krypton	–157
Xenon	–108

It should be noted that the noble gases neon, argon, krypton and xenon are obtained by this method; however, helium is more profitably obtained from natural gas.

QUESTION

a Use information given in the text to construct a flow chart to show the processes involved in the extraction of gases from air.

Figure 10.6 BOC produces large amounts of gases obtained from the fractional distillation of liquid air

Uses of the gases

Oxygen

- Large quantities are used in industry to convert pig iron into steel (Chapter 9, p. 122) and for producing very hot flames for welding by mixing with gases such as ethyne (acetylene).
- It is used in hospitals to help with breathing difficulties (Figure 10.7).

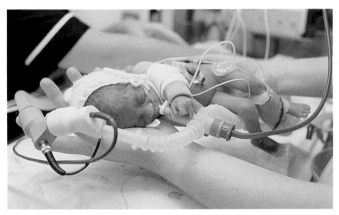

Figure 10.7 The incubator has its own oxygen supply

- People such as mountaineers and divers use oxygen (Figure 10.8).
- It is carried in space rockets so that the hydrogen and kerosene fuels can burn.
- Space shuttles carry oxygen gas to use in fuel cells which convert chemical energy into electrical energy.
- Astronauts must carry their own supply of oxygen, as do fire-fighters.
- It is used to restore life to polluted lakes and rivers and in the treatment of sewage.

Nitrogen

- Nitrogen is used in very large quantities in the production of ammonia gas (Chapter 15, p. 192), which is used to produce nitric acid. Nitric acid is used in the manufacture of dyes, explosives and fertilisers.
- Liquid nitrogen is used as a refrigerant. Its low temperature (-196 °C) makes it useful for freezing food quickly.
- Because of its unreactive nature, nitrogen is used as an inert atmosphere for some processes and chemical reactions. For example, empty oil tankers are filled with nitrogen to prevent fires.
- It is used in food packaging, for example in crisp packets, to keep the food fresh and in this case to prevent the crisps being compressed.

Noble gases

Argon is used:

- to fill ordinary and long-life light bulbs to prevent the tungsten filament from reacting with oxygen in the air and forming the oxide (Figure 10.9)
- to provide an inert atmosphere in arc welding and in the production of titanium metal.

Figure 10.8 These divers carry their own supply of oxygen

Figure 10.9 This long-life bulb contains argon

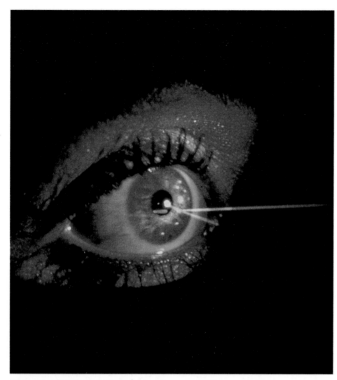

Figure 10.10 A helium–neon laser used in eye surgery

Figure 10.11 Helium is used to fill this airship as it has a low density and is unreactive

132

Neon is used:

- in advertising signs, because it glows red when electricity is passed through it
- in the helium–neon gas laser (Figure 10.10)
- in Geiger–Müller tubes, which are used for the detection of radioactivity.

Helium is used:

- to provide an inert atmosphere for welding
- as a coolant in nuclear reactors
- with 20% oxygen as a breathing gas used by deep-sea divers
- to inflate the tyres of large aircraft
- to fill airships and weather balloons (Figure 10.11)
- in the helium–neon laser
- in low-temperature research, because of its low boiling point.

Krypton and xenon are used:

- in lamps used in photographic flash units, in strobo-scopic lamps and in lamps used in lighthouses.

> **QUESTIONS**
>
> **a** How does oxygen help to restore life to polluted lakes?
> **b** Why is it important to have nitrogen in fertilisers?
> **c** Why is helium needed to produce an inert atmosphere for welding?

Resources in the oceans

In some hot, arid countries the sea is the main source of pure water for drinking. The water is obtained by a process known as **desalination** (Chapter 2, p. 20).

Sea water contains about 35 g of dissolved solids in each kilogram. A typical analysis of the elements present in sea water is shown in Table 10.4. Some elements are extracted from sea water on a commercial basis.

Table 10.4 Elements present in sea water

Element	g dm^{-3} of sea water
Bromine	0.07
Calcium	0.4
Chlorine	19.2
Magnesium	1.3
Potassium	0.4
Sodium	10.7
Sulphur	0.9
Other elements	1.4

Bromine

Eighty per cent of the world's supply of bromine is extracted from sea water. Bromine is extracted in this way in a chemical plant in Anglesey, Wales, but one of the biggest factories for extracting bromine is to be found on the south-west shores of the Dead Sea in Israel. The Dead Sea is a particularly rich source of bromine as it contains $5.2g$ of the element per dm^3 of sea water.

The extraction of bromine from sea water begins with the evaporation of the sea water using energy from the Sun, forming a more concentrated solution. The bromine is present in solution as bromide ions ($Br^-(aq)$), which can be converted to bromine via a displacement reaction with chlorine gas. The process is as follows.

- Chlorine gas is bubbled through the warm, partially evaporated Dead Sea water and displacement takes place.

$$\text{chlorine} + \text{bromide} \rightarrow \text{bromine} + \text{chloride}$$
$$\text{gas} \qquad \text{ions} \qquad \text{gas} \qquad \text{ions}$$
$$Cl_2(g) + 2Br^-(aq) \rightarrow Br_2(aq) + 2Cl^-(aq)$$

- Steam is blown through the resulting solution and the volatile bromine is given off as bromine vapour mixed with steam.
- Bromine is not very soluble in water and two separate layers form when the bromine and steam condense. The less dense water floats on top of the denser bromine layer. The bromine layer is then run off.
- The impure bromine is then purified by distillation and dried.

Half a tonne of chlorine is required for each tonne of bromine manufactured. Bromine is used in the manufacture of the fuel additive tetraethyl lead(IV) (TEL), although to a decreasing extent. Bromine compounds are also used in photography ($AgBr$), in medicines (KBr) and in herbicides.

Magnesium

Magnesium can also be extracted from sea water. The sea water is first treated with calcium hydroxide, and magnesium hydroxide is precipitated and can be filtered off.

$$\text{magnesium} + \text{hydroxide} \rightarrow \text{magnesium}$$
$$\text{ions} \qquad \text{ions} \qquad \text{hydroxide}$$
$$Mg^{2+}(aq) + 2OH^-(aq) \rightarrow Mg(OH)_2(s)$$

The magnesium hydroxide is then dissolved in hydrochloric acid and the resulting solution is evaporated to produce magnesium chloride. This is then fused, at about $700\,°C$, with additives to lower its melting point and is electrolysed with a graphite anode and a

Figure 10.12 Magnesium (as the element) is used in alloys to build space shuttles (top). Magnesium (in a compound) is used in the manufacture of toothpaste (bottom)

steel cathode. The magnesium produced is liberated at the cathode and is 99.9% pure.

$$\text{magnesium ions} + \text{electrons} \rightarrow \text{magnesium}$$
$$Mg^{2+}(aq) + 2e^- \rightarrow Mg(s)$$

Magnesium is used in alloys which are used in the space and aviation industries (Figure 10.12). Compounds of magnesium are also used in medicines ($Mg(OH)_2$ and $MgSO_4.7H_2O$) and toothpastes (Figure 10.12).

133

Sodium chloride

This is the most abundant resource in sea water. There is about 25 g of sodium chloride per dm³ of sea water. It is extracted in several areas of the world by evaporation, for example in France, Saudi Arabia and Australia.

The sea water is kept in shallow ponds until all the water has been evaporated by the heat of the Sun. The salt is then harvested.

Sodium chloride is used to flavour food, in the manufacture of sodium carbonate and sodium hydrogencarbonate and as the raw material for the chlor-alkali industry (Chapter 6, p. 75).

QUESTIONS

a Use the information in the text and any other information you can obtain from other sources to construct a flow diagram to show the processes involved in the extraction of bromine from sea water.

b In the extraction of magnesium from sea water, the magnesium hydroxide precipitate is treated with dilute hydrochloric acid. Write both a word and a balanced chemical equation to describe this reaction. In addition, write an electrode equation to show the production of chlorine gas at the anode.

The water cycle

Figure 10.13 illustrates the water cycle, which shows how water circulates around the Earth. The driving force behind the water cycle is the heat of the Sun.

- Heat from the Sun causes evaporation from oceans, seas and lakes. Water vapour is also formed from the evaporation of water from leaves (transpiration), through respiration and through combustion. The water vapour rises and cools, and condenses to form tiny droplets of water. These droplets form clouds.
- The clouds are moved along by air currents. As they cool, the tiny droplets join to form larger droplets, which fall as rain when they reach a certain size.
- The water that falls as rain runs into streams and rivers and then on into lakes, seas and oceans.

QUESTION

a Construct a simplified version of the water cycle using 'key words' in boxes and the 'processes involved' over linking arrows.

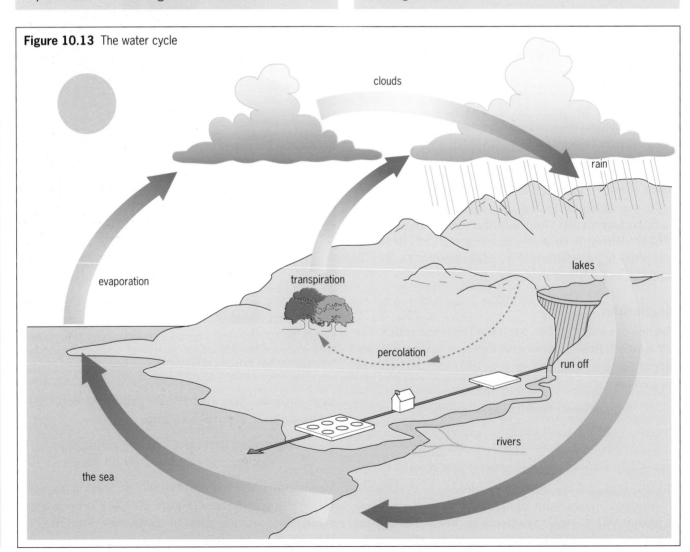

Figure 10.13 The water cycle

Pollution

The two major resources considered in this chapter, water and air, are essential to our way of life and our very existence. However, we are continually guilty of polluting these resources. We now look at the effects of the various sources of pollution and the methods used to control or eliminate them.

Water pollution

Water is very good at dissolving substances. It is, therefore, very unusual to find really pure water on this planet. As water falls through the atmosphere and down onto and through the surface of the Earth, it dissolves a tremendous variety of substances. Chemical fertilisers washed off surrounding land will add nitrate ions (NO_3^-) and phosphate ions (PO_4^{3-}) to the water, owing to the use of artificial fertilisers such as ammonium nitrate and ammonium phosphate. It may also contain human waste as well as insoluble impurities such as grit and bacteria and oil and lead 'dust' (to a decreasing extent) from the exhaust fumes of lorries and cars (Figure 10.14).

All these artificial as well as natural impurities must be removed from the water before it can be used. Recent European Union (EU) regulations have imposed strict guidelines on the amounts of various substances allowed in drinking water (Figure 10.15).

Most drinking water in the UK is obtained from lakes and rivers where pollution levels are low. The process of water treatment involves both filtration and chlorination and is summarised in Figure 10.16.

1 Impure water is first passed through screens to filter out floating debris.
2 Filtration through coarse sand traps larger, insoluble particles. The sand also contains specially grown microbes which remove some of the bacteria.

Figure 10.14 A badly polluted river

Figure 10.15 A water treatment works

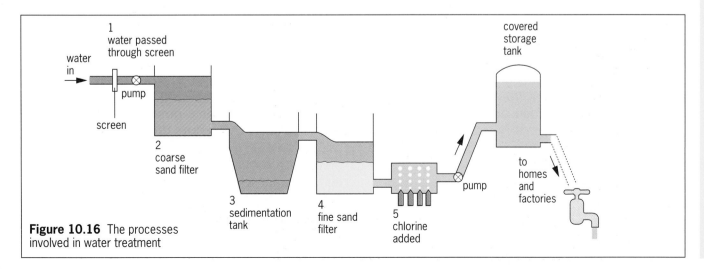

Figure 10.16 The processes involved in water treatment

3 A sedimentation tank has chemicals known as flocculants, for example alum, added to it to make the smaller particles, which remain in the water, stick together and sink to the bottom of the tank.

4 These particles are removed by further filtration through fine sand.

5 Finally, a little chlorine gas is added, which kills any remaining bacteria. This sterilises the water.

Sewage treatment

After we have used water it must be treated again before it can be returned to rivers, lakes and seas. The process known as sewage treatment is shown in Figure 10.17.

Used water, sewage, contains waste products such as human waste and washing-up debris as well as everything else that we put down a drain or a sink. The processes involved are as follows.

1 Large screens remove large pieces of rubbish.

2 Sand and grit are separated in large sedimentation tanks. The sand and grit often contain large amounts of useful chemicals which, by the action of selected microbes, can be used as fertilisers.

3 The impure water is then removed and sent to a trickling filter, where it is allowed to drain through gravel on which microbes have been deposited. These kill off any remaining bacteria in the water.

4 Finally, the treated water is chlorinated and returned to a river.

Atmospheric pollution

Air pollution is all around us. Concentrations of gases in the atmosphere such as carbon monoxide, sulphur dioxide and nitrogen oxides are increasing with the increasing population. As the population increases there is a consequent increase in the need for energy, industries and motor vehicles. These gases are produced primarily from the combustion of the fossil fuels coal, oil and gas, but they are also produced by the smoking of cigarettes.

Motor vehicles are responsible for much of the air pollution in large towns and cities. They produce four particularly harmful pollutants: carbon monoxide, sulphur dioxide, hydrocarbons and oxides of nitrogen. Concern about pollution due to cars has led to the introduction of strict regulations by the European Union (EU) and now all new cars must have a device known as a catalytic converter fitted to eliminate the production of some of these gases.

The catalytic converter acts as a device to speed up reactions which involve the pollutant gases, converting them to less harmful products, such as nitrogen and carbon dioxide. It should be noted that

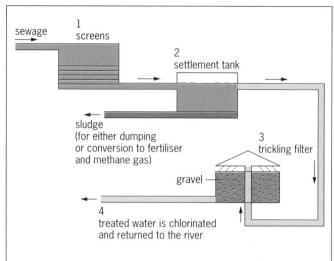

Figure 10.17 The processes involved in sewage treatment

catalytic converters can only be used with unleaded petrol as the lead 'poisons' the catalyst, preventing it from catalysing the reactions. For a further discussion of catalytic converters see Chapter 11, p. 145.

Another method that has been introduced to reduce the amount of pollutants is that of the 'lean burn' engine. Although this type of engine reduces the amounts of carbon monoxide and oxides of nitrogen produced, it actually increases the amount of hydrocarbons in the exhaust gases.

Power stations are a major source of sulphur dioxide, a pollutant formed by the combustion of coal, oil and gas, which contain small amounts of sulphur.

$$\text{sulphur} + \text{oxygen} \rightarrow \text{sulphur dioxide}$$
$$S(s) \ + \ O_2(g) \ \rightarrow \ SO_2(g)$$

This sulphur dioxide gas dissolves in rainwater to form the weak acid, sulphurous acid (H_2SO_3).

$$\text{sulphur dioxide} + \text{water} \rightleftharpoons \text{sulphurous acid}$$
$$SO_2(g) \ + H_2O(l) \rightleftharpoons \ H_2SO_3(aq)$$

A further reaction occurs in which the sulphurous acid is oxidised to sulphuric acid.

Solutions of these acids are the principal contributors to **acid rain**. For a further discussion of acid rain, see Chapter 16, p. 201.

Units called flue gas desulphurisation (FGD) units are being fitted to some power stations to prevent the emission of sulphur dioxide gas. Here, the sulphur dioxide gas is removed from the waste gases by passing it through calcium hydroxide slurry. This not only removes the sulphur dioxide but also creates calcium sulphate, which can be sold to produce plasterboard. The FGD units are very expensive and therefore the sale of the calcium sulphate is an important economic part of the process.

Figure 10.18 Sulphur dioxide is a major pollutant produced by industry

QUESTIONS

a Make a list of four major water pollutants and explain where they come from. What damage can these pollutants do?
b Write a balanced chemical equation to represent the reaction which takes place between sulphur dioxide and calcium hydroxide slurry in the FGD unit of a power station.
c State the environmental problems associated with the quarrying of limestone for FGD units.

Checklist

After studying Chapter 10 you should know and understand the following terms.

Primary atmosphere   The original thick layer of gases, mainly hydrogen and helium, that surrounded the Earth's core soon after the planet was formed 4500 million years ago.

Secondary atmosphere   Early volcanic activity created this mixture of gases. The mixture that formed this atmosphere included ammonia, nitrogen, methane, carbon monoxide, carbon dioxide and sulphur dioxide gases.

Ozone (trioxygen)   A colourless gas produced in the stratosphere by the action of high-energy ultra-violet radiation on oxygen gas, producing oxygen atoms. These oxygen atoms then react with further oxygen molecules to produce ozone. Its presence in the stratosphere acts as a screen (ozone layer) against dangerous ultraviolet radiation.

Troposphere   A layer of the atmosphere closest to the Earth which contains about 75% of the mass of the atmosphere. The composition of dry air is relatively constant in this layer of the atmosphere.

Stratosphere   A layer of the atmosphere above the troposphere in which the temperature increases with increasing height.

Mesosphere   A section of the atmosphere above the stratosphere in which the temperature falls with increasing height.

Fractional distillation of air   Air is a major raw material. The mixture of gases is separated by first liquefying the mixture at low temperature and high pressure. The temperature is then allowed to rise and the gases collected as they boil off. The gases so produced have many and varied uses.

Sea water as a resource   Sea water is a major resource. By evaporation (desalination) we can produce water for drinking or irrigation. Also, bromine, magnesium and salt are extracted in large quantities from sea water.

Water cycle   This cycle shows how water circulates around the Earth. The driving force behind the water cycle is the Sun.

Pollution   The modification of the environment caused by human influence. It often renders the environment harmful and unpleasant to life. Water pollution is caused by many substances, such as those found in fertilisers and in industrial effluent. Atmospheric pollution is caused by gases such as sulphur dioxide, carbon monoxide and nitrogen oxides being released into the atmosphere by a variety of industries and also by the burning of fossil fuels.

QUESTIONS

1 The apparatus shown on p. 129, Figure 10.4, was used to estimate the proportion of oxygen in the atmosphere. A volume of dry air (90 cm³) was passed backwards and forwards over heated copper until no further change in volume took place. The apparatus was then allowed to cool down to room temperature and the final volume reading was then taken. Some typical results are shown below.

Volume of gas before passing over heated copper = 90.0 cm³
Volume of gas after passing over heated copper = 70.7 cm³

During the experiment the copper slowly turned black.
a Why was the apparatus allowed to cool back to room temperature before the final volume reading was taken?
b (i) Using the information given above, calculate the volume reduction which has taken place.
   (ii) Calculate the percentage reduction in volume.
c Explain briefly why there is a change in volume.
d What observation given above supports your explanation in (c)? Write a balanced chemical equation for any reaction which has occurred.
e Give the name of the main residual gas at the end of the experiment.
f Would you expect the copper to have increased or decreased in mass during the experiment? Explain your answer.

2 Explain the following.
a Air is a mixture of elements and compounds.
b The percentage of carbon dioxide in the atmosphere does not significantly vary from 0.03%.
c When liquid air has its temperature slowly raised from −270 °C, helium is the first gas to boil off.
d Eighty per cent of the world's supply of bromine is extracted from sea water.
e Power stations are thought to be a major cause of acid rain.

3 Air is a raw material from which several useful substances can be separated. They are separated in the following process.
Dry and 'carbon dioxide free' air is cooled under pressure. Most of the gases liquefy as the temperature falls below −200 °C. The liquid mixture is separated by fractional distillation. The boiling points of the gases left in the air after removal of water vapour and carbon dioxide are given in the table below:

Gas	Boiling point (°C)
Argon	−186
Helium	−269
Krypton	−157
Neon	−246
Nitrogen	−196
Oxygen	−183
Xenon	−108

a Why is the air dried and carbon dioxide removed before it is liquefied?
b Which of the gases will not become liquid at −200 °C?
c Which of the substances in the liquid mixture will be the first to change from liquid to gas as the temperature is slowly increased?
d Give a use for each of the gases shown in the table.
e Use the data given in Table 10.1, p. 129, to calculate the volume of each of the gases found in 1 dm³ of air.

4 Explain what is meant by the term 'pollution' with reference to air and water.
a (i) Name an air pollutant produced by the burning of coal.
   (ii) Name a different air pollutant produced by the combustion of petrol in a car engine.
   (iii) Power stations are beginning to fit FGD units. What gas are they being fitted to remove? Briefly describe what happens in an FGD unit.
b Some of our drinking water is obtained by purifying river water.
   (i) Would distillation or filtration produce the purest water from river water? Give a reason for your answer.
   (ii) Which process, distillation or filtration, is actually used to produce drinking water from river water? Comment on your answer in comparison to your answer in (c)(i).
c Power stations produce warm water. This causes thermal pollution as this warm water is pumped into nearby rivers.
   (i) Why do power stations produce such large quantities of warm water?
   (ii) What effect does this warm water have on aquatic life?

5 In the final stage of the extraction of magnesium from sea water, molten magnesium chloride is electrolysed. Substances are added to bring the working temperature of the electrolysis cell down to 700 °C.
a Write equations to represent the reactions taking place at the cathode and anode. State clearly whether oxidation or reduction is taking place.
b Why are substances added to the molten magnesium chloride before electrolysis takes place?
c The industrial production of magnesium by electrolysis uses a current of 14 000 amps. Calculate the time required to produce 50 kg of magnesium from molten magnesium chloride.
(1 Faraday = 96 500 coulombs. A_r: Mg = 24)
d A large proportion of the magnesium produced (43%) is used to make alloys.
   (i) Why has it such a use in the production of magnesium alloys?
   (ii) The world production of magnesium is approximately 250 000 tonnes. Calculate the number of tonnes used in the production of magnesium alloys.

6 The Western world production of bromine is about 270 000 tonnes per year. Eighty per cent of this production is from sea water.

  a Calculate the amount of bromine obtained in the Western world from sea water.

  b Sea water contains 0.07 g of bromine per dm^3. Using your answer to (a), calculate the volume of sea water required to obtain that amount of bromine.

  c The main reaction in the extraction process is that involving displacement of bromine using chlorine gas.

    (i) Write an ionic equation for this reaction.

    (ii) Explain why you could not use iodine instead of chlorine in the displacement reaction.

  d Chlorine and bromine are hazardous substances. Describe some of the precautions that have to be taken to ensure the safety of members of the workforce who deal with these two substances.

7 France obtains some of its sodium chloride by evaporation of sea water.

  a If France produces 1 million tonnes of salt per year by this method and sea water contains 25 g of sodium chloride per dm^3, calculate the volume of sea water required to produce the annual salt production.

  b Give four important uses of sodium chloride.

  c Sodium chloride is an ionic substance.

    (i) Draw a diagram to show the bonding which takes place within sodium chloride.

    (ii) What are the properties of ionic substances such as sodium chloride?

8 In plants, during photosynthesis, carbon dioxide and water are converted into carbohydrates such as glucose ($C_6H_{12}O_6$), and oxygen is released.

  a Write a balanced chemical equation for the reaction taking place during photosynthesis.

  b What conditions are essential for this reaction to take place?

  c Why are animals unable to photosynthesise?

  d Which process do we obtain if we reverse photosynthesis?

  e Some of the oxygen released during photosynthesis is broken up by ultraviolet radiation. Some of the oxygen atoms produced combine with further oxygen molecules to produce an important allotrope of oxygen.

    (i) Name and give the formula of this allotrope of oxygen.

    (ii) Write a balanced chemical equation for the reaction in which this allotrope is produced.

    (iii) Why is this allotrope of oxygen so important to us?

11 RATES OF REACTION

Factors that affect the rate of a reaction	**Enzymes**
Surface area	**Checklist**
Concentration	**Questions**
Temperature	
Light	
Catalysts	

Figure 11.1 shows some slow and fast reactions. The two photographs on the left show examples of slow reactions. The ripening of apples takes place over a number of weeks, and the making and maturing of cheese may take months. The burning of solid fuels, such as coal, can be said to involve chemical reactions taking place at a medium speed or rate. The other example shows a fast reaction. The chemicals inside explosives, such as TNT, react very rapidly in reactions which are over in seconds or fractions of seconds.

As new techniques have been developed, the processes used within the chemical industry have become more complex. Therefore, chemists and chemical engineers have increasingly looked for ways to control the rates at which chemical reactions take place. In doing so, they have discovered that there are five main ways in which you can alter the rate of a chemical reaction. These ideas are not only very useful, but can also be applied to reactions which occur in the school laboratory.

Figure 11.1 Some slow (ripening fruit and cheesemaking), medium (coal fire) and fast (explosion) reactions

Factors that affect the rate of a reaction

- Surface area of the reactants.
- Concentration of the reactants.
- Temperature at which the reaction is carried out.
- Light.
- Use of a catalyst.

Surface area

In Chapter 8, we discussed the use of limestone (calcium carbonate) as a substance which can be used to neutralise soil acidity. Powdered limestone is used as it neutralises the acidity faster than if lumps of limestone are used. Why do you think this is the case?

In the laboratory, the reaction between acid and limestone in the form of lumps or powder can be observed in a simple test tube experiment. Figure 11.2 shows the reaction between dilute hydrochloric acid and limestone in lump and powdered form.

hydrochloric+ calcium → calcium + carbon + water
    acid   carbonate   chloride dioxide
$2HCl_{(aq)}$ + $CaCO_{3(s)}$ → $CaCl_{2(aq)}$ + $CO_{2(g)}$ + $H_2O_{(l)}$

The rates at which the two reactions occur can be found by measuring either:

- the volume of the carbon dioxide gas which is produced, or
- the loss in mass of the reaction mixture with time.

Figure 11.2 The powdered limestone (left) reacts faster with the acid than the limestone in the form of lumps

These two methods are generally used for measuring the rate of reaction for processes involving the formation of a gas as one of the products.

The apparatus shown in Figure 11.3 is used to measure the loss in mass of the reaction mixture. The

Figure 11.3 After 60 seconds the mass has fallen by 1.24 g

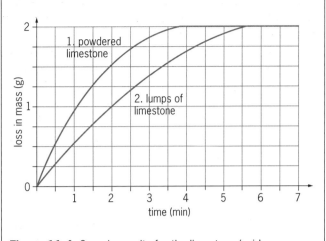

Figure 11.4 Sample results for the limestone/acid experiment

mass of the conical flask plus the reaction mixture is measured at regular intervals. The total loss in mass is calculated for each reading of the balance, and this is plotted against time. Some sample results from experiments of this kind have been plotted in Figure 11.4.

The reaction is generally at its fastest in the first minute. This is indicated by the slopes of the curves during this time. The steeper the slope, the faster the rate of reaction. You can see from the graphs in Figure 11.4 that the rate of reaction is greater with the powdered limestone.

The surface area has been increased by powdering the limestone (Figure 11.5). The acid particles now have an increased amount of surface of limestone with which to collide. The products of a reaction are formed when collisions occur between reactant particles. Therefore, the increase in surface area of the limestone increases the rate of reaction.

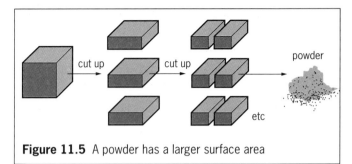

Figure 11.5 A powder has a larger surface area

Concentration

A yellow precipitate is produced in the reaction between sodium thiosulphate and hydrochloric acid.

sodium + hydrochloric → sodium + sulphur + sulphur + water
thiosulphate   acid   chloride   dioxide

$$Na_2S_2O_3(aq) + 2HCl(aq) \rightarrow 2NaCl(aq) + S(s) + SO_2(g) + H_2O(l)$$

The rate of this reaction can be followed by recording the time taken for a given amount of sulphur to be precipitated. This can be done by placing a conical flask containing the reaction mixture onto a cross on a piece of paper (Figure 11.6). As the precipitate of sulphur forms, the cross is obscured and finally disappears from view. The time taken for this to occur is a measure of the rate of this reaction. To obtain sufficient information about the effect of changing the concentration of the reactants, several experiments of this type must be carried out, using different concentrations of sodium thiosulphate or hydrochloric acid.

Some sample results of experiments of this kind have been plotted in Figure 11.7. You will note from the graph that when the most concentrated sodium thiosulphate solution was used, the reaction was at its fastest. This is shown by the shortest time taken for the cross to be obscured.

QUESTIONS

a What apparatus would you use to measure the rate of reaction of the limestone with dilute hydrochloric acid by measuring the volume of carbon dioxide produced?

b The following results were obtained from an experiment of the type you were asked to design in (a):

Time (min)	0	0.5	1.0	1.5	2.0	2.5	3.0	3.5	4.0	4.5	5.0
Total volume of CO₂ gas (cm³)	0	15	24	28	31	33	35	35	35	35	35

(i) Plot a graph of the total volume of CO_2 against time.

(ii) At which point is the rate of reaction fastest?

(iii) What volume of CO_2 was produced after 1 minute 15 seconds?

(iv) How long did it take to produce 30 cm³ of CO_2?

Figure 11.6 The precipitate of sulphur obscures the cross

As discussed earlier, the products of the reaction are formed as a result of the collisions between reactant particles. There are more particles in a more concentrated solution and collisions occur more often. The more often they collide, the more chance they have of reacting. This means that the rate of a chemical reaction will increase if the concentration of reactants is increased.

In reactions involving only gases, for example the Haber process (Chapter 15, p. 193), an increase in the overall pressure at which the reaction is carried out increases the rate of the reaction. The increase in pressure results in the gas particles being pushed closer together. This means that they collide more often and so react faster.

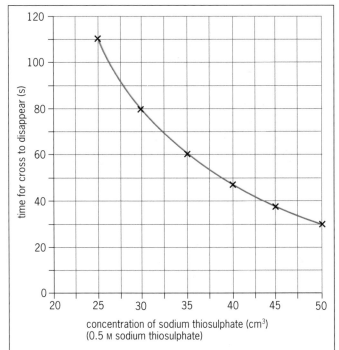

Figure 11.7 Sample data for the sodium thiosulphate/acid experiment at different concentrations of sodium thiosulphate

QUESTION

a Devise an experiment to show the effect of changing the concentration of dilute acid on the rate of reaction between magnesium and hydrochloric acid.

Temperature

Why do you think food is stored in a refrigerator? The reason is the rate of decay is slower at lower temperatures. This is a general feature of the majority of chemical processes.

The reaction between sodium thiosulphate and hydrochloric acid can also be used to study the effect of temperature on the rate of a reaction. Figure 11.8 shows some sample results of experiments with sodium thiosulphate and hydrochloric acid carried out at different temperatures.

You can see from the graph that the rate of the reaction is fastest at high temperatures. When the temperature at which the reaction is carried out is increased, the energy that the particles have also increases — the particles move faster. This has the result of increasing the number of collisions of sodium thiosulphate and hydrochloric acid particles, and the collisions which occur are more energetic and so more likely to form products. Therefore, if the temperature at which a reaction takes place is increased then the rate of reaction will increase.

QUESTIONS

a Explain why potatoes cooked in oil cook faster than those cooked in water.
b Devise an experiment to study the effect of temperature on the reaction between magnesium and hydrochloric acid.
c Explain why food cooks faster in a pressure cooker.

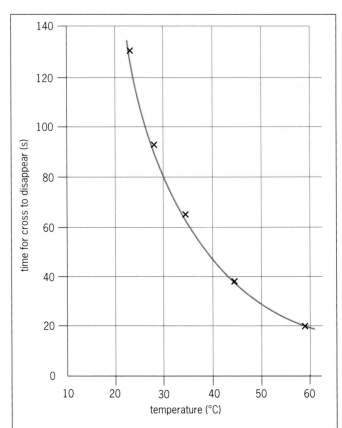

Figure 11.8 Sample data for the sodium thiosulphate/acid experiment at different temperatures

Light

Some chemical reactions are affected by light. Photosynthesis is a very important reaction (Chapter 8, p. 105) which occurs only when sunlight falls on leaves containing the green pigment chlorophyll.

Another chemical reaction which takes place only when exposed to light is that which occurs when you take a photograph. Photographic film is a transparent plastic strip coated with emulsion: a layer of gelatin throughout which are spread many millions of tiny crystals of silver halides, in particular silver bromide (AgBr). The emulsion used is similar for both black-and-white and colour film. In the case of colour film there are three layers of emulsion with each layer of emulsion containing a different dye.

When light hits a silver bromide crystal, silver cations (Ag^+) accept an electron from the bromide ions (Br^-) and silver atoms are produced.

$$\text{silver ion} + \text{electron} \rightarrow \text{silver atom}$$
$$Ag^+ \quad + \quad e^- \quad \rightarrow \quad Ag$$

The bromine atom produced in the process is trapped in the gelatin. The more light which falls on the photographic film then the greater the amount of silver deposited.

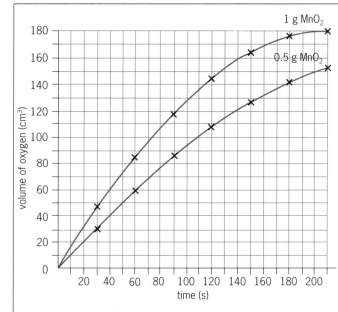

Figure 11.9 Sample data for differing amounts of MnO_2 catalyst

QUESTION

a Devise an experiment to show how sunlight affects the rate of formation of silver from the silver salts, silver chloride and silver bromide.

Catalysts

Over 90% of industrial processes use **catalysts**. A catalyst is a substance which can alter the rate of a reaction without being chemically changed at the end of it. In the laboratory, the effect of a catalyst can be observed using the decomposition of hydrogen peroxide as an example.

$$\text{hydrogen peroxide} \rightarrow \text{water} + \text{oxygen}$$
$$2H_2O_2(aq) \quad \rightarrow 2H_2O(l) + O_2(g)$$

The rate of decomposition at room temperature is very slow. There are substances, however, which will speed up this reaction, one of these being manganese(IV) oxide. When black manganese(IV) oxide powder is added to hydrogen peroxide solution, oxygen is produced rapidly. The rate at which this occurs can be followed by measuring the volume of oxygen gas produced with time.

Some sample results from experiments of this type have been plotted in Figure 11.9. At the end of the reaction, the manganese(IV) oxide can be filtered off and used again. The reaction can be made even faster by increasing the amount and surface area of the catalyst used. This is because the activity of a catalyst involves its surface. It should be noted that, in gaseous reactions, if dirt or impurities are present on the surface of the catalyst, it will not act as efficiently;

it is said to have been 'poisoned'. Therefore, the gaseous reactants must be pure.

Chemists have found that:

- a small amount of catalyst will produce a large amount of chemical change
- catalysts remain unchanged chemically after a reaction has taken place, but they can change physically. For example, a finer manganese(IV) oxide powder is left behind after the decomposition of hydrogen peroxide
- catalysts are very specific to a particular chemical reaction.

Some examples of chemical processes and the catalysts used are shown in Table 11.1.

A catalyst increases the rate by providing an alternative reaction path with a lower **activation energy**. The activation energy is the energy barrier which reactants must overcome, when their particles collide, to successfully react and form products (Figure 11.10).

Table 11.1 Examples of catalysts

Process	Catalyst
Haber process —for the manufacture of ammonia	Iron
Contact process —for the manufacture of sulphuric acid	Vanadium(V) oxide
Oxidation of ammonia to give nitric acid	Platinum
Fermentation of sugars to produce alcohol	Enzymes (in yeast)
Hydrogenation of unsaturated oils to form fats in the manufacture of margarines	Nickel

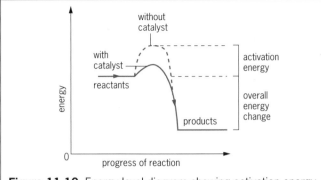

Figure 11.10 Energy level diagram showing activation energy

Catalytic converters

In the previous chapter you saw that European regulations state that all new cars have to be fitted with catalytic converters as part of their exhaust system (Figure 11.11). Car exhaust fumes contain pollutant gases such as carbon monoxide (CO) and nitrogen(II) oxide (NO). The following reactions proceed of their own accord but very slowly under the conditions inside an exhaust.

carbon monoxide + oxygen → carbon dioxide
$$2CO(g) \quad + \quad O_2(g) \quad \rightarrow \quad 2CO_2(g)$$

nitrogen(II) + carbon → nitrogen + carbon
oxide   monoxide   dioxide
$$2NO(g) \quad + \quad 2CO(g) \quad \rightarrow \quad N_2(g) \quad + 2CO_2(g)$$

The catalyst in the converter speeds up these reactions considerably. In these reactions, the pollutants are converted to carbon dioxide and nitrogen, which are naturally present in the air. It should be noted, however, that the catalytic converter can only be used with unleaded petrol and that, due to impurities being deposited on the surface of the catalyst, it becomes poisoned and has to be replaced every 5 or 6 years.

QUESTIONS

a Using a catalysed reaction of your choice, devise an experiment to follow the progress of the reaction and determine how effective the catalyst is.

b Why do some people consider catalytic converters not to be as environmentally friendly as suggested in their advertising material?

c Unreacted hydrocarbons such as octane, C_8H_{18} (from petrol), also form part of the exhaust gases. These gases are oxidised in the converter to carbon dioxide and water vapour. Write an equation for the oxidation of octane.

Enzymes

Enzymes are protein molecules produced in living cells. These substances are used by living organisms as catalysts to speed up hundreds of different chemical reactions going on inside them. These biological catalysts are very specific in that each chemical

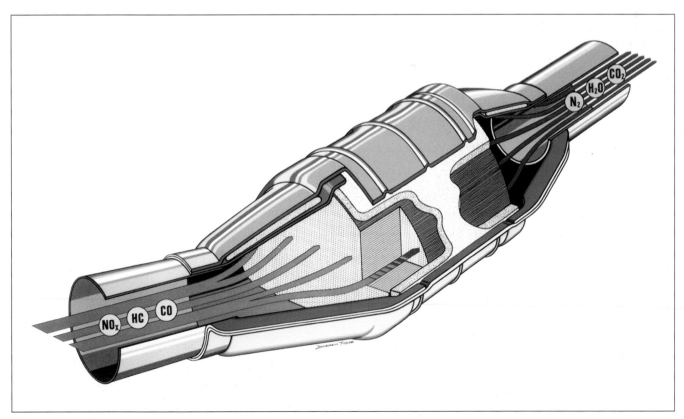

Figure 11.11 A cut through of a catalytic converter

Figure 11.12 These biological washing powders contain enzymes

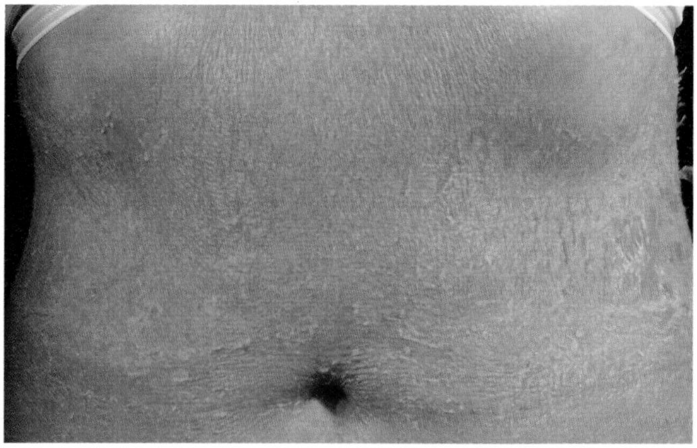

Figure 11.13 An allergic reaction to a biological detergent

reaction taking place has a different enzyme catalyst. You can imagine, therefore, that there are literally hundreds of different kinds of enzyme. For example, hydrogen peroxide is a substance naturally produced within our bodies (a natural metabolic product). However, it is extremely damaging and must be decomposed very rapidly. Catalase is the enzyme which converts hydrogen peroxide into harmless water and oxygen within our livers:

$$\text{hydrogen peroxide} \xrightarrow{\text{catalase}} \text{water} + \text{oxygen}$$

$$2H_2O_2(aq) \xrightarrow{\text{catalase}} 2H_2O(l) + O_2(g)$$

Although many chemical catalysts can work under various conditions of temperature and pressure as well as alkalinity or acidity, biological catalysts operate only under very particular conditions. For example, they operate over a very narrow temperature range and if the temperature becomes too high, they become inoperative. At temperatures above about 45 °C, they denature.

A huge multimillion-pound industry has grown up around the use of enzymes. For example, biological washing powders (Figure 11.12) contain enzymes to break down stains such as sweat, blood and egg, and they do this at the relatively low temperature of 40 °C. This reduces energy costs, because the washing water does not need to be heated as much.

There were problems associated with the early biological washing powders. Some customers suffered from skin rashes, because they were allergic to the enzymes (Figure 11.13). This problem has been overcome to a certain extent by advising that extra rinsing is required. Also, many manufacturers have placed warnings on their packets, indicating that the powder contains enzymes which may cause skin rashes.

QUESTIONS

a When using biological washing powders what factors have to be taken into consideration?

b Enzymes in yeast are used in the fermentation of glucose. Why, when the temperature is raised to 45 °C, is very little ethanol actually produced compared with the amount formed at room temperature?

Checklist

After studying Chapter 11 you should know and understand the following terms.

Reaction rate   A measure of the change which happens during a reaction in a single unit of time. It may be affected by the following factors:
- surface area of the reactants
- concentration of the reactants
- the temperature at which the reaction is carried out
- light
- use of a catalyst.

Activation energy   The excess energy (E_A) that a reaction must acquire to permit the reaction to occur.

Catalyst   A substance which alters the rate of a chemical reaction without itself being chemically changed.

Catalytic converter   A device for converting dangerous exhaust gases from cars into less dangerous emissions. For example, carbon monoxide gas is converted to carbon dioxide gas.

Enzymes   Protein molecules produced in living cells. They act as biological catalysts and are specific to certain reactions. They operate only within narrow temperature and pH ranges.

QUESTIONS

1 Explain the following statements.
   a A car exhaust pipe will rust much faster if the car is in constant use.
   b Carrots cook faster when they are chopped up.
   c Industrial processes become more economically viable if a catalyst can be found for the reactions involved.
   d In fireworks it is usual for the ingredients to be powdered.
   e Tomatoes ripen faster in a greenhouse.
   f The reaction between zinc and dilute hydrochloric acid is slower than the reaction between zinc and concentrated hydrochloric acid.

2 A student performed two experiments to establish how effective manganese(IV) oxide was as a catalyst for the decomposition of hydrogen peroxide.
   The results below were obtained by carrying out these experiments with two different quantities of manganese(IV) oxide. The volume of the gas produced was recorded against time.

Time (s)	0	30	60	90	120	150	180	210
Volume for 0.3g (cm³)	0	29	55	79	98	118	133	146
Volume for 0.5g (cm³)	0	45	84	118	145	162	174	182

   a Draw a diagram of the apparatus you could use to carry out these experiments.
   b Plot a graph of the results.
   c Is the manganese(IV) oxide acting as a catalyst in the decomposition of hydrogen peroxide? Explain your answer.
   d (i)   At which stage does the reaction proceed most quickly?
      (ii) How can you tell this from the graph you have drawn?
      (iii) In terms of particles, explain why the reaction is quickest at the point you have chosen in (i).
   e Why does the slope of the graph become less steep as the reaction proceeds?

   f What volume of gas has been produced when using 0.3g of manganese(IV) oxide after 50s?
   g How long did it take for 60 cm³ of gas to be produced when the experiment was carried out using 0.5g of the manganese(IV) oxide?
   h Write a balanced chemical equation for the decomposition of hydrogen peroxide.

3 a Which of the following reaction mixtures will produce hydrogen more quickly at room temperature:
      (i) magnesium ribbon + dilute nitric acid?
      (ii) magnesium powder + dilute nitric acid?
   b Give an explanation of your answer to (a).
   c Suggest two other methods by which the speed of this reaction can be altered.

4 A flask containing dilute hydrochloric acid was placed on a digital balance. An excess of limestone chippings was added to this acid, a plug of cotton wool was placed in the neck of the flask and the initial mass was recorded. The mass of the apparatus was recorded every 2 minutes. At the end of the experiment the loss in mass of the apparatus was calculated and the following results were obtained.

Time (min)	0	2	4	6	8	10	12	14	16
Loss in mass (g)	0	2.1	3.0	3.1	3.6	3.8	4.0	4.0	4.0

   a Plot the results of the experiment.
   b Which of the results would appear to be incorrect? Explain your answer.
   c Write a balanced chemical equation to represent the reaction taking place.
   d Why did the mass of the flask and its contents decrease?
   e Why was the plug of cotton wool used?
   f How does the rate of reaction change during this reaction? Explain this using particle theory.
   g How long did the reaction last?

QUESTIONS (continued)

5 a What is a catalyst?
  b List the properties of catalysts.
  c Name the catalyst used in the following processes:
   (i) the Contact process
   (ii) the Haber process
   (iii) the hydrogenation of unsaturated fats.
  d Which series of metallic elements, in the periodic table, do the catalysts you have named in (c) belong to?
  e What are the conditions used in the industrial processes named in (c)? The following references will help you: Chapters 12, 14 and 15.

6 This question concerns the reaction of copper(II) carbonate with dilute hydrochloric acid. The equation for the reaction is:

$$CuCO_3(s) + 2HCl(aq) \rightarrow CuCl_2(aq) + CO_2(g) + H_2O(l)$$

  a Sketch a graph to show the rate of production of carbon dioxide when an excess of dilute hydrochloric acid is added. The reaction lasts 40 s and produces 60 cm^3 of gas.
  b Find on your graph the part which shows:
   (i) where the reaction is at its fastest
   (ii) when the reaction has stopped.
  c Calculate the mass of copper(II) carbonate used to produce 60 cm^3 of carbon dioxide. (A_r: Cu = 63.5; C = 12; O = 16. One mole of a gas occupies 24 dm^3 at room temperature and pressure (rtp).)
  d Sketch a further graph using the same axes to show what happens to the rate at which the gas is produced if:
   (i) the concentration of the acid is decreased
   (ii) the temperature is increased.

7 European regulations state that all new cars have to be fitted with catalytic converters as part of their exhaust system.
  a Why are these regulations necessary?
  b Which gases are removed by catalytic converters?
  c Which metals are often used as catalysts in catalytic converters?
  d What does the term 'poisoned' mean with respect to catalysts?
  e The latest converters will also remove unburnt petrol. An equation for this type of reaction is:

$$2C_7H_{14}(g) + 21O_2(g) \rightarrow 14CO_2(g) + 14H_2O(g)$$

   (i) Calculate the mass of carbon dioxide produced by 1.96 g of unburnt fuel.
   (ii) Convert this mass of carbon dioxide into a volume measured at rtp.
   (iii) If the average car produces 7.84 g of unburnt fuel a day, calculate the volume of carbon dioxide produced by the catalytic converter measured at rtp. (A_r: C = 12; H = 1; O = 16. One mole of any gas occupies 24 dm^3 at rtp.)

8 a Give examples of chemical reactions which happen:
   (i) very slowly
   (ii) at a moderate rate
   (iii) very quickly.
  b How could you speed up the reaction named in (a)(i)?
  c How could you slow down the reaction named in (a)(ii)?

12 THE PETROLEUM INDUSTRY

Substances from oil

What do the modes of transport shown in Figure 12.1 have in common? They all use liquids obtained from **crude oil** as fuels.

Oil refining

Crude oil is a complex mixture of compounds known as **hydrocarbons** (Figure 12.2a). Hydrocarbons are molecules which contain only the elements carbon and hydrogen bonded together covalently (Chapter 4, p. 49). These carbon compounds form the basis of a

Figure 12.2a Crude oil is a mixture of hydrocarbons

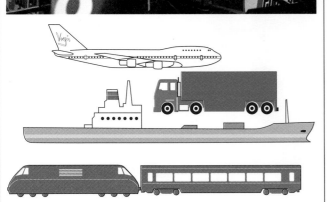

Figure 12.1 Modes of transport

Figure 12.2b The objects above are made from substances obtained from oil

group called **organic compounds**. All living things are made from organic compounds based on chains of carbon atoms similar to those found in crude oil. Crude oil is not only a major source of fuel but is also a raw material of enormous importance. It supplies a large and diverse chemical industry to make dozens of products (Figure 12.2b).

Crude oil is not very useful to us until it has been processed. The process, known as **refining**, is carried out at an oil refinery (Figure 12.3).

Refining involves separating crude oil into various batches or **fractions**. Chemists use a technique called **fractional distillation** to separate the different fractions. This process works in the same way as that discussed in Chapter 2, p. 21, for separating ethanol (alcohol) and water. The different components (fractions) separate because they have different boiling points. The crude oil is heated to about 400 °C to vaporise all the different parts of the mixture. The mixture of vapours passes into the fractionating column near the bottom (Figure 12.4). Each

Figure 12.3 An oil refinery

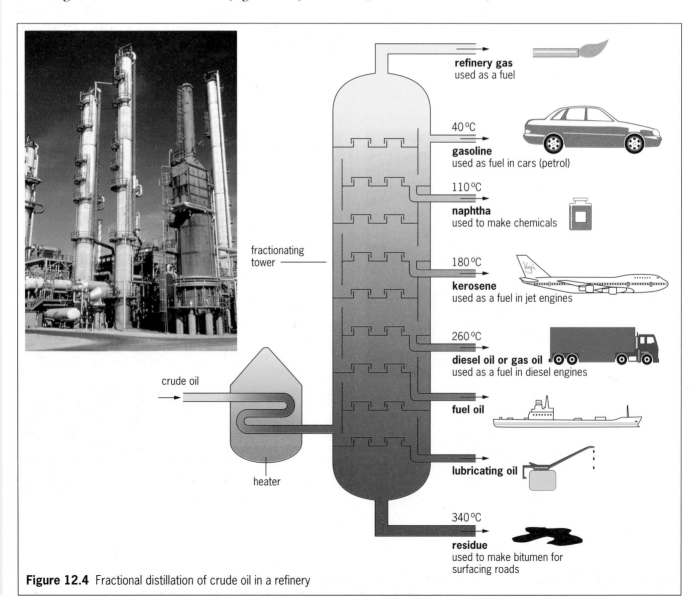

refinery gas
used as a fuel

40 °C
gasoline
used as fuel in cars (petrol)

110 °C
naphtha
used to make chemicals

180 °C
kerosene
used as a fuel in jet engines

260 °C
diesel oil or gas oil
used as a fuel in diesel engines

fuel oil

lubricating oil

340 °C
residue
used to make bitumen for surfacing roads

fractionating tower

crude oil

heater

150   **Figure 12.4** Fractional distillation of crude oil in a refinery

fraction is obtained by collecting hydrocarbon molecules which have a boiling point in a given range of temperatures (Figure 12.4). For example, the fraction we know as petrol contains molecules which have boiling points between 30 °C and 110 °C. The molecules in this fraction contain between five and ten carbon atoms. These smaller molecules with lower boiling points condense higher up the tower. The bigger hydrocarbon molecules which have the higher boiling points condense in the lower half of the tower.

The liquids condensing at different levels are collected on **trays**. In this way the crude oil is separated into different fractions. These fractions usually contain a number of different hydrocarbons. The individual single hydrocarbons can then be obtained by further refining the fraction by further distillation.

It is important to realise that the uses of the fractions depend on their properties. For example, one of the lower fractions, which boils in the range 250–350 °C, is quite thick and sticky and makes a good lubricant. However, the petrol fraction burns very easily and this therefore makes it a good fuel for use in engines.

QUESTIONS

a What do you understand by the term hydrocarbon?
b All organisms are composed of compounds which contain carbon. Why do you think carbon chemistry is often called 'organic chemistry'?
c List the main fractions obtained by separating the crude oil mixture and explain how they are obtained in a refinery.

Alkanes

Most of the hydrocarbons in crude oil belong to the family of compounds called **alkanes**. The molecules within the alkane family contain carbon atoms covalently bonded to four other atoms by single bonds. Because these molecules possess only single bonds they are said to be **saturated**, as no further atoms can be added (Figure 12.5). The physical properties of the first six members of the alkane family are shown in Table 12.1.

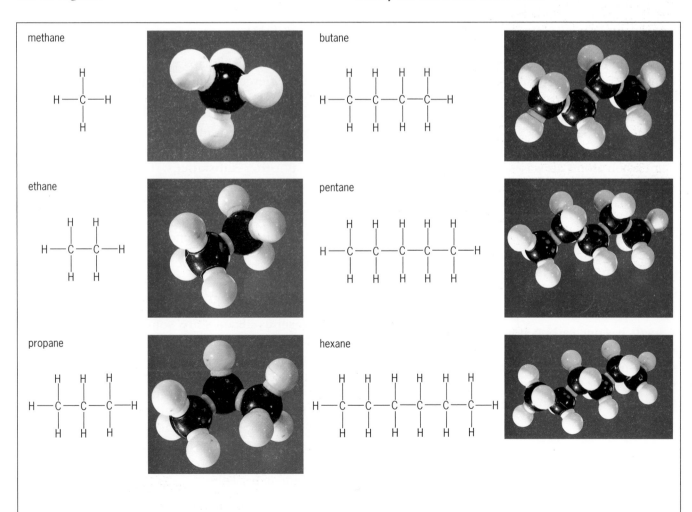

Figure 12.5 The alkane molecules look like the models in the photographs

Table 12.1 Some alkanes and their physical properties

Alkane	Formula	Melting point (°C)	Boiling point (°C)	Physical state at room temperature
Methane	CH_4	–182	–164	Gas
Ethane	C_2H_6	–183	–87	Gas
Propane	C_3H_8	–190	–42	Gas
Butane	C_4H_{10}	–138	0	Gas
Pentane	C_5H_{12}	–129	36	Liquid
Hexane	C_6H_{14}	–95	69	Liquid

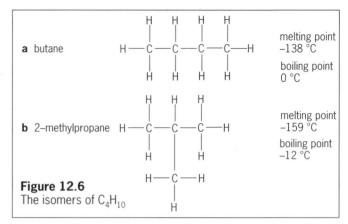

Figure 12.6
The isomers of C_4H_{10}

You will notice from Figure 12.5 and Table 12.1 that the compounds have a similar structure and similar name endings. They also behave chemically in a similar way and the family of compounds can be represented by a general formula. In the case of the alkanes the general formula is:

$$C_nH_{(2n+2)}$$

where n is the number of carbon atoms present.

A family with the above factors in common is called a **homologous series**. As you go up a homologous series, in order of increasing number of carbon atoms, the physical properties of the compounds gradually change. For example, the melting and boiling points of the alkanes shown in Table 12.1 gradually increase. This is due to an increase in the intermolecular forces (van der Waals' forces) as the size and mass of the molecule increases (Chapter 4, p. 52).

Under normal conditions molecules with up to four carbon atoms are gases, those with between five and 16 carbon atoms are liquids, while those with greater than 16 carbon atoms are solids.

Naming of the alkanes

All the alkanes have names ending in *-ane*. The rest of the name tells you the number of carbon atoms present in the molecule. For example, the compound whose name begins with:

- *meth-* has one carbon atom
- *eth-* has two carbon atoms
- *prop-* has three carbon atoms
- *but-* has four carbon atoms
- *pent-* has five carbon atoms

and so on.

Isomerism

Sometimes it is possible to write more than one structural formula to represent a molecular formula. The structural formula of a compound shows how the atoms are joined together by the covalent bonds.

For example, there are two different compounds with the molecular formula C_4H_{10}. The structural formulae of these two substances along with their names and physical properties are shown in Figure 12.6.

Compounds such as those in Figure 12.6 are known as **isomers**. Isomers are substances which have the same molecular formula but different structural formulae. The different structures of the compounds shown in Figure 12.6 have different melting and boiling points. Molecule (b) contains a branched chain and has a lower melting point than molecule (a), which has no branched chain. All the alkane molecules with four or more carbon atoms possess isomers. Perhaps you can see why there are so many different organic compounds!

The chemical behaviour of alkanes

Alkanes are rather unreactive compounds. For example, they are generally not affected by alkalis, acids or many other substances. Their most important property is that they burn easily.

Gaseous alkanes, such as methane, will burn in a good supply of air, forming carbon dioxide and water as well as plenty of heat energy.

$$\text{methane} + \text{oxygen} \rightarrow \text{carbon dioxide} + \text{water} + \text{energy}$$

$$CH_4(g) + 2O_2(g) \rightarrow CO_2(g) + 2H_2O(g)$$

The gaseous alkanes are some of the most useful fuels. Methane, better known as natural gas, is used for cooking as well as for heating our offices, schools and homes (Figure 12.7a). Propane and butane burn with very hot flames and they are sold as liquefied petroleum gas (LPG). In rural areas where there is no supply of natural gas, central heating systems can be run on propane gas (Figure 12.7b). Butane, sometimes mixed with propane, is used in portable blow-lamps and in gas lighters.

Another useful reaction worth noting is that between the alkanes and the halogens. For example, methane and chlorine react in the presence of sun-

Figure 12.7 (**a**, top) This is burning methane. (**b**, bottom) Central heating systems can be run on propane

light (or ultraviolet light). The ultraviolet light splits the chlorine molecules into atoms. When this type of reaction takes place, these atoms are called **free radicals** and they are very reactive.

$$\text{chlorine gas} \xrightarrow{\text{sunlight}} \text{chlorine atoms}$$
$$\text{(free radicals)}$$
$$Cl_2(g) \longrightarrow 2Cl(g)$$

The chlorine atoms then react further with methane molecules, and a hydrogen chloride molecule is produced along with a methyl free radical.

$$\begin{array}{ccc} \text{chlorine} + \text{methane} & \rightarrow & \text{methyl} + \text{hydrogen} \\ \text{atom} & & \text{radical} \quad \text{chloride} \\ Cl(g) + CH_4(g) & \rightarrow & CH_3(g) + HCl(g) \end{array}$$

The methyl free radical reacts further.

$$\begin{array}{ccc} \text{methyl} + \text{chlorine} & \rightarrow & \text{chloromethane} + \text{chlorine} \\ \text{radical} \quad \text{gas} & & \text{atom} \\ CH_3(g) + Cl_2(g) & \rightarrow & CH_3Cl(g) + Cl(g) \end{array}$$

This chlorine free radical, in turn, reacts further and the process continues until all the chlorine and the methane have been used up. This type of process is known as a **chain reaction** and it is very rapid. The overall chemical equation for this process is:

$$\begin{array}{ccc} \text{methane} + \text{chlorine} & \rightarrow & \text{chloromethane} + \text{hydrogen} \\ & & \text{chloride} \\ CH_4(g) + Cl_2(g) & \rightarrow & CH_3Cl(g) + HCl(g) \end{array}$$

We can see from this final equation that one hydrogen atom of the methane molecule is **substituted** by a chlorine atom. This type of reaction is known as a **substitution reaction**.

Because we cannot control the chlorine free radicals produced in this reaction, we also obtain small amounts of other 'substituted' products — CH_2Cl_2 (dichloromethane), $CHCl_3$ (trichloromethane or chloroform) and CCl_4 (tetrachloromethane). Many of these so-called **halogenoalkanes** are used as solvents. For examples, 1,1,1-trichloroethane is used as a solvent in some correction fluids (Figure 12.8).

Early anaesthetics relied upon trichloromethane, $CHCl_3$, or chloroform. Unfortunately, this anaesthetic had a severe problem since the lethal dose was only slightly higher than that required to anaesthetise the patient. In 1956, halothane was discovered by chemists working at ICI. This is a compound containing chlorine, bromine and fluorine. Its formula is $CF_3CHBrCl$. However, even this is not the perfect anaesthetic since evidence suggests that prolonged exposure to this substance may cause liver damage. The search continues for even better anaesthetics.

A group of compounds were discovered in the 1930s and were called the chlorofluorocarbons or

Figure 12.8 1,1,1-trichloroethane is used as the solvent in this correction fluid

153

Figure 12.9 These cans used to have CFC propellants in them

CFCs for short. Because of their inertness they found many uses, especially as a propellant in aerosol cans. CFC-12 or dichlorodifluoromethane, CF_2Cl_2, was one of the most popular CFCs in use in aerosol cans (Figure 12.9). Scientists now believe that the CFCs released from aerosols are destroying the ozone layer.

The ozone hole problem

Our atmosphere protects us from harmful ultraviolet radiation from the Sun. This damaging radiation is absorbed by the relatively thin ozone layer found in the stratosphere some 25–50 km above the Earth's surface (Figure 12.10).

Large holes have recently been discovered in the ozone layer over Antarctica and Europe. Scientists think that these holes have been produced by CFCs such as CFC-12. CFCs escape into the atmosphere and, because of their inertness, remain without further reaction until they reach the stratosphere and the ozone layer. In the stratosphere the high-energy ultraviolet radiation causes a chlorine atom to split off from the CFC molecule. This chlorine atom, or free radical, then reacts with the ozone.

$$Cl(g) + O_3(g) \rightarrow OCl(g) + O_2(g)$$

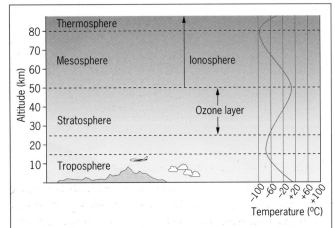

Figure 12.10 The ozone layer is between 25 km and 50 km above sea level

This is not the only problem with CFCs. They are also powerful 'greenhouse gases' (Chapter 8, p. 105). The ozone depletion and greenhouse effects have become such serious problems that an international agreement known as the *Montreal Protocol on Substances which Deplete the Ozone Layer* was agreed in 1987. The proposed controls were tightened in 1990 by the second meeting of the parties to the *Montreal Protocol*.

Research is now going ahead, with some success, to produce safer alternatives to CFCs. At present, better alternatives called hydrochlorofluorocarbons (HCFCs) have been developed. These substances have lower ozone-depletion effects and are not very effective 'greenhouse gases'.

Other uses of alkanes

Besides their major use as fuels (Chapter 13, p. 162), some of the heavier alkanes are used as waxes (Figure 12.11), as lubricating oils and in the manufacture of alkenes.

Figure 12.11 Candles contain a mixture of heavier alkanes

QUESTIONS

a Estimate the boiling points for the alkanes with formulae:
   (i) C_8H_{18}
   (ii) $C_{12}H_{26}$.

b Name the alkanes which have the following formulae:
   (i) C_7H_{16}
   (ii) $C_{10}H_{22}$.

c Draw the structural formulae for the isomers of:
   (i) C_5H_{12}
   (ii) C_6H_{14}.

d Write a balanced chemical equation to represent the combustion of propane.

e In what mole proportions should chlorine and methane be mixed to produce:
   (i) mainly chloromethane?
   (ii) mainly tetrachloromethane?

f Describe a method you would use to separate chloromethane from the other possible reaction products when methane reacts with chlorine.

g Why do you think that CFCs release chlorine atoms into the stratosphere and not fluorine atoms?

Alkenes

Alkenes form another homologous series of hydrocarbons of the general formula:

$$C_nH_{2n}$$

where n is the number of carbon atoms. The alkenes are more reactive than the alkanes because they each contain a double covalent bond between the carbon atoms (Figure 12.12). Molecules that possess a double covalent bond of this kind are said to be **unsaturated**, because it is possible to break this double bond and add extra atoms to the molecule.

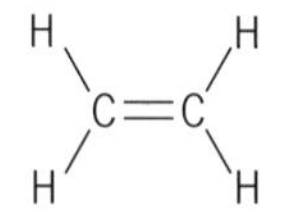

Figure 12.12
The covalent bonding in ethene, the simplest alkene

Table 12.2 The first three alkenes and their physical properties

Alkene	Formula	Melting point (°C)	Boiling point (°C)	Physical state at room temperature
Ethene	C_2H_4	–169	–104	Gas
Propene	C_3H_6	–185	–47	Gas
Butene	C_4H_8	–184	–6	Gas

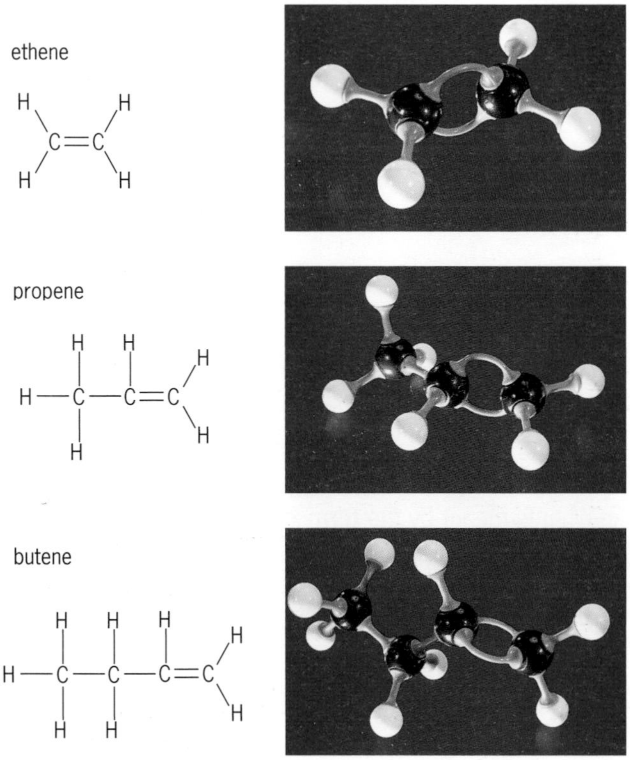

ethene

propene

butene

Figure 12.13 Structure and shape of the first three alkenes

All alkenes have names ending in *-ene*. Alkenes, especially ethene, are very important industrial chemicals. They are used extensively in the plastics industry and in the production of alcohols such as ethanol and propanol. Table 12.2 gives the names and formulae and some physical properties of the first three members of the alkene family. Figure 12.13 shows the structure of these first three members as well as models of their shape.

Where do we get alkenes from?

Very few alkenes are found in nature. Most of the alkenes used by the petrochemical industry are obtained by breaking up larger, less useful alkane molecules obtained from the fractional distillation of crude oil. This is usually done by a process called **catalytic cracking**. In this process the alkane molecules to be 'cracked' (split up) are passed over a mixture of aluminium and chromium oxides heated to about 500 °C.

dodecane   →   decane   + ethene

$$C_{12}H_{26}(g) \quad \rightarrow \quad C_{10}H_{22}(g) + C_2H_4(g)$$

(found in kerosene)   shorter alkane   alkene

Figure 12.14 shows the simple apparatus that can be used to carry out cracking reactions in the laboratory. You will notice that in the laboratory we may use a catalyst of broken, unglazed pottery.

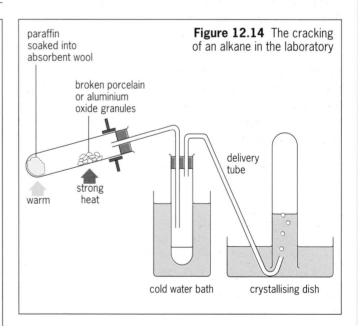

paraffin soaked into absorbent wool

Figure 12.14 The cracking of an alkane in the laboratory

broken porcelain or aluminium oxide granules

delivery tube

warm

strong heat

cold water bath   crystallising dish

The chemical behaviour of alkenes

The double bond makes alkenes more reactive than alkanes since it tends to break open during chemical reactions and join onto other atoms (Figure 12.15). For example, hydrogen will add across the double bond, under suitable conditions, forming ethane.

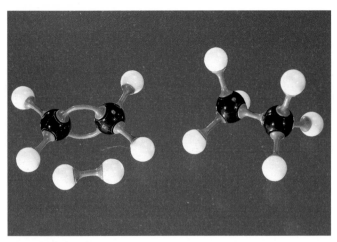

Figure 12.15 The addition of hydrogen to ethene using molecular models

$$\text{ethene} \quad + \quad \text{hydrogen} \xrightarrow[\text{Ni/Pt catalyst}]{150-300°C} \text{ethane}$$

$$C_2H_4(g) \quad + \quad H_2(g) \longrightarrow C_2H_6(g)$$

This reaction is called **hydrogenation**. Hydrogenation reactions like the one shown with ethene are used in the manufacture of margarines from vegetable oils. Vegetable oils contain fatty acids, such as linoleic acid ($C_{18}H_{32}O_2$). These are unsaturated molecules, containing several double bonds. These double bonds make the molecule less flexible. By hydrogenation, it is possible to convert these molecules into more saturated ones. Now the molecules are less rigid and can flex and twist more easily, and hence pack more closely together. This in turn causes an increase in the intermolecular forces and so raises the melting point. The now solid margarines can be spread on bread more easily than liquid oils.

There is another side to this process. Many doctors now believe that unsaturated fats are more healthy than saturated ones. Because of this, many margarines are left partially unsaturated. They do not have all the C=C taken out of the fat molecules. However, the matter is far from settled and the debate continues.

Another important **addition reaction** is the one used in the manufacture of ethanol. Ethanol has important uses as a solvent and a fuel (p. 165). It is formed when water (as steam) is added across the double bond in ethene. For this reaction to take place, the reactants have to be passed over a catalyst of phosphoric(v) acid (absorbed on silica pellets) at a temperature of 300°C and pressure of 60 atmosphere (1 atmosphere = 1×10^5 Pascals).

$$\text{ethene} \quad + \quad \text{steam} \underset{\substack{\text{phosphoric(v)}\\ \text{acid catalyst}}}{\overset{300°C, 60 \text{ atm}}{\rightleftharpoons}} \text{ethanol}$$

$$C_2H_4(g) \quad + \quad H_2O(g) \rightleftharpoons C_2H_5OH(g)$$

This reaction is reversible as is shown by the ⇌ sign. The conditions have been chosen to ensure the highest possible yield of ethanol. In other words, the conditions have been chosen which favour the forward reaction.

A test for unsaturated compounds

The addition reaction between bromine dissolved in an organic solvent, for example 1,1,1-trichloroethane, and alkenes is used as a chemical test for the presence of a double bond between two carbon atoms. When a few drops of this bromine solution are shaken with the hydrocarbon, if it is an alkene, such as ethene, a reaction takes place in which bromine joins to the alkene double bond. This results in the bromine solution losing its colour. If an alkane such as hexane is shaken with a bromine solution of this type, no colour change takes place (Figure 12.16). This is because there are no double bonds between the carbon atoms of alkanes.

$$\text{ethene} \quad + \text{bromine} \longrightarrow \text{dibromoethane}$$

$$C_2H_4(g) \quad + Br_2(\text{in solution}) \longrightarrow C_2H_4Br_2(\text{in solution})$$

QUESTIONS

a Using the information in Table 12.2, make an estimate of the boiling point of hexene.

b Write a balanced chemical equation to represent the process that takes place when decane is cracked.

c What is meant by the term 'addition reaction'?

d Write a word and balanced chemical equation for the reaction between ethene and hydrogen chloride.

e Write the structural formula for pentene.

Figure 12.16 An alkene decolorises bromine in 1,1,1-trichloroethane

Checklist

After studying Chapter 12 you should know and understand the following terms.

Organic chemistry   The branch of chemistry concerned with compounds of carbon found in living organisms.

Hydrocarbon   A substance which contains atoms of carbon and hydrogen only.

Oil refining   The general process of converting the mixture that is collected as crude oil into separate fractions. These fractions, known as petroleum products, are used as fuels, lubricants, bitumens and waxes. The fractions are separated from the crude oil mixture by fractional distillation.

Alkanes   A family of saturated hydrocarbons with the general formula C_nH_{2n+2}. The term 'saturated', in this context, is used to describe molecules that have only single bonds. The alkanes can only undergo substitution reactions in which there is replacement of one atom in the molecule by another atom.

Isomers   Compounds which have the same molecular formula but different structural arrangements of the atoms.

Free radicals   These are atoms or groups of atoms with unpaired electrons and are therefore highly reactive. They can be produced by high-energy radiation such as ultraviolet light in photochemical reactions.

Chain reaction   A reaction which is self-sustaining owing to the products of one step of the reaction assisting in promoting further reaction.

Halogenoalkanes   Organic compounds in which one or more hydrogen atoms of an alkane have been substituted by halogen atoms such as chlorine.

CFC Abbreviation for chlorofluorocarbon, which is a type of organic compound in which some or all of the hydrogen atoms of an alkane have been replaced by fluorine and chlorine atoms. These substances are generally unreactive but they can diffuse into the stratosphere where they break down under the influence of ultraviolet light. The products of this photochemical process then react with ozone (in the ozone layer). Because of this, their use has been discouraged. They are now being replaced by hydrochlorofluorocarbons (HCFCs).

Alkenes   A family of unsaturated hydrocarbons with the general formula C_nH_{2n}. The term 'unsaturated', in this context, is used to describe molecules which contain one or more double carbon–carbon bonds. Unsaturated compounds undergo addition reactions across the carbon–carbon double bonds and so produce saturated compounds. The addition of hydrogen across the carbon–carbon double bonds is used to reduce the amount of unsaturation during the production of margarines.

Catalytic cracking   The decomposition of higher alkanes into alkenes and alkanes of lower relative molecular mass. The process involves passing the larger alkane molecules over a catalyst of aluminium and chromium oxides, heated to 500 °C.

Test for unsaturation   A few drops of bromine dissolved in an organic solvent are shaken with the hydrocarbon. If it is decolorised, the hydrocarbon is unsaturated.

157

QUESTIONS

1 Explain the following.
  a Ethene is called an unsaturated hydrocarbon.
  b The cracking of larger alkanes into simple alkanes and alkenes is important to the petrochemical industry.
  c The conversion of ethene to ethanol is an example of an addition reaction.

2 The following question is about some of the reactions of ethene.

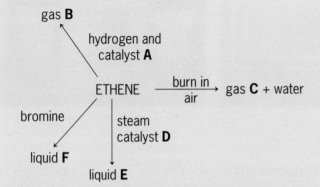

  a Give the names and formulae for the substances **A** to **F**.
  b **(i)** Write a word and balanced chemical equation to represent the reaction in which liquid **E** is formed.
  (ii) What reaction conditions are required for the process to take place?
  (iii) Hydrogen is used in the production of margarine to remove unsaturation. Explain what you understand by this statement.
  c Name the homologous series that gas **B** belongs to.
  d Describe a chemical test which would allow you to identify gas **C**.

3 a Crude oil is a mixture of *hydrocarbons* which belong to the *homologous series* called the *alkanes*. This mixture can be separated into fractions by the process of *fractional distillation*. Some of the fractions obtained are used as *fuels*. Some of the other fractions are subjected to *catalytic cracking* in order to make *alkenes*.
  Explain the meaning of the terms in italics.
  b Alkanes can be converted into substances which are used as solvents. To do this the alkane is reacted with a halogen, such as chlorine, in the presence of ultraviolet light.
  (i) Write a word and balanced chemical equation for the reaction between methane and chlorine.
  (ii) Name the type of reaction taking place.
  (iii) Highly reactive chlorine atoms are produced in the presence of ultraviolet light. When atoms are produced in this way, what are they called?
  (iv) Write a balanced chemical equation for the reaction which takes place between CHF_3 and Cl_2 to produce a chlorofluorocarbon (CFC).
  (v) Why are CFCs such a problem?

4 Crude oil is a mixture of hydrocarbons. The refining of crude oil produces fractions which are more useful to us than crude oil itself. Each fraction is composed of hydrocarbons which have boiling points within a specific range of temperature. The separation is carried out in a fractionating column, as shown below.

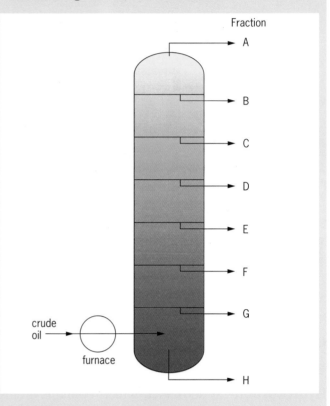

  a Which separation technique is used to separate the fractions?
  b Name each of the fractions **A** to **H** and give a use for each.
  c Why do the fractions come from the fractionating column in this order?
  d What is the connection between your answer to (c) and the size of the molecules in each fraction?
  e Which of the fractions will be the most flammable?

5 Alkanes and alkenes are hydrocarbons. They are composed of molecules which contain covalent bonds. For each of the molecules below, use a dot and cross diagram to show the bonding it contains.
  a Methane, CH_4.
  b Propene, C_3H_6.
  c Ethane, C_2H_6.
  d Ethene, C_2H_4.

6 Crude oil is an important source of organic chemical fuels. It is refined by fractional distillation. Use the information in the table below to answer the questions which follow.

Fraction	Boiling point (°C)
A	40
B	80
C	200
D	350
E	above 350

a For each of the questions that follow, give the letter of the fraction which is most appropriate as an answer. You should also give a reason for your answer in each case.
   (i) Which fraction would contain the most volatile substances?
   (ii) Which of the fractions would collect at the bottom of the fractionating column?
   (iii) Which fraction could be used as a fuel for cars?

(iv) Which fraction would contain the largest molecules?

b Some of the fractions undergo a further process called cracking to produce further substances.
   (i) Explain what you understand by the term 'cracking'. What conditions are employed when cracking occurs?
   (ii) Write a word and balanced chemical equation to show how octane can be produced by the cracking of $C_{15}H_{32}$.

7 a A hydrocarbon contains 82.8% by mass of carbon. Work out the empirical formula of this hydrocarbon.
  b The relative molecular mass of this hydrocarbon was found to be 58 by mass spectrometry. Work out its molecular formula. (A_r: C = 12, H = 1)

8 a Which of the following formulae represent alkanes, which represent alkenes and which represent neither? CH_3, C_6H_{12}, C_5H_{12}, C_6H_6, C_9H_{20}, $C_{12}H_{24}$, $C_{20}H_{42}$, C_2H_4, C_8H_{18}, C_3H_7
  b Draw all the possible isomers which have the molecular formula C_6H_{14}.

13 ENERGY SOURCES

Fossil fuels

Coal, oil and natural gas are all examples of **fossil fuels**. The term fossil fuels is derived from the fact that they are formed from dead plants and animals which were fossilised over 200 million years ago during the Carboniferous era.

Coal was produced by the action of pressure and heat on dead wood from ancient forests which once grew in the swamp land in many parts of the world under the prevailing weather conditions of that time. When dead trees fell into the swamps they were buried by mud. This prevented aerobic decay (which takes place in the presence of air). Over millions of years, due to movement of the Earth's crust (Chapter 18, p. 217) as well as to changes in climate, the land sank and the decaying wood became covered by even more layers of mud and sand. Anaerobic decay (which takes place in the absence of air) occurred, and as time passed the gradually forming coal became more and more compressed as other material was

Figure 13.1 Piece of coal showing a fossilised leaf

Figure 13.2 Cutting of coal is extremely mechanised

laid down above it (Figure 13.1). Over millions of years, as the layers of forming coal were pushed deeper and the pressure and temperature increased, the final conversion to coal took place (Figure 13.2).

Different types of coal were formed as a result of different pressures being applied during its formation. For example, anthracite is a hard coal with a high carbon content, typical of coal produced at greater depths. Table 13.1 shows some of the different types of coal along with their carbon contents.

Oil and gas were formed during the same period as coal. It is believed that oil and gas were formed from the remains of plants, animals and bacteria that once lived in seas and lakes. This material sank to the bottom of these seas and lakes and became covered in mud, sand and silt which thickened with time. Anaerobic decay took place, and, as the mud layers built up, high temperatures and pressures were created which converted the material slowly into oil and gas. As rock formed, earth movements caused it to buckle and split, and the oil and gas was trapped in folds beneath layers of non-porous rock or cap-rock (Figures 13.3 and 13.4).

Table 13.1 Types of coal

Type of coal	Carbon content (%)
Anthracite	90
Bituminous coal	60
Lignite	40
Peat	20

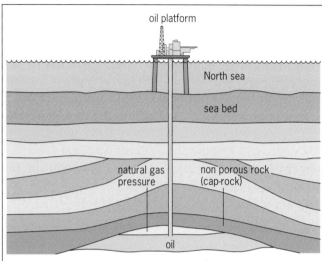

Figure 13.4 Natural gas and oil are trapped under non-porous rock

QUESTIONS

a Coal, oil and natural gas are all termed 'fossil fuels'. Why is the word 'fossil' used in this context?

b (i) Name the process by which plants convert carbon dioxide and water into glucose.

(ii) What conditions are necessary for this process to occur?

c Draw a flow diagram to represent the formation of coal, oil or gas.

Figure 13.3 Oil production in the North Sea

What is a fuel?

A fuel is a substance which can be conveniently used as a source of energy. Fossil fuels produce energy when they undergo **combustion**.

$$\text{fossil fuel} + \text{oxygen} \rightarrow \text{carbon dioxide} + \text{water} + \text{energy}$$

For example, natural gas burns readily in air (Chapter 12, p. 152).

$$\text{methane} + \text{oxygen} \rightarrow \text{carbon dioxide} + \text{water} + \text{energy}$$
$$CH_4(g) + 2O_2(g) \rightarrow CO_2(g) + 2H_2O(l)$$

It should be noted that natural gas, like crude oil, is a mixture of hydrocarbons such as methane, ethane and propane, and may also contain some sulphur. The sulphur content varies from source to source (Chapter 16, p. 199). Natural gas obtained from the North Sea is quite low in sulphur.

The perfect fuel would be:

- cheap
- available in large quantities
- safe to store and transport
- easy to ignite and burn, causing no pollution
- capable of releasing large amounts of energy.

Solid fuels are safer than volatile liquid fuels like petrol and gaseous fuels like natural gas.

How are fossil fuels used?

A major use of fossil fuels is in the production of electricity. Coal, oil and natural gas are burned in power stations (Figure 13.5) to heat water to produce steam which is then used to drive large turbines (Figure 13.6). At least 80% of the electricity generated in the UK is generated using fossil fuels. However, it should be noted that the relative importance of the three major fossil fuels is changing. Coal and oil are becoming less important while natural gas is increasingly important (DASH FOR GAS).

In a power station, the turbine drives a generator to produce electricity which is then fed into the National Grid (Figure 13.6). The National Grid is a system for distributing electricity throughout the country.

Other major uses of the fossil fuels are:

- as a major feedstock (raw material) for the chemicals and pharmaceuticals industries
- for domestic and industrial heating and cooking
- as fuels for various forms of transport vehicle.

QUESTIONS

a 'We have not yet found the perfect fuel.' Discuss this statement.

b 'Fossil fuels are a major feedstock for the chemical and pharmaceutical industries.' By reference to Chapter 12, give examples which support this statement.

Alternative sources of energy

Fossil fuels are an example of **non-renewable** resources, so called because they are not being replaced at the same rate as they are being used up. For example, we have approximately 60 years' supply of crude oil remaining from known reserves if we continue to use it at the current rate as a source of energy and chemicals (Table 13.2). It is important to use non-renewable fuels carefully and to consider alternative, **renewable** sources of energy for use in the future.

Nuclear power

Calder Hall power station in Cumbria, on the site of the present-day nuclear power complex at Sellafield (Figure 13.7), opened in 1956 and was the first nuclear reactor in the world to produce electricity on an industrial scale.

Figure 13.5 A power station

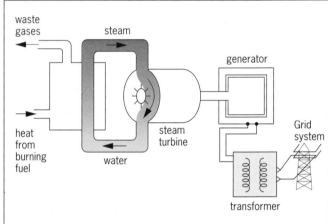

Figure 13.6 The way in which fuels are used to produce electricity

Table 13.2 Estimates of how long our fossil fuels will last

Fossil fuel	Estimated date it is expected to run out
Gas	2045
Oil	2055
Coal	2500

Figure 13.7 The nuclear power complex at Sellafield, Cumbria

Nuclear reactors obtain energy from the **fission** of uranium-235. **Nuclear fission** occurs when the unstable nucleus of a radioactive isotope splits up, forming smaller atoms and producing a large amount of energy (Chapter 17, p. 211). Scientists believe that the energy comes from the conversion of some of the mass of the isotope into energy.

This fission process begins when a neutron hits an atom of uranium-235, causing it to split and produce three further neutrons. These three neutrons split three more atoms of uranium-235, which produces nine neutrons and so on. This initiates a **chain reaction** (Figure 13.8).

In a reactor the fission process cannot be allowed to get out of control as it does in an atomic bomb. To prevent this, boron control rods can be pushed into different positions in the reactor to absorb some of the neutrons which are produced and so slow down the chain reaction. If this is done, the energy released from the reaction is obtained in a more controlled way. The energy is used to produce steam, which in turn is used to generate electricity (Figure 13.6).

However, there are problems. The main problem associated with a nuclear power station is that the reactor produces highly radioactive waste materials. These waste materials are difficult to store and cannot be disposed of very easily. Also, leaks of radioactive material have occurred at various sites throughout the world. Accidents at a small number of nuclear power stations, such as Chernobyl, in the Ukraine (Figure 13.9), and Three Mile Island, in the US, have led to a great deal of concern about their safety.

In the UK the safety record of the nuclear power industry is relatively good because it is subject to strict controls.

Hydroelectric power (HEP)

Hydroelectric power is electricity generated from the energy of falling water (Figure 13.10). It is an excellent energy source and electricity has been generated in this way in the mountainous areas of Scotland and Wales for some time. It is a very cheap source of electricity. Once you have built the power station, the energy is absolutely free. In some mountainous areas of the world, such as the Alps, HEP is the main source of electricity. One of the main advantages of this system is that it can be quickly used to supplement the National Grid at times of high demand. A disadvantage of HEP schemes is that they often require valleys to be flooded and communities to be moved.

Geothermal energy

Water is pumped into hot rocks in the Earth's crust far below ground level. The internal heat of the rocks

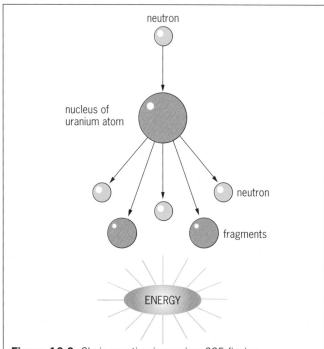

Figure 13.8 Chain reaction in uranium-235 fission

Figure 13.9 A nuclear accident happened at the Chernobyl power station in 1986

Figure 13.10 Hydroelectric power station

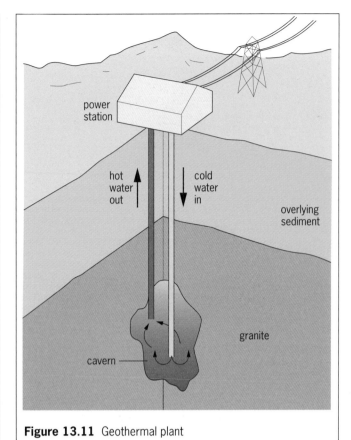

Figure 13.11 Geothermal plant

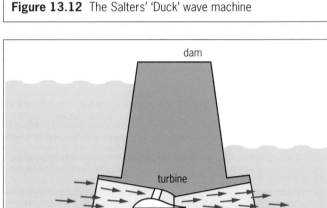

Figure 13.12 The Salters' 'Duck' wave machine

Figure 13.13 Tidal power — the ebb and flow of the tide drives the turbine set into the barrage or dam

converts the water to steam, which is used to drive turbines and hence generate electricity (Figure 13.11). This is a major source of electrical energy in Iceland.

Geothermal energy is a natural, non-polluting source of energy. Experiments on the viability of geothermal energy are being conducted in Cornwall.

Wave power

In this method the energy of moving waves is used to generate electricity. Figure 13.12 shows the Salters' 'Duck' wave machine in operation. The vertical motion of the waves is converted to rotary motion, which is used to drive a generator producing electricity. A disadvantage of this method for generating large amounts of electricity is that vast numbers of strings of these ducks would be required.

Wave power is a non-polluting source of energy.

Tidal power

The ebb and flow of the tides drives turbines built into a dam or barrage across an estuary where the height difference between high and low tides is large (Figure 13.13). There is a successful tidal power station across the river Rance near St Malo in northern France. The most likely place in Britain to put a tidal barrage is across the Severn Estuary near Bristol, but there would be environmental disadvantages with such a scheme. For example, there would be a threat to the wildlife around the estuary, since the mud

banks are one of the few remaining sites in Britain for wildfowl and wading birds to nest.

Wind power

This method uses the force of the wind to turn generators to produce electricity. Wind machines 24 m high are capable of generating 200 kW of electricity per machine. Large 'wind farms' have been developed in many parts of the world, for example in the US and on the Pennines in England (Figure 13.14).

Figure 13.14 Wind farm at Ovenden Moor, Yorkshire

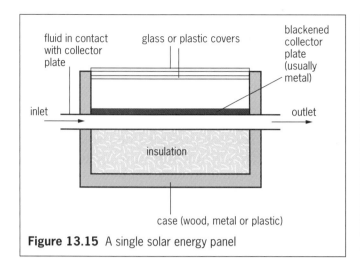

Figure 13.15 A single solar energy panel

Figure 13.16 A bank of photocells

The disadvantages of wind power are that the farms are unsightly, they require vast amounts of land, as the machines have to be carefully arranged so that their operation is not impaired, and they produce a lot of noise.

Solar energy

Two possible methods exist to use the energy of the Sun. In the first, the Sun's energy can be absorbed onto black-painted collector plates and used to heat water and homes (Figure 13.15). This method is relatively cheap. The second method involves the use of photovoltaic cells or photocells to generate electricity (Figure 13.16).

Disadvantages include the initial cost of cells as well as the fact that the Sun does not shine all the time. To solve this problem, solar cells are commonly linked to storage batteries.

Biomass and biogas

When any biological material, whether plant or animal, can be converted into energy, this energy is called **biomass** energy. It can be taken from animal or plant materials in different ways:

- by burning it, for example wood (Figure 13.17)
- by pressing out oils that can be burned
- by fermenting it to produce fuels such as ethanol or methane.

At least 50% of the world's population rely on wood as their main energy source.

In India there are millions of methane generators. Methane generated by the digestion of animal waste is called **biogas**. The biogas produced is used for cooking, heating and lighting. The by-product of this process is an excellent fertiliser.

Some countries have already experimented with ethanol as a fuel for cars. Up to 20% of ethanol can be added to petrol without the need to adjust the carburettor. Brazil, which has few oil reserves, pro-

Figure 13.17 Biomass energy is produced by burning wood

duces ethanol by fermentation (breakdown by enzymes) of sugar cane and grain, and uses it as a petrol additive (Figure 13.18). The Brazilian government has cut down its petrol imports by up to 60% through using this alcohol/petrol mixture.

Figure 13.18 In Brazil cars use an ethanol/petrol mixture

Chemical energy

We obtain our energy needs from the combustion of fuels, such as hydrocarbons, from the combustion of foods and from many other chemical reactions.

Combustion

When natural gas burns in a plentiful supply of air it produces a large amount of energy.

methane + oxygen → carbon + water + heat
                    dioxide   energy

$$CH_4(g) + 2O_2(g) \rightarrow CO_2(g) + 2H_2O(l) + \text{heat energy}$$

During this process, the **complete combustion** of methane, heat is given out. It is an **exothermic** reaction. If only a limited supply of air is available then the reaction is not as exothermic and the poisonous gas carbon monoxide is produced.

methane + oxygen → carbon + water + heat
                    monoxide   energy

$$2CH_4(g) + 3O_2(g) \rightarrow 2CO(g) + 4H_2O(l) + \text{heat energy}$$

This process is known as the **incomplete combustion** of methane.

The energy changes that take place during a chemical reaction can be shown by an **energy level diagram**. Figure 13.19 shows the energy level diagram for the complete combustion of methane.

When any chemical reaction occurs, the chemical bonds in the reactants have to be broken — this requires energy. When the new bonds in the products are formed, energy is given out (Figure 13.20). The **bond energy** is defined as the amount of energy in kilojoules (kJ) associated with the breaking or making of one mole of chemical bonds in a molecular element or compound.

Using the bond energy data from Table 13.3, which tells us how much energy is needed to break a chemi-

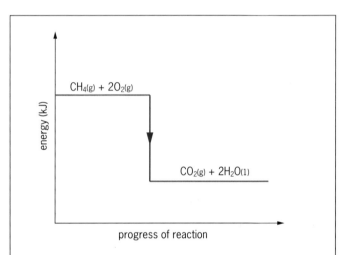

Figure 13.19 Energy level diagram for the complete combustion of methane

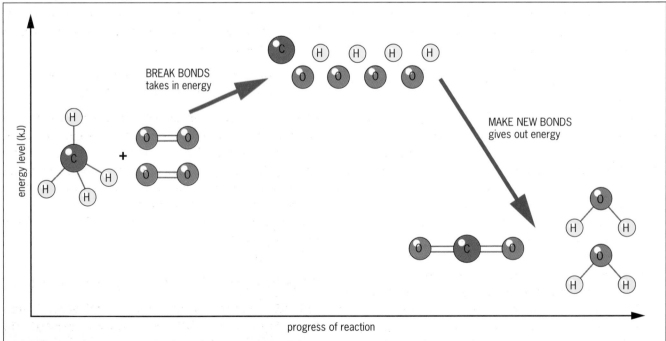

Figure 13.20 Breaking and forming bonds during the combustion of methane

Table 13.3 Bond energy data

Bond	Bond energy (kJ mol⁻¹)
C–H	435
O=O	497
C=O	803
H–O	464
C–C	347

cal bond and how much is given out when it forms, we can calculate how much energy is involved in each stage.

Bond breaking

Breaking 4 C–H bonds in methane requires 4×435
= 1740 kJ

Breaking 2 O=O bonds in oxygen requires 2×497
= 994 kJ

Total = 2734 kJ of energy

Making bonds

Making 2 C=O bonds in carbon dioxide gives out 2×803
= 1606 kJ

Making 4 O–H bonds in water gives out 4×464
= 1856 kJ

Total = 3462 kJ of energy

Energy difference
= energy required   –   energy given out
  to break bonds   when bonds are made
= 2734 – 3462
= –728 kJ

The negative sign shows that the chemicals are losing energy to the surroundings, that is, it is an exothermic reaction. A positive sign would indicate that the chemicals are gaining energy from the surroundings. This type of reaction would be called an **endothermic** reaction.

The energy stored in the bonds is called the **enthalpy** and is given the symbol H. The change in energy going from reactants to products is called the **change in enthalpy** and is shown as ΔH (pronounced 'delta H'). ΔH is called the **heat of reaction**.

For an exothermic reaction ΔH is negative and for an endothermic reaction ΔH is positive.

When fuels, such as methane, are burned they require energy to start the chemical reaction. This is known as the **activation energy**, E_A (Figure 13.21). In the case of methane reacting with oxygen, it is the energy involved in the initial bond breaking (Figure 13.20). The value of the activation energy will vary from fuel to fuel.

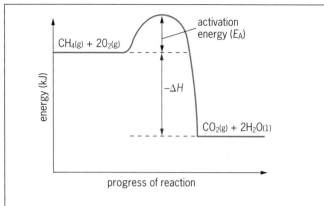

Figure 13.21 Energy level diagram for methane/oxygen

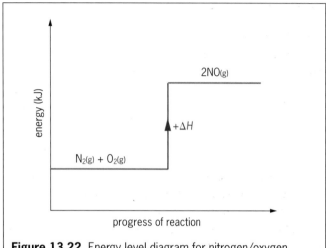

Figure 13.22 Energy level diagram for nitrogen/oxygen

Endothermic reactions are much less common than exothermic ones. In this type of reaction energy is absorbed from the surroundings so that the energy of the products is greater than that of the reactants. The reaction between nitrogen and oxygen gases is endothermic (Figure 13.22).

nitrogen + oxygen → nitrogen(II) oxide
$N_2(g)$ + $O_2(g)$ → $2NO(g)$

Dissolving is often an endothermic process. For example, when ammonium nitrate dissolves in water the temperature of the water falls, indicating that energy is being taken from the surroundings. Photosynthesis and thermal decomposition are other examples of endothermic processes.

In equations it is usual to express the ΔH value in units of kJ mol⁻¹.

For example:

$CH_4(g) + 2O_2(g) \rightarrow CO_2(g) + 2H_2O(l)$ $\Delta H = -728$ kJ mol⁻¹

This ΔH value tells us that when 1 mol of methane is burned in oxygen, 728 kJ of energy are released. This value is called the **enthalpy of combustion** of methane (or **molar heat of combustion** of methane).

Enthalpy of neutralisation (molar heat of neutralisation)

This is the enthalpy change that takes place when 1 mol of hydrogen ions ($H^+(aq)$) is neutralised.

$$H^+(aq) + OH^-(aq) \rightarrow H_2O(l) \quad \Delta H = -57\,kJ\,mol^{-1}$$

This process occurs, for example, in the titration of an alkali by an acid to produce a neutral solution (Chapter 7, p. 87).

QUESTIONS

a Using the bond energy data given in Table 13.3:
   (i) calculate the enthalpy of combustion of ethane
   (ii) draw an energy level diagram to represent this combustion process.
b How much energy is released if:
   (i) 0.5 mole of methane are burned?
   (ii) 5 moles of methane are burned?
   (iii) 4 g of methane are burned?
   (A_r: C = 12, H = 1)
c How much energy is released if:
   (i) 2 moles of hydrogen ions are neutralised?
   (ii) 0.25 mole of hydrogen ions are neutralised?
   (iii) 1 mole of sulphuric acid is completely neutralised?

Change of state

In Chapter 1, p. 3, we discussed the melting and boiling of a substance. The heating curve for water is shown in Figure 1.9. For ice to melt to produce liquid water, it must absorb energy from its surroundings. This energy is used to break down the weak forces between the water molecules (intermolecular forces) in the ice. This energy is called the **enthalpy of fusion** and is given the symbol ΔH_{fusion}. Similarly, when liquid water changes into steam, the energy required for this process to occur is called the **enthalpy of vaporisation** and is given the symbol ΔH_{vap}. Figure 13.23 shows the energy level diagrams representing both the fusion and the vaporisation processes.

QUESTIONS

a Describe the energy changes which take place when the processes described in this section, with water, are reversed.
b Using the knowledge you have obtained from Chapter 1, p. 4, give a full definition of the enthalpy of fusion and enthalpy of vaporisation for water.

Cells and batteries

A simple type of chemical cell is that shown in Figure 13.24 (top). In this cell the more reactive metal zinc dissolves in the dilute sulphuric acid, producing zinc ions ($Zn^{2+}(aq)$) and releasing two electrons.

$$Zn(s) \rightarrow Zn^{2+}(aq) + 2e^-$$

The electrons produced at the zinc electrode flow through the external circuit via the bulb and the bulb glows.

Bubbles of hydrogen are seen when the electrons arrive at the copper electrode. The hydrogen gas is produced from the hydrogen ions in the acid, which collect the electrons appearing at the copper electrode.

$$2H^+(aq) + 2e^- \rightarrow H_2(g)$$

Slowly, the zinc electrode dissolves in the acid and the bulb will then go out. If the zinc is replaced by a more reactive metal, such as magnesium, then the

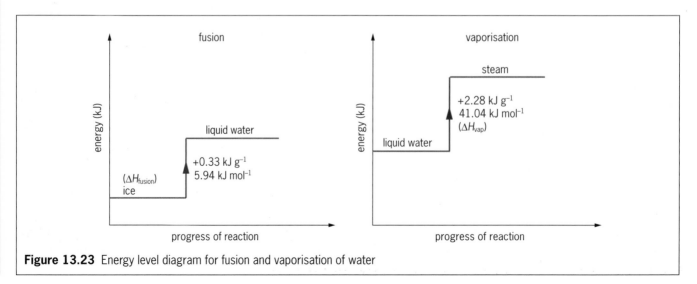

Figure 13.23 Energy level diagram for fusion and vaporisation of water

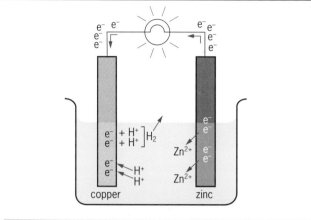

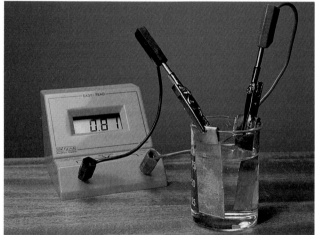

Figure 13.24 A chemical cell in operation

bulb glows more brightly. This happens because the magnesium loses electrons more easily as it reacts faster with the dilute acid.

The difference in the reactivity between the two metals used in the cell creates a particular voltage reading on the voltmeter shown in Figure 13.24 (bottom). The more the two metals differ in reactivity, the larger is the voltage shown and delivered by the cell. This method can be used to confirm the order of reactivity of the metals (Chapter 9, p. 111). Other types of chemical cell in common use are:

- dry cells used in radios, torches etc.
- lead–acid accumulators used in motor vehicles.

Fuel cells

Scientists have found a much more efficient way of changing chemical energy into electrical energy, using a fuel cell (Figure 13.25). Fuel cells are like the chemical cells in the previous section, except that the reagents are supplied continuously to the electrodes. The reagents are usually hydrogen and oxygen. The fuel cell principle was first discovered by Sir William Grove in 1839.

Figure 13.25 The space shuttle computers use electricity produced by fuel cells

When he was electrolysing water and he switched off the power supply, he noticed that a current still flowed but in the reverse direction. Subsequently, the process was explained in terms of the reactions at the electrodes' surfaces of the oxygen and hydrogen gases which had been produced during the electrolysis.

a Describe how simple chemical cells can be used to confirm the order of reactivity of the metals in the reactivity series.
b The fuel cell was discovered during electrolysis experiments with water. It is the reverse process which produces the electricity. Write a balanced chemical equation to represent the overall reaction taking place in a fuel cell.

Checklist

After studying Chapter 13 you should know and understand the following terms.

Fossil fuels   Fuels, such as coal, oil and natural gas, formed from the remains of plants and animals.

Anaerobic decay   Decay which takes place in the absence of air.

Aerobic decay   Decay which takes place in the presence of air.

Nuclear fission   The disintegration of a radioactive nucleus into two or more lighter fragments. The energy released in the process is called nuclear energy.

Chain reaction   A nuclear reaction which is self-sustaining as a result of one of the products causing further reactions.

Non-renewable energy sources   Sources of energy, such as fossil fuels, which take millions of years to form and which we are using up at a rapid rate.

Renewable energy   Sources of energy which cannot be used up or which can be made at a rate faster than the rate of use.

Combustion   A chemical reaction in which a substance reacts rapidly with oxygen with the production of heat and light.

Exothermic reaction   A chemical reaction that releases heat energy into its surroundings.

Endothermic reaction   A chemical reaction which absorbs heat energy from its surroundings.

Bond energy   An amount of energy associated with a particular bond in a molecular element or compound.

Enthalpy   Energy stored in chemical bonds, given the symbol H.

Enthalpy change   Given the symbol ΔH, it represents the difference between energies of reactants and products.

Enthalpy of combustion   The enthalpy change which takes place when one mole of a substance is completely burned in oxygen.

Enthalpy of neutralisation   The enthalpy change which takes place when one mole of hydrogen ions is completely neutralised.

Enthalpy of fusion   The enthalpy change that takes place when one mole of a solid is changed to one mole of liquid at the same temperature.

Enthalpy of vaporisation   The enthalpy change that takes place when one mole of liquid is changed to one mole of vapour at the same temperature.

Chemical cell   A system for converting chemical energy to electrical energy.

QUESTIONS

1 a State which of the following processes is endothermic and which is exothermic.
   (i) The breaking of a chemical bond.
   (ii) The forming of a chemical bond.
 b The table on the right shows the bond energy data for a series of covalent bonds.
   (i) Use the information given in the table to calculate the overall enthalpy change for the combustion of ethanol producing carbon dioxide and water.
   (ii) Is the process in (i) endothermic or exothermic?

Bond	Bond energy (kJ mol^{-1})
C–H	435
O=O	397
C=O	803
H–O	464
C–C	347
C–O	358

2 Explain the following.
  a Hydroelectric power is a relatively cheap source of electricity.
  b Geothermal energy is a non-polluting form of energy.
  c A disadvantage of wind power is that it causes noise pollution.
  d The by-product from the process by which methane is generated by the digestion of animal waste is an excellent fertiliser.
  e The fission of uranium-235 in a nuclear reactor is an example of a chain reaction.
  f Tidal- and wave-generated electricity has a major environmental disadvantage.

3 One of the first practical chemical cells was the Daniell cell invented by John Daniell in 1836. A diagram of this type of cell is shown below.

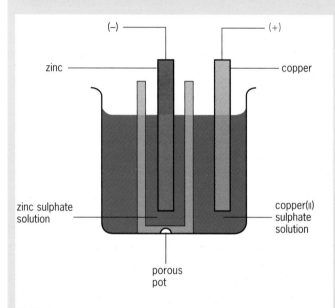

It is capable of generating about 1.1 volts and was used to operate small electrical items such as doorbells.
  a The electrode reaction taking place at a copper anode is:
$$Cu^{2+}(aq) + 2e^- \rightarrow Cu(s)$$
  Write an electrode equation for the process taking place at the cathode.
  b Which way would the electrons flow in the wire connected to the voltmeter—from 'copper to zinc' or 'zinc to copper'?
  c Why should copper(II) sulphate crystalise at the bottom of the outer container?
  d What is the function of the porous pot?
  e There are problems associated with the Daniell cell which has led to it being replaced by other types of cell. Give two reasons why Daniell cells are no longer in use today.

4 This question is about endothermic and exothermic reactions.
  a Explain the meaning of the terms endothermic and exothermic.

b (i) Draw an energy level diagram for the reaction:
$$NaOH(aq) + HCl(aq) \rightarrow NaCl(aq) + H_2O(l) \quad \Delta H = -57\,kJ\,mol^{-1}$$
  (ii) Is this reaction endothermic or exothermic?
  (iii) Calculate the energy change associated with this reaction if 2 moles of sodium hydroxide were neutralised by excess hydrochloric acid.

c (i) Draw an energy level diagram for the reaction:
$$2H_2O(l) \rightarrow 2H_2(g) + O_2(g) \quad \Delta H = +575\,kJ\,mol^{-1}$$
  (ii) Is this reaction endothermic or exothermic?
  (iii) Calculate the energy change for this reaction if only 9 g of water were converted into hydrogen and oxygen.

5 The following results were obtained from an experiment carried out to measure the enthalpy of combustion (heat of combustion) of ethanol. The experiment involved heating a known volume of water with the flame from an ethanol burner. The burner was weighed initially and after the desired temperature rise had been obtained.

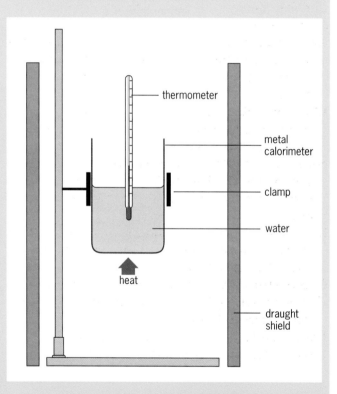

Volume of water in glass beaker = 200 cm³
Mass of ethanol burner at start = 85.3 g
Mass of ethanol burner at end = 84.8 g
Temperature rise of water = 12 °C
(Density of water = 1 g cm⁻³)

Heat energy = mass of water × 4.2 × temperature
given to water   (g)   (Jg⁻¹°C⁻¹)   rise (°C)

  a Calculate the mass of ethanol burned.
  b Calculate the amount of heat produced, in joules, in this experiment by the ethanol burning.
  c Convert your answer to (b) into kilojoules.

QUESTIONS (continued)

d Calculate the amount of heat produced by 1 g of ethanol burning.

e What is the mass of 1 mole of ethanol (C_2H_5OH)? (A_r: C = 12; O = 16; H = 1)

f How much heat would be produced if 1 mole of ethanol had been burned? (This is the heat of combustion of ethanol.)

g Compare your value with the actual value of 1371 kJ mol^{-1} and suggest two reasons for the difference in values.

h Write a balanced chemical equation to represent the combustion of ethanol.

6 The following results were obtained from a neutralisation reaction between 1 M hydrochloric acid and 1 M sodium hydroxide. This experiment was carried out to measure the heat of neutralisation of hydrochloric acid. The temperature rise which occurred during the reaction was recorded.

> Volume of sodium hydroxide used = 50 cm^3
> Volume of acid used = 50 cm^3
> Temperature rise = 5 °C
> (Density of water = 1 g cm^{-3})

Heat energy given = mass of water × 4.2 × temperature
out during reaction   (g)   (J g^{-1} °C^{-1})   rise (°C)

a Write a balanced chemical equation for the reaction.

b What mass of solution was warmed during the reaction?

c How much heat energy was produced during the reaction?

d How many moles of hydrochloric acid were involved in the reaction?

e How much heat would be produced if 1 mole of hydrochloric acid had reacted? (This is the heat of neutralisation of hydrochloric acid.)

f The heat of neutralisation of hydrochloric acid is −57 kJ mol^{-1}. Suggest two reasons why there is a difference between this and your calculated value.

7 Write down which factors are most important when deciding on a particular fuel for the purpose given:

a fuel for a cigarette lighter

b fuel for a camping stove

c fuel for an aeroplane

d fuel for an underground transport system

e fuel for the space shuttle

f fuel for domestic heating.

8 'Propagas' is used in some central heating systems where natural gas is not available. It burns according to the following equation:

$$C_3H_8(g) + 5O_2(g) \rightarrow 3CO_2(g) + 4H_2O(l) \quad \Delta H = -2220 \text{ kJ mol}^{-1}$$

a What are the chemical names for 'propagas' and natural gas?

b Would you expect the heat generated per mole of 'propagas' burned to be greater than that for natural gas? Explain your answer.

c What is 'propagas' obtained from?

d Calculate:

  (i) the mass of 'propagas' required to produce 5550 kJ of energy

  (ii) the heat energy produced by burning 0.5 mole of 'propagas'

  (iii) the heat energy produced by burning 11 g of 'propagas'

  (iv) the heat energy produced by burning 2000 dm^3 of 'propagas'.

(A_r: C = 12; H = 1; O = 16. One mole of any gas occupies 24 dm^3 at room temperature and pressure.)

14 THE (WIDER) ORGANIC MANUFACTURING INDUSTRY

Plastics and polymers

In 1933, Gibson and Fawcett carried out a reaction involving ethene and another organic chemical called benzaldehyde (C_6H_5CHO) using a pressure of about 2000 atmospheres. They were hoping to make these two chemicals react to produce another organic substance from a homologous series called ketones. The reaction vessel leaked and some air (oxygen) got in and much more ethene had to be added. Upon opening the reaction vessel they found a white waxy solid instead of what they expected. They subsequently discovered that this solid was in fact a new type of substance. It was found to consist of very large molecules, each one made up of many thousands of ethene molecules (Figure 14.1). 'Many' in Greek is 'poly', and so the new substance was called 'polyethene' (or **polythene**). The modern plastics industry was really born with this accidental discovery.

Polythene has many useful properties.

- It is easily moulded.
- It is an excellent electrical insulator.
- It does not corrode.
- It is tough.
- It is not affected by the weather.
- It is durable.

It was first used to insulate telephone cables and its unique electrical properties were essential during the

Figure 14.1 Molecules of ethene in this polythene chain are represented by individual poppet beads

development of radar. Its properties continued to be exploited, and today it can be found as a substitute for natural materials in plastic bags, sandwich boxes, washing-up bowls, wrapping film, milk-bottle crates and squeezy bottles.

We now manufacture polythene by heating ethene to a relatively high temperature under a high pressure in the presence of a catalyst.

$$n\begin{pmatrix} H & & H \\ & \diagdown & \diagup \\ & C & = & C \\ & \diagup & & \diagdown \\ H & & H \end{pmatrix} \longrightarrow \begin{pmatrix} H & H \\ | & | \\ -C & - & C- \\ | & | \\ H & H \end{pmatrix}_n$$

where n is a very large number. In polythene the

173

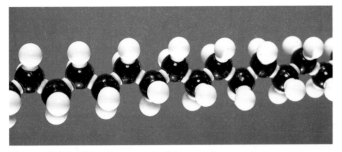

Figure 14.2 This shows part of the polythene polymer chain

ethene molecules have joined together to form a very long hydrocarbon chain (Figure 14.2). The ethene molecules are able to form chains like this because they possess carbon–carbon double bonds.

Other alkene molecules can also produce substances like polythene; for example, propene produces polypropene, which is used to make ropes and packaging. When small molecules like ethene join together to form long chains of atoms, the process is called **polymerisation**. The small molecules, like ethene, which join together in this way are called **monomers**. A polymer chain often consists of many thousands of monomer units and in any piece of plastic there will be many millions of polymer chains. Since in this polymerisation process the monomer units add together to form the polymer, the process is called **addition polymerisation**.

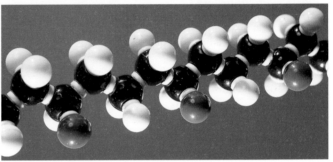

Figure 14.3 Models of the PVC monomer and part of the PVC polymer chain

Other addition polymers

Many other addition polymers have been produced. Often the plastics are produced with particular properties in mind, for example PTFE (polytetrafluoroethene) and PVC (polyvinyl chloride or polychloroethene). Both of these plastics have monomer units similar to ethene.

PVC monomer
(vinyl chloride or
chloroethene)

PTFE monomer
(tetrafluoroethene)

If we start from chloroethene (Figure 14.3), the polymer we make is slightly stronger and harder than polythene and is therefore particularly good for making pipes for plumbing.

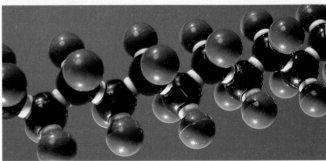

Figure 14.4 Models of the PTFE monomer and part of the PTFE polymer chain

$$n\left(\begin{array}{c} H \qquad H \\ \diagdown \quad \diagup \\ C = C \\ \diagup \quad \diagdown \\ H \qquad Cl \end{array}\right) \longrightarrow \left(\begin{array}{c} H \quad H \\ | \quad\ | \\ -C-C- \\ |\quad\ | \\ H \quad Cl \end{array}\right)_n$$

monomer

polymer chain

If we start from tetrafluoroethene (Figure 14.4) the polymer we make has some slightly unusual properties:

- it will withstand very high temperatures
- it forms a very slippery surface.

174

These properties make PTFE an ideal 'non-stick' coating for frying pans.

$$n \begin{pmatrix} \begin{array}{ccc} F & & F \\ \diagdown & & \diagup \\ & C = C & \\ \diagup & & \diagdown \\ F & & F \end{array} \end{pmatrix} \longrightarrow \begin{pmatrix} \begin{array}{cc} F & F \\ | & | \\ -C - C- \\ | & | \\ F & F \end{array} \end{pmatrix}_n$$

monomer   polymer chain

The properties of some addition polymers along with their uses are given in Table 14.1.

Condensation polymers

Wallace Carothers discovered a different sort of plastic when he developed nylon in 1935. Nylon is made by reacting two different chemicals together, unlike polythene which is made only from monomer units of ethene. Polythene is formed by addition polymerisation which can be represented by:

-A-A-A-A-A-A-A-A-A-A-

where A = monomer.

The starting molecules for nylon are more complicated than those for polythene and are called 1,6-diaminohexane and hexanedioic acid.

1,6–diaminohexane   hexanedioic acid
$H_2N(CH_2)_6NH_2$ + $HOOC(CH_2)_4COOH$

↓

$H_2N(CH_2)_6\mathbf{NHOC}(CH_2)_4COOH$ + H_2O

amide link

The polymer chain is made up from the two starting molecules arranged alternately (Figure 14.5) as these molecules react and therefore link up. Each time a reaction takes place a molecule of water is lost, and because of this it is called **condensation polymerisation**. Because an amide link is formed during the polymerisation, nylon is known as a **polyamide**. This type of polymerisation, in which two kinds of monomer unit react, results in a chain of the type:

-A-B-A-B-A-B-A-B-A-B-

When nylon is made in industry, it forms as a solid which is melted and forced through small holes (Figure 4.20, p. 52). The long filaments cool and solid nylon fibres are produced which are stretched to align the polymer molecules and then dried. The resulting yarn can be woven into fabric to make

Table 14.1 Some addition polymers

Plastic	Monomer	Properties	Uses
Polyethene	$CH_2{=}CH_2$	Tough, durable	Carrier bags, bowls buckets, packaging
Polypropene	$CH_3CH{=}CH_2$	Tough, durable	Ropes, packaging
PVC	$CH_2{=}CHCl$	Strong, hard (less flexible than polyethene)	Pipes, electrical insulation, guttering
PTFE	$CF_2{=}CF_2$	Non-stick surface, withstands high temperature	Non-stick frying pans, soles of irons
Polystyrene	$CH_2{=}CHC_6H_5$	Light, poor conductor of heat	Insulation, packaging (especially as foam)
Perspex	$CH_2{=}C(CO_2CH_3)CH_3$	Transparent	Used as a glass substitute

Figure 14.5 A nylon polymer chain is made up from the two molecules arranged alternately just like the two different coloured poppet beads in the photo

shirts, ties, sheets and parachutes or turned into ropes or racket strings for tennis and badminton rackets.

We can obtain different polymers with different properties if we carry out condensation polymerisation reactions between other monomer molecules. For example, if we react ethane-1,2-diol with benzene-1,4-dicarboxylic acid, then we produce a polymer called terylene.

ethane-1,2-diol   benzene-1,4-dicarboxylic acid
$HO(CH_2)_2OH$ + $HOOC(C_6H_4)COOH$

↓

$HO(CH_2)_2\mathbf{OCO}(C_6H_4)COOH$ + H_2O

ester link

Figure 14.6 Blow moulding in progress in the recycling of polythene

Figure 14.7 These objects are made from compression-moulded plastic

Like nylon, terylene can be turned into yarn, which can then be woven. Terylene clothing is generally softer than that made from nylon but both are hard wearing. Because an ester link is formed during the polymerisation, terylene is known as a **polyester**.

Thermosoftening and thermosetting plastics

Plastics can be put into one of two categories. If they melt or soften when heated (like polyethene, PVC and polystyrene), then they are called **thermoplastics** or **thermosoftening plastics**. If they do not soften on heating but only char and decompose on further heating, they are known as **thermosetting plastics**.

Thermoplastics are easily moulded or **formed** into useful articles. Once they are molten they can be injected or blown into moulds, and a variety of different-shaped items can be produced (Figure 14.6).

Thermosetting plastics can be heated and moulded only once, usually by compression moulding (Figure 14.7).

Figure 14.8 shows the different molecular structures for thermosetting and thermosoftening plastics. Thermosetting plastics have polymer chains which are linked or bonded to each other to give a **cross-linked** structure, and so the chains are held firmly in place and no softening takes place on heating. Thermosoftening plastics do not have polymer chains joined in this way. When thermosoftening plastics are heated their polymer chains flow over one another and the plastic softens.

Disposal of plastics

In the last 30 years plastics have taken over as replacement materials for metals, glass, paper and wood as well as for natural fibres such as cotton and wool. This is not surprising since they are light, cheap, relatively unreactive, can be easily moulded and can be dyed bright colours. However, this situation has contributed significantly to the household

Figure 14.8 a In thermosetting plastic the chains are cross-linked. **b** In thermosoftening plastic there is no cross-linking

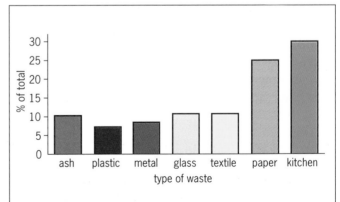

Figure 14.9 Amount of plastic waste compared with other household waste in EU countries

waste problem. Figure 14.9 shows some EU average figures for solid household waste.

In the recent past much of our plastic waste has been used to landfill disused quarries. However, these sites are getting harder to find and it is becoming more and more expensive. The alternatives to dumping plastic waste are certainly more economical and more satisfactory.

- Incineration — schemes have been developed to use the heat generated for heating purposes.
- Recycling — large quantities of black plastic bags and sheeting are produced for resale.
- Biodegradable plastics, as well as those polymers that degrade in sunlight (**photodegradable**, Figure 14.10a), have been developed. Other common categories of degradable plastics include synthetic biodegradable plastics which are broken down by bacteria as well as plastics which dissolve in water (Figure 14.10b).

QUESTIONS

a Write the general equation to represent the formation of polystyrene from its monomer.
b Suggest other uses for the polymers shown in Table 14.1.
c Draw the structure of the repeating units found in:
  (i) nylon
  (ii) terylene.
d Explain the differences between an addition polymer and a condensation polymer.
e Give two advantages and two disadvantages of plastic waste (rubbish).

Figure 14.10 (**a**, top) This plastic bag is photodegradable. (**b**, bottom) This plastic dissolves in water

Some biopolymers

Starch is a **biopolymer** or **natural polymer**. It is a condensation polymer of glucose. It is often produced as a way of storing energy and is formed as a result of photosynthesis in green plants.

$$\text{carbon} + \text{water} \xrightarrow[\substack{\text{sunlight/chlorophyll} \\ \text{(in green plants)}}]{\text{photosynthesis}} \text{glucose} + \text{oxygen}$$

$$6CO_2(g) + 6H_2O(l) \longrightarrow C_6H_{12}O_6(aq) + 6O_2(g)$$

$$\text{glucose} \rightarrow \text{starch} + \text{water}$$

$$nC_6H_{12}O_6(aq) \rightarrow (C_6H_{10}O_5)_n(s) + nH_2O(l)$$

Both starch and glucose are **carbohydrates**. Carbohydrates are a class of naturally occurring organic compounds which can be represented by the general formula $(CH_2O)_x$. Starch occurs in potatoes, rice and wheat. Glucose, from which starch is polymerised, belongs to a group of simple carbohydrates known as **monosaccharides**. They are sweet to taste and soluble in water. Starch belongs to the more complicated group of carbohydrates known as **polysaccharides**. Starch does not form a true solution and it does not have a sweet taste. With iodine it gives an intense blue colour (nearly black), which is used as a test for starch or iodine itself.

Hydrolysis of starch

Starch can be broken down in two ways, both of which take place in the presence of water. Hence the reactions are known as **hydrolysis** reactions. Hydrolysis of starch is the key reaction that enables us to use this energy source. If starch is boiled for about one hour with dilute hydrochloric acid, it is broken down into its monomers, glucose molecules.

$$\text{starch} + \text{water} \xrightarrow[\text{heat}]{\text{dilute acid}} \text{glucose}$$

$$(C_6H_{10}O_5)_n(s) + nH_2O(l) \longrightarrow nC_6H_{12}O_6(aq)$$

If starch is mixed with saliva and left to stand for a few minutes, it will break down to maltose, a disaccharide. The enzyme present in the saliva, called amylase, catalyses this hydrolysis reaction.

$$\text{starch} + \text{water in saliva} \xrightarrow{\text{amylase}} \text{maltose}$$
$$2(C_6H_{10}O_5)_n(s) + nH_2O(l) \longrightarrow nC_{12}H_{22}O_{11}(aq)$$

Enzymes are very efficient natural catalysts present in plants and animals. They do not require high temperatures to break down the starch to maltose (Chapter 11, p. 145). In humans, a salivary amylase breaks down the starch in our food. If you chew on a piece of bread for several minutes, you will notice a sweet taste in your mouth. The above hydrolysis reactions are summarised in Figure 14.11.

QUESTIONS

a In the hydrolysis of starch, how, using a chemical test, could you tell whether all the starch had been hydrolysed?

b Describe a method you could possibly use to identify the products of the different types of hydrolysis.

Ethanol (C_2H_5OH)—an alcohol

The alcohols form another homologous series with the general formula $C_nH_{2n+1}OH$. All the alcohols possess an –OH as the **functional group**. The functional group is the group of atoms responsible for the characteristic reactions of a compound. Table 14.2 shows the first three members along with their melting and boiling points. Figure 14.12 shows the actual arrangement of the atoms in these first three members of this family.

Ethanol is by far the most important of the alcohols and is often just called 'alcohol'. Ethanol can be produced by fermentation (p. 179) as well as by the hydration of ethene (Chapter 12, p. 156). It is a neutral, colourless, volatile liquid which does not conduct electricity. Ethanol burns quite readily with a clean, hot flame.

$$\text{ethanol} + \text{oxygen} \rightarrow \text{carbon} + \text{water} + \text{energy}$$
$$\text{dioxide}$$
$$CH_3CH_2OH + 3O_2(g) \rightarrow 2CO_2(g) + 3H_2O(g) + \text{energy}$$

As methylated spirit, or meths, it is used in spirit (camping) stoves. Methylated spirit is ethanol with small amounts of poisonous substances added to stop you drinking it. Also, some countries, like Brazil, are already using ethanol mixed with petrol as a fuel for cars (Chapter 13, p. 165).

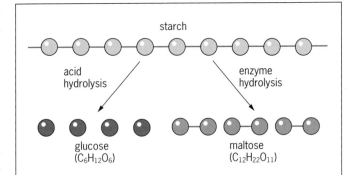

Figure 14.11 Starch produces glucose or maltose depending on the type of hydrolysis used

Table 14.2 The first three members of the alcohol family

Alcohol	Formula	Melting point (°C)	Boiling point (°C)
Methanol	CH_3OH	–94	64
Ethanol	CH_3CH_2OH	–117	78
Propanol	$CH_3CH_2CH_2OH$	–126	97

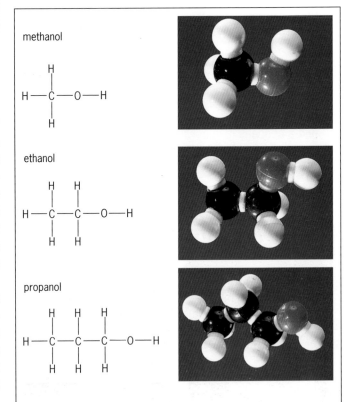

Figure 14.12 The molecules look like the models in the photographs

Many other materials, such as food flavourings, are made from ethanol. Ethanol is a very good solvent and evaporates easily. It is therefore used extensively as a solvent for paints, glues, aftershave and many other everyday products.

Ethanol can be oxidised to ethanoic acid (an organic acid also called acetic acid) by powerful oxi-

Figure 14.13 Orange potassium dichromate(VI) slowly turns green as it oxidises ethanol to ethanoic acid

dising agents, such as warm acidified potassium dichromate(VI). During the reaction the orange colour of potassium dichromate(VI) changes to a dark green (Figure 14.13) as the ethanol is oxidised to ethanoic acid.

$$\text{ethanol} + \text{oxygen} \xrightarrow[\substack{\text{(from potassium} \\ \text{dichromate(VI))}}]{\text{heat}} \text{ethanoic acid} + \text{water}$$

$$CH_3CH_2OH(l) + 2[O] \longrightarrow CH_3COOH(aq) + H_2O(l)$$

carboxylic acid group

Figure 14.14 Fermenting glucose and yeast to produce ethanol. The bag is inflated during the experiment by CO_2

A similar oxidation process takes place, but more slowly, if wine or beer is left open. It will eventually turn to 'vinegar' (ethanoic acid) due to bacterial oxidation of the ethanol in the alcoholic drink.

QUESTIONS

a Write the structural formula for butanol.
b Write a word and balanced chemical equation for:
  (i) the combustion of butanol
  (ii) the oxidation of butanol.

Biotechnology

Biotechnology involves making use of micro-organisms or their components, such as enzymes, for the benefit of humans to produce, for example, foods such as yoghurt and bread. One of the oldest bio-technologies is that of **fermentation**. It involves a series of biochemical reactions brought about by micro-organisms or enzymes. Fermentation is the basic process behind the baking, wine- and beer-making industries.

Fermentation in the laboratory can be carried out using sugar solution. A micro-organism called yeast is added to the solution. The yeast uses the sugar for energy during **anaerobic respiration** (respiration without air), and so the sugar is broken down to give carbon dioxide and ethanol. The best temperature for this process to be carried out at is 37 °C.

$$\text{glucose} \xrightarrow{\text{yeast}} \text{ethanol} + \text{carbon dioxide}$$
$$C_6H_{12}O_6(aq) \longrightarrow 2C_2H_5OH(l) + 2CO_2(g)$$

Figure 14.14 shows a simple apparatus for obtaining ethanol from glucose in the laboratory.

Alcoholic drinks like beer and wine are made on a large scale in vast quantities. Beer is made from barley to which hops have been added to produce the distinctive flavour (Figure 14.15a), whereas wine is made by fermenting grape juice, which contains glucose (Figure 14.15b). The micro-organisms in yeast will carry out the fermentation quite success-fully in both cases. Beer normally contains only about 4% by volume of ethanol, whereas wine contains about 11%. The yeast is killed off if there is much more ethanol than this, so it is not possible to make stronger drinks by fermentation. Some of the stronger drinks in Table 14.3, like sherry, are made up to 'strength' by adding pure ethanol.

Figure 14.15 (**a**, left) Beer is produced on a large scale. (**b**, right) Grape picking for the wine-making industry

Table 14.3 Ethanol content of different drinks

Drink	Percentage ethanol by volume
Whisky, brandy	40
Sherry	20
Martini	14
Wine	11
Beer	4

Figure 14.16 A whisky distillery

Spirits such as whisky and brandy are the more concentrated forms of alcoholic drink. These higher concentrations of ethanol are produced by distillation after the fermentation process is complete (Figure 14.16).

Treat alcohol carefully

As you will know, it is the alcohol (ethanol) in these drinks that can make people drunk. It is also the alcohol which causes headaches (the 'hangover'), dizziness and vomiting associated with being drunk. Alcohol can damage vital organs in your body. Excessive intake over an extended period of time can cause irreparable damage to the liver (cirrhosis) and can result in death.

Alcohol is a drug and is addictive. There are about half a million alcoholics in the UK (Figure 14.17). Because of their addiction, alcoholics often have health, family and work problems.

Even small amounts of alcohol in the bloodstream can reduce your judgement and skill. Up to a third of all road accidents are associated with alcohol. The introduction of the breathalyser had a marked effect on road accident figures. In its first year of use there were 40 000 fewer accidents on the roads of the UK. The 'Alcotest 80' breathalyser tubes contain potassium dichromate(VI) crystals (Figure 14.18). When you breathe into a breathalyser, if your exhaled air contains alcohol vapour some of the potassium dichromate(VI) crystals turn green. How far the green colour extends along the tube is a measure of the amount of alcohol in your breath. The extent of the green colour is taken as a measure of the blood alcohol concentration (BAC). The legal limit on BAC is 80 mg of alcohol per 100 cm^3 of blood. If you 'fail'

Table 14.4 Effects of different blood alcohol levels

Number of drinks	BAC level	Effects
1 pint of beer	30 mg	Increased chance of an accident
3 single whiskies	50 mg	Cheerful, judgement affected, less inhibited
5 single whiskies	80 mg	This is the legal limit for driving. You are four times more likely to have an accident

Figure 14.17 Alcohol abuse is a very serious social problem in our society

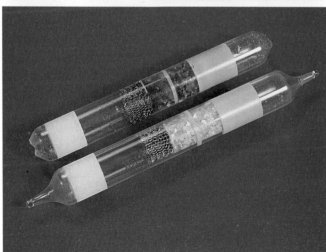

Figure 14.18 (**a**, top) A driver's alcohol level being assessed using a breathalyser. (**b**, bottom) 'Alcotest 80' breathalyser tubes (before, orange, and after, green)

on the 'Alcotest 80' breathalyser, a blood sample is taken to confirm the BAC result.

Table 14.4 shows the effects of different blood alcohol levels on an average person. It is important to appreciate that young people may experience these effects after less alcohol than the table shows.

Baking

A **baker** is involved with biotechnology. To make bread, fresh yeast is mixed with warm sugar solution and the mixture added to the flour. This dough mixture is then put into a warm place to rise. The dough rises due to the production of carbon dioxide from **aerobic respiration** (respiration with air) by the yeast. The products of this style of respiration are different to those of anaerobic respiration.

$$\text{sugar} + \text{oxygen} \xrightarrow{\text{yeast}} \text{carbon dioxide} + \text{water} + \text{energy}$$

$$C_6H_{12}O_6(aq) + 6O_2(g) \longrightarrow 6CO_2(g) + 6H_2O(l) + \text{energy}$$

After the dough has 'risen', it is baked and the heat kills the yeast and the bread stops rising.

New applications

A large number of firms throughout the world are investing large sums of money in the newer biotechnology applications now in use.

- Enzymes can be isolated from micro-organisms and used to catalyse reactions in other processes. For example, proteases are used in biological detergents to digest protein stains such as blood and food. Also, catalase is used in the rubber industry to help convert latex into foam rubber.
- Our ability to manipulate an organism's genes to make it useful to us is called **genetic engineering**. This is being used, for example, to develop novel plants for agriculture as well as making important human proteins such as insulin and growth hormone.

However, a word of caution is necessary. The new biotechnologies may not be without dangers. For example, new pathogens (organisms that cause disease)

181

might be created accidentally. Also, new pathogens may be created deliberately for use in warfare. As you can imagine, there are very strict guidelines covering these new biotechnologies, especially in the area of research into genetic engineering.

Table 14.5 The first two members of the carboxylic acid series

Carboxylic acid	Formula	Melting point (°C)	Boiling point (°C)
Methanoic acid	HCOOH	9	101
Ethanoic acid	CH$_3$COOH	17	118

QUESTIONS

a What do you understand by the term 'biotechnology'? In your answer make reference to the making of beer and bread.

b With reference to the data shown in Table 14.3, answer the following.

(i) Calculate the amount of ethanol (alcohol) in 1 litre of:
whisky
brandy.

(ii) An advert for Campari suggested it contained 15% more alcohol by volume than sherry. Calculate the percentage by volume of alcohol in Campari and also calculate the amount of alcohol in a 1 litre bottle of Campari.

(iii) Low-alcohol beer contains about 0.5% of alcohol by volume. How many litres of this beer contains the same volume of alcohol as 2 litres of normal-strength beer?

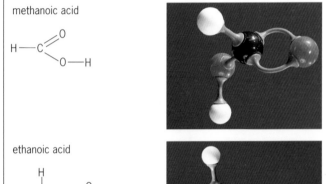

methanoic acid

ethanoic acid

Figure 14.19 The molecules look like the models in the photographs

Ethanoic acid (CH$_3$COOH)—a carboxylic acid

The carboxylic acids form another homologous series with the general formula C$_n$H$_{2n+1}$COOH. All the carboxylic acids possess –COOH as their functional group.

Table 14.5 shows the first two members of this homologous series along with their melting and boiling points. Figure 14.19 shows the actual arrangement of the atoms in these two members of this family.

Methanoic acid is present in stinging nettles and ant stings. Ethanoic acid, however, is the most well known as the main constituent of vinegar. Like other acids, ethanoic acid affects indicators and will react with metals such as magnesium. However, whereas the mineral acids such as hydrochloric acid are called strong acids (Chapter 7, p. 83), ethanoic acid is a weak acid. Even though it is a weak acid, it will still react with bases to form salts. For example, the salt sodium ethanoate is formed when ethanoic acid reacts with dilute sodium hydroxide.

ethanoic + sodium → sodium + water
acid   hydroxide   ethanoate

CH$_3$COOH(aq) + NaOH(aq) → CH$_3$COONa(aq) + H$_2$O(l)

Figure 14.20 Perfumes contain esters

Ethanoic acid will react with ethanol, in the presence of a few drops of concentrated sulphuric acid, to produce ethyl ethanoate—an **ester**.

conc H$_2$SO$_4$

ethanoic + ethanol ⇌ ethyl + water
acid   ethanoate

CH$_3$COOH(l) + CH$_3$CH$_2$OH(l) ⇌ CH$_3$COOCH$_2$CH$_3$(aq) + H$_2$O(l)

This reaction is called **esterification**.

Members of the 'ester' family have strong and pleasant smells. Many of them occur naturally and are responsible for the flavours in fruits and the smells of flowers. They are used, therefore, in some food flavourings and in perfumes (Figure 14.20).

Fats and oils are naturally occurring esters which are used as energy storage compounds by plants and animals (Figure 14.21).

Figure 14.21 Olive oil is obtained from a plant whereas lard is made from animal fat

Soaps and detergents

Millions of tonnes of soaps and soapless detergents are manufactured worldwide every year. Soap is manufactured by heating natural fats and oils of either plants or animals with a strong alkali. These fats and oils are called triglycerides and are complicated ester molecules.

Fat is boiled with aqueous sodium hydroxide to form soap. The esters are broken down in the presence of water — hydrolysed. This type of reaction is called **saponification**. The equation given below is that for the saponification of glyceryl stearate (a fat).

glyceryl + sodium → sodium + glycerol
stearate   hydroxide   stearate (soap)

$$H_{35}C_{17} - COOCH_2$$
$$H_{35}C_{17} - COOCH + 3NaOH_{(aq)} \longrightarrow 3C_{17}H_{35}COO^-Na^+_{(l)} + HO - CH$$
$$H_{35}C_{17} - COOCH_{2(l)} \qquad\qquad HO - CH_{2(l)}$$
$$HO - CH_2$$

The cleaning properties of the soap depend on its structure and bonding. Sodium stearate consists of a long hydrocarbon chain which is hydrophobic (water hating) attached to an ionic 'head' which is hydrophilic (water loving) (Figure 14.22).

Covalent compounds are generally insoluble in water but they are more soluble in organic solvents. Ionic compounds are generally water soluble but tend to be insoluble in organic solvents. When a soap is put into water which has a greasy dish (or a greasy cloth) in it, the hydrophobic hydrocarbon chain on each soap molecule becomes attracted to the grease and becomes embedded in it (Figure 14.23). On the other hand, the hydrophilic ionic head group is not attracted to the grease but is strongly attracted to the water molecules. When the water is stirred, the

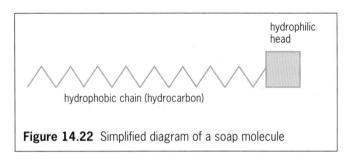

Figure 14.22 Simplified diagram of a soap molecule

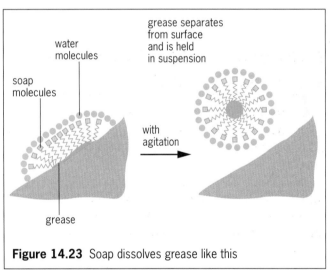

Figure 14.23 Soap dissolves grease like this

grease is slowly released and is completely surrounded by the soap molecules. The grease is, therefore, 'solubilised' and removed from the dish. The soap is able to remove the grease because of the combination of the covalent and ionic bonds present.

Soapless detergents

In Chapter 8, p. 103, we discussed the way in which, in hard water areas, an insoluble scum forms when soap is used. This problem has been overcome by the development of synthetic **soapless detergents**. These new substances do not form scum with hard

183

water since they do not react with Ca^{2+} and Mg^{2+} present in such water. Also, these new soapless detergent molecules have been designed so that they are biodegradable. Bacteria readily break down these new molecules so that they do not persist in the environment.

Sodium alkyl benzene sulphonates were developed in the early 1970s. The structure of sodium 3-dodecylbenzene sulphate, $C_{18}H_{29}SO_3Na$, is given below.

$SO_3^-Na^+$

The calcium and magnesium salts of this detergent molecule are water soluble, so the problem of scum is solved. Very many of our washing powders (and liquids) contain this type of substance. The manufacture of soapless detergents is explained in Chapter 16, p. 204.

QUESTIONS

a What class of organic compound do substances like glyceryl stearate belong to?
b What do you understand by the terms:
  (i) hydrophobic?
  (ii) hydrophilic?
  (iii) saponification?
c What is the main advantage of detergents over soaps?

Amino acids

There are 20 different amino acids and they each possess two functional groups. One is the carboxylic acid group, -COOH. The other is the amine group, -NH$_2$. The two amino acids shown below are glycine and alanine.

General structure   Glycine   Alanine

Amino acids are the building blocks of proteins. Proteins are polyamides, as they contain the -CONH- group, which is called the amide or peptide link. Proteins are formed by condensation polymerisation.

Glycine   Alanine

$$H_2N - C - CONH - C - COOH + H_2O$$

A dipeptide
(composed of two amino acids joined together)

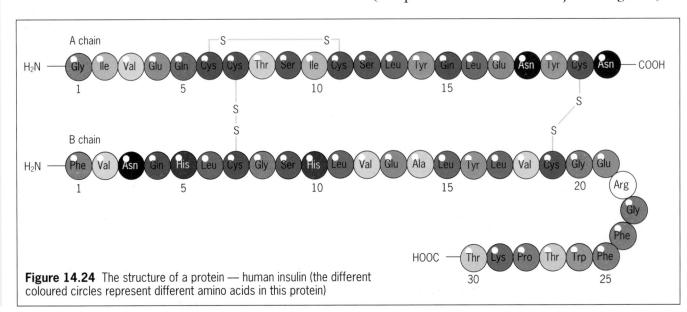

Figure 14.24 The structure of a protein — human insulin (the different coloured circles represent different amino acids in this protein)

Further reaction with many more amino acids takes place at opposite ends of each molecule to produce the final protein (Figure 14.24). For a molecule to be a protein, there must be at least 100 amino acids involved. Below this number, they are called polypeptides. Proteins make up some 15% of our body weight.

Proteins fall broadly into two groups: fibrous and globular.

- Fibrous proteins — these have linear molecules and are insoluble in water and resistant to alkalis and acids. Collagen (in tendons and muscles), keratin (in nails, hair, horn and feathers), and elastin (in arteries) are all fibrous proteins.
- Globular proteins — these have complicated three-dimensional structures and are soluble in water. They are easily affected by acids, alkalis and temperature increase, when they are said to be denatured. Casein (in milk), albumen (in egg white) and enzymes are examples of globular proteins.

Protein analysis

How can you determine which amino acids are present in a particular protein? This involves hydrolysis of the peptide (amide) bonds in the protein so that the individual amino acids are released. This can only be done by heating the protein with dilute hydrochloric acid. The mixture of amino acids is then separated by thin layer chromatography (TLC) or electrophoresis (Figure 14.25). In both cases, a locating agent (Chapter 2, p. 23), such as ninhydrin, is used. This ensures that the spots of acid are visible.

If you are only trying to show the presence of a protein, a quick test to carry out is known as the **Biuret test**. A mixture of dilute sodium hydroxide and 1% copper(II) sulphate solution is shaken with a sample of the material under test. If a protein is present, a purple colour appears after about 3 minutes (Figure 14.26).

DNA

Deoxyribonucleic acid (DNA) belongs to a group of chemicals called the nucleic acids (Figure 14.27). They are also biopolymers. DNA controls the protein synthesis within your cells. When you eat a food containing proteins, such as meat or cheese, your digestive enzymes break down the proteins present into individual amino acids. The DNA in your cells controls the order in which the amino acids are repolymerised to make the proteins you need!

Figure 14.25 Amino acids can be separated and identified by TLC

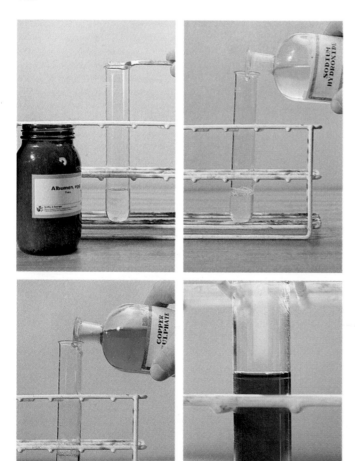

Figure 14.26 Biuret test (the sequence is left to right, top to bottom). The purple colour shows the presence of a protein

Genes are the units of heredity that control the characteristics of organisms. A gene is made of DNA. No two individuals have the same DNA sequence. DNA 'fingerprinting' has become a very powerful forensic science tool in the investigation of crime (Figure 14.28).

QUESTIONS

a Which two functional groups do amino acids possess?
b How many amino acids have to be involved before the biopolymer is called a protein?
c Name the process by which the individual amino acids in a protein are released by reaction with a dilute acid.
d Explain how DNA fingerprinting may be used in paternity suits.

Pharmaceuticals

Pharmaceuticals are chemicals called **drugs** that are prepared and sold with the intention of treating illness. A drug is any substance, natural or synthetic, which alters the way in which the body works. There are many categories of drugs. The following are some examples.

- Anaesthetics — these induce loss of feeling and/or consciousness, for example fluothane.
- Analgesics — these relieve pain, for example aspirin.
- Antibiotics — these are substances, originally produced by micro-organisms, which are used to kill other types of microbe, for example penicillin. However, many of these antibiotics are now made in chemical laboratories, for example carbenicillin.
- Sedatives — these induce sleep, for example barbiturates.
- Tranquillisers — these will give relief from anxiety, for example Valium.

There are, of course, many other types of drug available which have very specific uses. For example, methyldopa was developed to relieve hypertension (high blood pressure), and Corgard was developed to treat heartbeat irregularity.

The pharmaceutical industry is one of the most important parts of the chemical industry and is an important consumer of the products of the petrochemical industry. It is a high-profit industry but with very high research and development costs. For example, it costs in excess of £100 million to discover, test and get a single drug on to the market.

Today, the pharmaceutical industry could be called the 'medicines by design' industry. Companies such as SmithKline Beecham have teams of chemists and

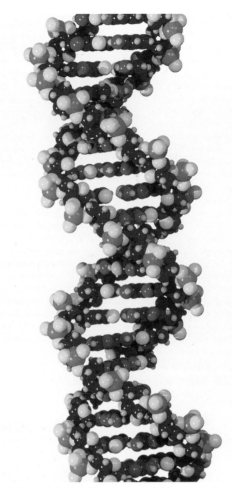

Figure 14.27 Crick and Watson based this model for DNA on X-ray studies and chemical analysis

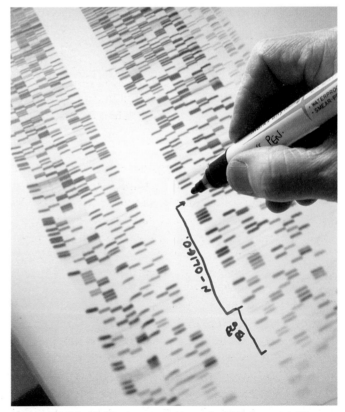

Figure 14.28 Identical patterns shown by DNA fingerprinting can identify the criminal

Table 14.6 Commonly available drugs

Name	Structure	For the treatment of
Aspirin	(structure of aspirin showing benzene ring with $C-O-H$ and $O-CH_3$ ester groups)	Headaches, mild pain, heart disease
Paracetamol	$HO-$(benzene ring)$-N(H)-C(=O)-CH_3$	Headaches, mild pain
Methyldopa	$HOOC-C(CH_3)(NH_2)-CH_2-$(benzene ring with two OH groups)	High blood pressure (hypertension)
Penicillin	$CH_3-C(=O)-N(H)-$ (penicillin ring structure with S, two CH_3, $C=O$, OH)	A variety of bacterial infections
Valium	(Valium structure with CH_3, N, $C=O$, Cl and phenyl ring)	Feelings of anxiety or depression

biochemists working almost around the clock to discover, test, check for safety and produce drugs that can deal with almost every known illness.

Table 14.6 shows the structures of a selection of some of the more common drugs available at the present time, along with their uses. The common names for these drugs are used, since their systematic, theoretical names are extremely complex.

Drug abuse

Some of the very useful drugs developed by chemists can be habit forming. For example, barbiturates (in sleeping tablets) and amphetamines (stimulants) fall into this category. Another drug that has recently created problems is Valium, which is not itself addictive but when used in the long term makes people dependent on it. Severe psychological and physiological problems can arise.

It should be noted, however, that it is the opiates which cause **addiction** (Figure 14.29). Cocaine and heroin are just two examples of such substances. Consequences of the addiction include personal neglect, both of nutritional needs and of hygiene. For a short-term feeling of well-being ('fix'), the addict is prepared to do almost anything. Addicts often turn to a life of crime to fulfil their cravings for the opiates.

Figure 14.29 Increasing public awareness of the risks associated with drug abuse is very important

Addicts, especially those injecting drugs, are at a high risk of HIV (human immunodeficiency virus) infection as they often share needles with other drug addicts, who may be HIV positive. Public awareness campaigns aim to educate everyone on the dangers in society, including drug abuse and its related risks (Figure 14.29).

QUESTIONS

a Using the data given in Table 14.6, suggest which of the pharmaceuticals contains:
  (i) sulphur
  (ii) an –OH group
  (iii) an –NH$_2$ group
  (iv) an ester group.
b Drug abuse is a rapidly growing problem. Using the information given in this section as well as your research skills, make a list of the addictive drugs. Also explain the problems that drug abuse can cause.

Checklist

After studying Chapter 14 you should know and understand the following terms.

Monomer　A simple molecule, such as ethene, which can be polymerised.

Polymer　A substance possessing very large molecules consisting of repeated units or monomers. Polymers therefore have a very large relative molecular mass.

Polymerisation　The chemical reaction in which molecules (monomers) join together to form a polymer.

Addition polymer　A polymer formed by an addition reaction. For example, polythene is formed from ethene.

Condensation polymer A polymer formed by a condensation reaction (one in which water is given out). For example, nylon is produced by the condensation reaction between 1,6-diaminohexane and hexanedioic acid.

Thermoplastics Plastics which soften when heated (for example polythene, PVC).

Thermosetting plastics Plastics which do not soften on heating but only char and decompose (for example melamine).

Cross-linking The formation of side covalent bonds linking different polymer chains and therefore increasing the rigidity of the plastic. Thermosetting plastics are usually heavily cross-linked.

Biodegradable plastics Plastics designed to degrade (decompose) under the influence of bacteria.

Photodegradable plastics Plastics designed to degrade under the influence of sunlight.

Biopolymers Natural polymers such as starch and proteins.

Carbohydrates A group of naturally occurring organic compounds which can be represented by the general formula $(CH_2O)_x$.

Monosaccharides A group of simple carbohydrates. They are sweet to taste and are water soluble (for example glucose).

Polysaccharides A group of more complicated carbohydrates. They generally do not form true solutions and do not have a sweet taste (for example starch).

Hydrolysis A chemical reaction involving the reaction of a compound with water. Acid hydrolysis usually involves dilute hydrochloric acid, and enzyme hydrolysis involves enzymes such as amylase.

Biotechnology This involves making use of microorganisms in industrial and commercial processes. For example, the process of fermentation is brought about by the enzymes in yeast.

Alcohols These are organic compounds containing the –OH group. They have the general formula $C_nH_{2n+1}OH$. Ethanol is by far the most important of the alcohols and is often just called 'alcohol'.

Functional group The atom or group of atoms responsible for the characteristic reactions of a compound.

Carboxylic acids A family of organic compounds containing the functional group –COOH. They have the general formula $C_nH_{2n+1}COOH$. The most important and well known of these acids is ethanoic acid, which is the main constituent in vinegar. Ethanoic acid is produced by the oxidation of ethanol.

Esters A family of organic compounds formed by the reaction of an alcohol with a carboxylic acid in the presence of concentrated H_2SO_4. This type of reaction is known as esterification. Esters are characterised by a strong and pleasant smell (many occur in nature and account for the smell of flowers).

Soaps Substances formed by saponification. In this reaction, the oil or fat (glyceryl ester) is hydrolysed by aqueous sodium hydroxide to produce the sodium salt of the fatty acid, particularly sodium stearate (from stearic acid). Soap will dissolve grease because of the dual nature of the soap molecule. It has a hydrophobic part (the hydrocarbon chain) and a hydrophilic part (the ionic head) and so will involve itself with both grease molecules and water. However, it forms a scum with hard water by reacting with the Ca^{2+} (or Mg^{2+}) present.

Soapless detergents These are soap-like molecules which do not form a scum with hard water. These substances have been developed from petrochemicals. Their calcium and magnesium salts are water soluble and they are biodegradable.

Amino acids These are naturally occurring organic compounds which possess both an –NH_2 group and a –COOH group on adjacent carbon atoms. There are 20 naturally occurring amino acids, of which glycine is the simplest.

Proteins Polymers of amino acids formed by condensation reactions. They fall broadly into the two categories: fibrous proteins (for example keratin and collagen) and globular proteins (for example casein and albumen).

Biuret test The test for proteins. A mixture of dilute sodium hydroxide and 1% copper(II) sulphate solution is shaken with the material under test. A purple colour appears after about 3 minutes if a protein is present.

DNA Abbreviation for deoxyribonucleic acid. It belongs to a group of biopolymers called the nucleic acids. It is involved in the polymerisation of amino acids in a specific order to form the particular protein required by a cell.

Pharmaceuticals These are chemicals called drugs that are prepared and sold with the intention of treating disease (for example methyldopa).

Drug This is any substance, natural or synthetic, that alters the way in which the body works.

Drug abuse This term usually applies to the misuse of addictive drugs, which include barbiturates and amphetamines, as well as the opiates, cocaine and heroin. These drugs create severe psychological and physiological problems through self-indulgence. This leads to a variety of personal problems for the user.

HIV Short for human immunodeficiency virus, from which AIDS (acquired immunodeficiency syndrome) can develop.

1 Explain the following.
   a The problem of plastic waste has been overcome.
   b The majority of detergents produced today are biodegradable.
   c In bread making, yeast is added to the mix and the dough left to stand for a period of time.
   d Drug abuse is a rapidly growing problem in our society.
   e When wine is left open it eventually turns to vinegar.
   f Polythene is a thermoplastic.

2 a A detergent molecule may be represented by the following simplified diagram.

   Use this representation of a detergent molecule in a series of labelled diagrams to show how detergents can remove grease from a piece of greasy cloth.
   b Explain why detergents do not form a scum with hard water, whereas soaps do.
   c The modern detergents are biodegradable.
   (i) Explain what this statement means.
   (ii) Why is it necessary for detergents to be biodegradable?

3 A piece of cheese contains protein. Proteins are natural polymers made up of amino acids. There are 20 naturally occurring amino acids. The structures of two amino acids are shown below.

Glycine

$$H_2N - \overset{\overset{\displaystyle H}{|}}{\underset{\underset{\displaystyle H}{|}}{C}} - COOH$$

Alanine

$$H_2N - \overset{\overset{\displaystyle H}{|}}{\underset{\underset{\displaystyle CH_3}{|}}{C}} - COOH$$

   a Name the type of polymerisation involved in protein formation.
   b Draw a structural formula to represent the part of the protein chain formed by the reaction between the amino acids shown above.
   c What is the name given to the common linkage present in protein molecules?
   d Why is there such a huge variety of proteins?
   e Name and describe the features of the two broad groups of proteins.

4

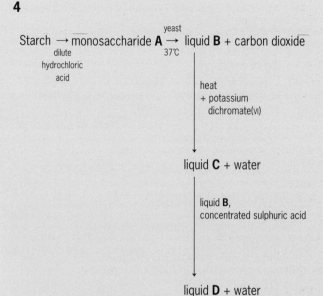

   a Name, give the formula and one use of each of the substances **A** to **D**.
   b Write word and balanced chemical equations for the reactions involved in the formation of liquids **B**, **C** and **D**.
   c Starch is classified as a natural polymer or 'biopolymer'. Explain the meaning of this statement.
   d Name the processes by which:
   (i) starch is broken down
   (ii) liquid **B** is formed
   (iii) liquid **C** is formed
   (iv) liquid **D** is formed.

5 a Copy the following table and complete it by writing the structural formulae for methanol and methanoic acid.

Methane	Methanol	Methanoic acid
$H - \overset{\overset{\displaystyle H}{\|}}{\underset{\underset{\displaystyle H}{\|}}{C}} - H$		

   b Describe a simple chemical test that could be used to distinguish methanol from methanoic acid.
   c (i) Name the class of compound produced when methanol reacts with methanoic acid.
   (ii) Name the type of reaction taking place.
   (iii) Write a word and balanced chemical equation for this reaction.
   (iv) Give two uses related to the class of compound formed in this reaction.
   d The following reaction takes place when methanol is burned:

$$2CH_3OH(l) + 3O_2(g) \rightarrow 2CO_2(g) + 4H_2O(g) \quad \Delta H = -1452 \, kJ \, mol^{-1}$$

(i) How much heat energy would be liberated by burning:
0.5 mol of methanol?
4.0 mol of methanol?
4 g of methanol?

(ii) Calculate the volume of carbon dioxide produced at room temperature and pressure (rtp) when 16 g of methanol is burned.
(A_r: C = 12; H = 1; O = 16. One mole of any gas at rtp occupies 24 dm^3.)

6 This question requires you to use the information in Table 14.6.

a Which grouping of atoms is common to both pain killers?

b Which of the pharmaceuticals contains:
(i) sulphur?
(ii) the greatest number of nitrogen atoms?
(iii) the carboxylic acid functional group?
(iv) chlorine?
(v) the greatest number of –CH$_3$ groups?

c Which of the pharmaceuticals is used to combat:
(i) bacterial infections?
(ii) depression?
(iii) heart disease?

d The pharmaceuticals industry has often been termed the 'medicines by design' industry. Explain the meaning of this statement.

7 When a bottle of wine is left open to the air it turns sour. This is due to the oxidation, by the oxygen in the air, of the ethanol in the wine to ethanoic acid.

a Write an equation to represent this oxidation process.

b What would you expect to happen to the pH of the wine over a number of days?

c Devise a method which would allow you to calculate the amount of ethanoic acid in the wine.

8 Why is it safe for us to use vinegar, which contains ethanoic acid, on food while it would be extremely dangerous for us to use dilute nitric acid for the same purpose?

15 NITROGEN

Nitrogen—an essential element

Nitrogen gas is all around us. It makes up 78% of the air. Each day each individual breathes thousands of litres of nitrogen into his/her lungs and then exhales it without any chemical change occurring.

Nitrogen gas has no smell, is colourless and is a very unreactive gas.

The nitrogen gas molecule comprises two nitrogen atoms covalently bonded together by a triple bond (Figure 15.1). This triple bond is very strong, requiring a large amount of energy to break it so that nitrogen can take part in chemical reactions. Only at very high temperatures or if an electrical spark is passed through nitrogen gas will it react with oxygen. This occurs in an internal combustion engine, where the spark from the spark plug is sufficient to convert some of the nitrogen to toxic nitrogen oxides (NO_x) which are then emitted as car exhaust fumes (Chapter 11, p. 145). Catalytic converters prevent the emission of nitrogen oxides by converting them back to nitrogen gas before they leave the exhaust pipe.

Nitrogen gas can be obtained industrially by the fractional distillation of liquid air. It is produced on a very large scale, since the gas has many very important industrial uses.

Nitrogen is an essential element necessary for the well-being of animals and plants. It is present in proteins, which are found in all living things. Proteins are essential for healthy growth. Animals obtain the nitrogen they need for protein production by feeding on plants and other animals. Most plants obtain the nitro-

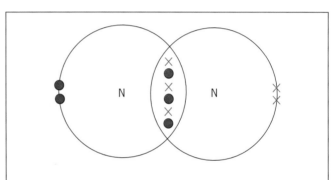

Figure 15.1 The nitrogen molecule

gen they require from the soil. Nitrogen is found in nitrogen-containing compounds called **nitrates**, which are produced as a result of the effect of lightning and as dead plants and animals decay. Nitrates are soluble compounds and can be absorbed by the plants through their roots. Nitrogen in the air is also converted into nitrates by some forms of bacteria. Some of these bacteria live in the soil. A group of plants called **leguminous** plants have these bacteria in nodules on their roots. These bacteria are able to take nitrogen from the atmosphere and convert it into a form in which it can be used to make proteins. This process is called **fixing nitrogen** and is carried out by plants such as beans and clover.

The vital importance of nitrogen to both plants and animals can be summarised by the **nitrogen cycle** (Figure 15.2). If farm crops are harvested from the land rather than left to decay, returning the nitrogen back to the soil, the soil becomes deficient in this important element. In addition, nitrates can be

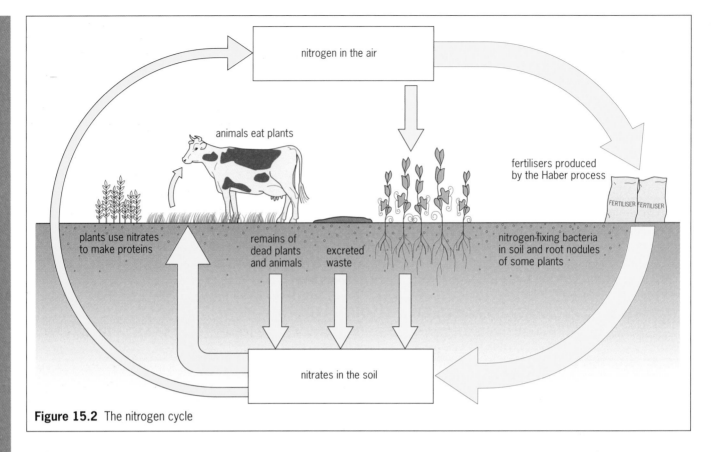

Figure 15.2 The nitrogen cycle

washed from the soil by the action of rain (leaching). For the soil to remain fertile for the next crop, the nitrates need to be replaced. The natural process is by decay or by the action of lightning on atmospheric nitrogen. Without the decay, however, the latter process is not efficient enough to produce nitrates on the scale required. Farmers often need to add substances containing these nitrates. Such substances include:

- farmyard manure — this is very rich in nitrogen-containing compounds which can be converted by bacteria in the soil to nitrates
- artificial fertilisers — these are manufactured compounds of nitrogen which are used on an extremely large scale to enable farmers to produce ever-bigger harvests for the increasing population of the world. One of the most commonly used artificial fertilisers is **ammonium nitrate**, which is made from ammonia gas and nitric acid, both nitrogen-containing compounds.

QUESTIONS

a Why can most plants not use nitrogen gas directly from the atmosphere?

b Nitrogen is essential in both plants and animals to form which type of molecule? Explain why these molecules are essential to life.

c Explain why natural sources of the element nitrogen are not sufficient to produce the world's annual crop requirement.

The Haber process

The Haber process forms the basis of the artificial fertiliser industry, as it is used to produce ammonia gas. The process was developed by a German scientist called Fritz Haber in 1913. He was awarded a Nobel Prize in 1918 for his work. The process involves the reaction of nitrogen and hydrogen to produce ammonia gas. It was initially developed to satisfy the need for explosives during World War I, as explosives can be made from ammonia.

The nitrogen for the process is obtained from the atmosphere and the hydrogen from the reaction between methane (natural gas) and steam.

$$\text{methane} + \text{steam} \rightleftharpoons \text{hydrogen} + \text{carbon monoxide}$$
$$CH_4(g) + H_2O(g) \rightleftharpoons 3H_2(g) + CO(g)$$

This process is known as **steam reforming**. This reaction is a **reversible** reaction and special conditions are employed to ensure that the reaction moves to the right (the forward reaction), producing hydrogen and carbon monoxide. The process is carried out at a temperature of 750°C, at a pressure of 30 atmospheres with a catalyst of nickel. These conditions enable the maximum amount of hydrogen to be produced at an economic cost.

The carbon monoxide produced is then allowed to reduce some of the unreacted steam to produce more hydrogen gas.

carbon + steam ⇌ hydrogen + carbon
monoxide   dioxide
$$CO(g) + H_2O(g) \rightleftharpoons H_2(g) + CO_2(g)$$

Nitrogen and hydrogen in the correct proportions (1:3) are then pressurised to approximately 200 atmospheres and passed over a catalyst of freshly produced, finely divided iron at a temperature of between 350 °C and 500 °C:

nitrogen + hydrogen ⇌ ammonia
$$N_2(g) + 3H_2(g) \rightleftharpoons 2NH_3(g) \quad \Delta H = -92\,kJ\,mol^{-1}$$

The reaction is exothermic.

Under these conditions the gas mixture leaving the reaction vessel contains about 15% ammonia, which is removed by cooling and condensing it as a liquid. The unreacted nitrogen and hydrogen are recirculated into the reaction vessel to react together once more to produce further quantities of ammonia.

The 15% of ammonia produced does not seem a great deal. The reason for this is the reversible nature of the reaction. Once the ammonia is made from nitrogen and hydrogen, it decomposes to produce nitrogen and hydrogen. There comes a point when the rate at which the nitrogen and hydrogen react to produce ammonia is equal to the rate at which the ammonia decomposes. This situation is called a **chemical equilibrium**. Because the processes continue to happen, the equilibrium is said to be **dynamic**. The conditions used ensure that the ammonia is made economically. The diagram below shows how the percentage of ammonia produced varies with the use of different temperatures and pressures (Figure 15.3).

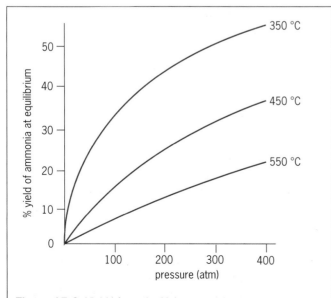

Figure 15.3 Yields from the Haber process

You will notice that the higher the pressure and the lower the temperature used, the more ammonia is produced. Relationships such as this were initially observed by Henri Le Chatelier, a French scientist, in 1888. He noticed that if the pressure was increased in reactions involving gases, the reaction which produced the fewest molecules of gas was favoured. If you look at the reaction for the Haber process you will see that, going from left to right, the number of molecules of gas goes from four to two. This is why the Haber process is carried out at high pressures. He also noticed that reactions which were exothermic produced more products if the temperature was low. Indeed, if the Haber process is carried out at room temperature you get a higher percentage of ammonia. However, in practice the rate of the reaction is lowered too much and the ammonia is not produced quickly enough for the process to be economical. An **optimum temperature** is used to produce enough ammonia at an acceptable rate. It should be noted, however, that the increased pressure used is very expensive in capital terms and so alternative, less expensive routes involving biotechnology (Chapter 14, p. 179) are being sought at the present time.

QUESTIONS

a Use the information given above and any other sources you may have to produce a flow diagram of the Haber process. Indicate the flow of the gases, and write equation(s) to show what happens at each stage. You may have to look up the boiling points of the gases involved for the stage in which the ammonia is separated from the reaction vessel mixture.

b What problems do the builders of a chemical plant to produce ammonia have to consider when they start to build such a plant?

c What problems are associated with building a plant which uses such high pressures as those required in the Haber process?

Ammonia gas

Making ammonia in a laboratory

Small quantities of ammonia gas can be produced by heating any ammonium salt, such as ammonium chloride, with an alkali, such as calcium hydroxide.

calcium + ammonium → calcium + water + ammonia
hydroxide   chloride   chloride
$$Ca(OH)_2(s) + 2NH_4Cl(s) \rightarrow CaCl_2(s) + 2H_2O(g) + 2NH_3(g)$$

Water vapour is removed from the ammonia gas by passing the gas formed through a drying tower containing calcium oxide (Figure 15.4).

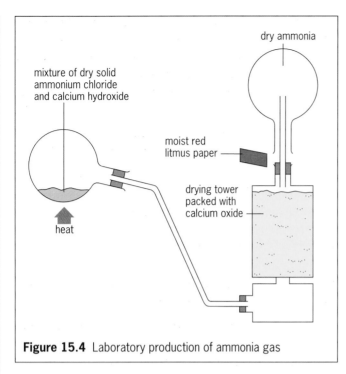

Figure 15.4 Laboratory production of ammonia gas

mixture of dry solid ammonium chloride and calcium hydroxide

moist red litmus paper

dry ammonia

drying tower packed with calcium oxide

heat

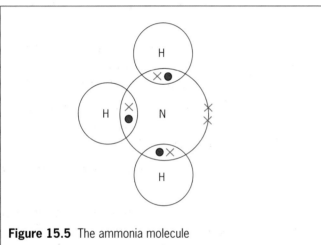

Figure 15.5 The ammonia molecule

This reaction forms the basis of a chemical test to show that a compound contains the ammonium ion (NH_4^+). If any compound containing the ammonium ion is heated with sodium hydroxide, ammonia gas will be given off which will turn damp red litmus paper blue.

Physical properties of ammonia

Ammonia (Figure 15.5):

- is a colourless gas
- is less dense than air
- has a sharp or pungent smell
- is very soluble in water with about $1200\,cm^3$ of ammonia in each $1\,cm^3$ of water.

Chemical properties of ammonia

The reason ammonia is so soluble in water is that some of it reacts with the water. The high solubility can be shown by the 'fountain flask experiment' (Figure 15.6). As the first drop of water reaches the top of the tube all the ammonia gas in the flask dissolves, creating a much reduced pressure. Water then rushes up the tube to fill the space once occupied by the dissolved gas. This creates the fountain. If the water initially contained some universal indicator, then you would also see a change from green to blue when it comes into contact with the dissolved ammonia. This shows that ammonia solution is a weak alkali, although dry ammonia gas is not. This is because a little of the ammonia gas has reacted with the water, producing ammonium ions and hydroxide ions. The hydroxide ions produced make a solution of ammonia alkaline.

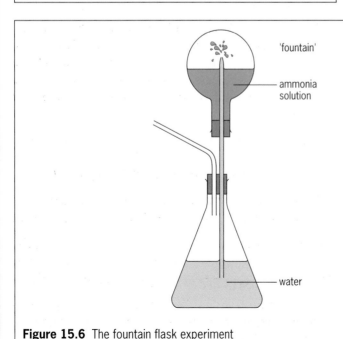

'fountain'

ammonia solution

water

Figure 15.6 The fountain flask experiment

ammonia + water $\rightleftharpoons$ ammonium + hydroxide
ions   ions

$$NH_3(g) + H_2O(l) \rightleftharpoons NH_4^+(aq) + OH^-(aq)$$

The solution is only weakly alkaline because of the reversible nature of this reaction, which results in a relatively low concentration of hydroxide ions. Ammonia gas dissolved in water is usually known as aqueous ammonia.

QUESTIONS

a Calcium oxide is used to dry ammonia gas in its laboratory preparation. Write a word and balanced chemical equation to show how calcium oxide can react with the water vapour to remove it from damp ammonia gas.

b Explain why ammonia gas only acts as a weak alkali in the presence of water.

Nitric acid

Manufacture of nitric acid

One of the largest uses of ammonia is in the production of nitric acid. The process, which was invented by Wilhelm Ostwald, has three stages.

1 A mixture of air and ammonia is heated to about 230 °C and is passed through a metal gauze made of platinum (90%) and rhodium (10%). The reaction produces a lot of heat energy. This energy is used to keep the reaction vessel temperature at around 800 °C. The reaction produces nitrogen monoxide (NO) and water.

ammonia + oxygen $\rightarrow$ nitrogen + water
monoxide

$$4NH_3(g) + 5O_2(g) \rightarrow 4NO(g) + 6H_2O(g)$$
$$\Delta H = -904 \, kJ \, mol^{-1}$$

2 The colourless nitrogen monoxide gas produced from the first stage is then reacted with oxygen from the air to form brown nitrogen dioxide gas (NO_2).

nitrogen monoxide + oxygen $\rightarrow$ nitrogen dioxide

$$2NO(g) + O_2(g) \rightarrow 2NO_2(g)$$
$$\Delta H = -114 \, kJ \, mol^{-1}$$

3 The nitrogen dioxide is then dissolved in water to produce nitric acid.

nitrogen + water $\rightarrow$ nitric + nitrogen
dioxide   acid   monoxide

$$3NO_2(g) + H_2O(l) \rightarrow 2HNO_3(aq) + NO(g)$$

A small amount of nitrogen dioxide cannot be turned into nitric acid and so it is expelled into the atmosphere through a tall chimney. This can cause pollution because it is an acidic gas.

The nitric acid produced is used in the manufacture of:

- artificial fertilisers, such as ammonium nitrate
- explosives, such as 2,4,6-trinitrotoluene (TNT)
- dyes
- artificial fibres, such as nylon.

It is also used in the treatment of metals.

Properties of nitric acid

Dilute nitric acid

Nitric acid is a strong acid. As a dilute acid it shows most of the properties of an acid.

- It forms an acidic solution in water and will turn pH paper red.
- It will react with bases such as sodium hydroxide and zinc oxide to give salts, called **nitrates**, and water.

sodium + nitric $\rightarrow$ sodium + water
hydroxide   acid   nitrate

$$NaOH(aq) + HNO_3(aq) \rightarrow NaNO_3(aq) + H_2O(l)$$

zinc + nitric $\rightarrow$ zinc + water
oxide   acid   nitrate

$$ZnO(s) + 2HNO_3(aq) \rightarrow Zn(NO_3)_2(aq) + H_2O(l)$$

- It reacts with carbonates to give a salt, water and carbon dioxide gas.

sodium + nitric $\rightarrow$ sodium + water + carbon
carbonate   acid   nitrate   dioxide

$$Na_2CO_3(s) + 2HNO_3(aq) \rightarrow 2NaNO_3(aq) + H_2O(l) + CO_2(g)$$

- Most acids will react with metals to give a salt and hydrogen gas. However, since nitric acid is an oxidising agent it behaves differently and hydrogen is rarely formed. The exceptions to this are the reactions of cold dilute nitric acid with magnesium and calcium.

magnesium + nitric $\rightarrow$ magnesium + hydrogen
acid   nitrate

$$Mg(s) + 2HNO_3(aq) \rightarrow Mg(NO_3)_2(aq) + H_2(g)$$

Dilute nitric acid reacts with copper, for example, to produce nitrogen monoxide instead of hydrogen.

copper + nitric $\rightarrow$ copper(II) + water + nitrogen
acid   nitrate   monoxide

$$3Cu(s) + 8HNO_3(aq) \rightarrow 3Cu(NO_3)_2(aq) + 4H_2O(l) + 2NO(g)$$

The nitrogen monoxide produced quickly reacts with oxygen from the air to give brown fumes of nitrogen dioxide gas.

Concentrated nitric acid

Concentrated nitric acid is a powerful **oxidising agent**. It will oxidise both metals and non-metals.

- It will oxidise carbon to carbon dioxide:

carbon + conc → carbon + nitrogen(IV) + water
nitric acid　　dioxide　　oxide

$$C(s) + 4HNO_3(l) \rightarrow CO_2(g) + 4NO_2(g) + 2H_2O(l)$$

- It will oxidise copper to copper(II) nitrate (Figure 15.7).

copper + conc → copper(II) + nitrogen + water
nitric acid　　nitrate　　dioxide

$$Cu(s) + 4HNO_3(l) \rightarrow Cu(NO_3)_2(aq) + 2NO_2(g) + 2H_2O(l)$$

QUESTIONS

a What do you think happens to the nitrogen monoxide produced in the third step of the industrial preparation of nitric acid?

b What is the maximum amount of nitric acid that could be produced from 500 tonnes of ammonia gas?

c In reality the amount of nitric acid produced would be less than the amount calculated above. Explain why this would be so.

d Write a word and balanced chemical equation to show the neutralisation reaction between nitric acid and ammonium hydroxide.

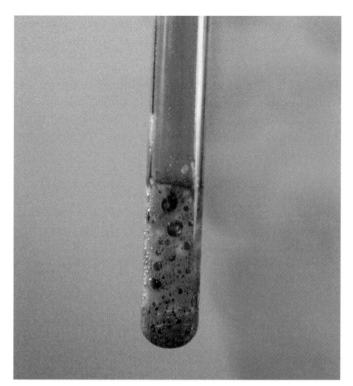

Figure 15.7 Copper reacting with concentrated nitric acid

Artificial fertilisers

The two processes so far described, the production of ammonia and nitric acid, are extremely important in the production of many artificial fertilisers. As was explained earlier, the use of artificial fertilisers is essential if farmers are to produce sufficient crops to feed the ever-increasing population. Crops remove nutrients from the soil as they grow; these include nitrogen, phosphorus and potassium. Artificial fertilisers are added to the soil to replace these nutrients and others, such as calcium, magnesium, sodium, sulphur, copper and iron. Examples of nitrogenous fertilisers (those which contain nitrogen) are shown in Table 15.1.

Table 15.1 Some nitrogenous fertilisers

Fertiliser	Formula
Ammonium nitrate	NH_4NO_3
Ammonium phosphate	$(NH_4)_3PO_4$
Ammonium sulphate	$(NH_4)_2SO_4$
Urea	$CO(NH_2)_2$

Artificial fertilisers can also make fertile land which was once unable to support crop growth. The fertilisers which add the three main nutrients (N, P and K) are called NPK fertilisers. They contain ammonium nitrate (NH_4NO_3), ammonium phosphate (($NH_4)_3PO_4$) and potassium chloride (KCl) in varying proportions.

Manufacture of ammonium nitrate

Ammonium nitrate (Nitram) is probably the most widely used nitrogenous fertiliser. It is manufactured by reacting ammonia gas and nitric acid.

ammonia + nitric acid → ammonium nitrate

$$NH_3(g) + HNO_3(aq) \rightarrow NH_4NO_3(aq)$$

Problems with fertilisers

If artificial fertilisers are not used correctly, problems can arise. If too much fertiliser is applied to the land, rain washes the fertiliser off the land and into rivers and streams. This process of leaching encourages the growth of algae and marine plants. As the algae die and decay oxygen is removed from the water, leaving insufficient amounts for fish and other organisms to survive.

QUESTIONS

a Calculate the percentage of nitrogen in each of the four fertilisers in Table 15.1.
(A_r: N = 14; H = 1; P = 31; O = 16; S = 32)

b Write down a method that you could carry out in a school laboratory to prepare a sample of ammonium sulphate fertiliser.

Checklist

After studying Chapter 15 you should know and understand the following terms.

Nitrogen fixation   The direct use of atmospheric nitrogen in the formation of important compounds of nitrogen. Bacteria present in root nodules of certain plants are able to take nitrogen directly from the atmosphere to form essential protein molecules.

Nitrogen cycle   The system by which nitrogen and its compounds, both in the air and in the soil, are interchanged.

Reversible reaction   A chemical reaction which is said to be reversible can go both ways. This means that once some of the products have been formed they will undergo a chemical change once more to re-form the reactants. The reaction from left to right, as the equation for the reaction is written, is known as the forward reaction and the reaction from right to left is known as the back reaction.

Chemical equilibrium   A chemical equilibrium is dynamic. The concentrations of the reactants and products remain constant because the rate at which the forward reaction occurs is the same as that of the back reaction.

Optimum temperature   A compromise temperature used in industry to ensure that the yield of product and the rate at which it is produced make the process as economical as possible.

Artificial fertiliser   A substance added to soil to increase the amount of elements such as nitrogen, potassium and phosphorus. This enables crops grown in the soil to grow more healthily and to produce higher yields.

QUESTIONS

1 Study the following reaction scheme.

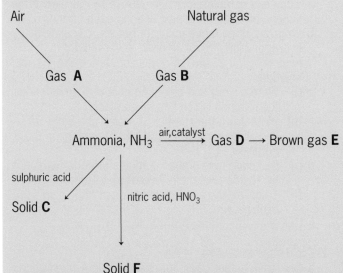

a Identify the substances **A** to **F** by giving their names and chemical formulae.
b How is gas **A** obtained from the air?
c Write a word and balanced chemical equation for the formation of ammonia gas from gases **A** and **B**.
d Write an equation for the formation of solid **C** from ammonia.
e Give a use for:
  (i) solid **C**
  (ii) nitric acid
  (iii) solid **F**.

2 a Write word and balanced chemical equations for the reaction of dilute nitric acid with:
  (i) sodium hydroxide
  (ii) copper(II) oxide
  (iii) magnesium carbonate.

b Describe how you would obtain dry crystals of potassium nitrate from dilute nitric acid and potassium hydroxide solution.

3 An international chemical firm has decided that it needs to build a new plant to produce the fertiliser ammonium sulphate, starting from the raw materials.
a What raw materials would be needed for this process?
b What factors would a chemical engineer look for before she decided where to build the new plant?
c At this type of plant there is always the danger of the leakage of some ammonia gas. What effect would a leak of ammonia gas have on:
  (i) people living in the surrounding area?
  (ii) the pH of the soil in the surrounding area?
  (iii) the growth of crops in the surrounding area?

4

Fertiliser	Formula	% nitrogen
Ammonia solution	NH_3	82.4
Calcium nitrate	$Ca(NO_3)_2$	17.1
Nitram	NH_4NO_3	35.0
Sodium nitrate	$NaNO_3$	
Potassium nitrate	KNO_3	

a Copy and complete the above table by calculating the percentage of nitrogen in the fertilisers sodium nitrate and potassium nitrate.
(A_r: N = 14; H = 1; O = 16; K = 39; Ca = 40; Na = 23)
b Including the data you have just calculated, which of the fertilisers contains:
  (i) the largest percentage of nitrogen?
  (ii) the smallest percentage of nitrogen?

c Give the chemical name for the fertiliser for which ICI uses the name Nitram.

d Ammonia can be used directly as a fertiliser but not very commonly. Think of two reasons why ammonia is not often used directly as a fertiliser.

e Nitram fertiliser is manufactured by the reaction of nitric acid with ammonia solution according to the equation:

$$NH_3(aq) + HNO_3(aq) \rightarrow NH_4NO_3(aq)$$

A bag of Nitram may contain 50 kg of ammonium nitrate. What mass of nitric acid would be required to make it?

5 Ammonia gas is made industrially by the Haber process, which involves the reaction between the gases nitrogen and hydrogen. The amount of ammonia gas produced from this reaction is affected by both the temperature and the pressure at which the process is run. The graph below shows how the amount of ammonia produced from the reaction changes with both temperature and pressure. The percentage yield of ammonia indicates the percentage of the nitrogen and hydrogen gases that are actually changed into ammonia gas.

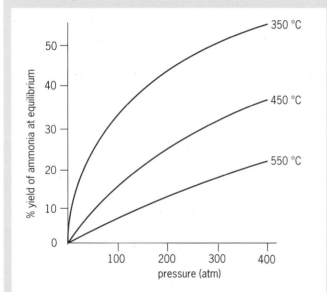

a Write a word and balanced chemical equation for the reversible reaction between nitrogen and hydrogen to produce ammonia using the Haber process.

b What is meant by the term a 'reversible reaction'?

c Use the graphs to say whether more ammonia is produced at:
   (i) higher or lower temperatures
   (ii) higher or lower pressures

d What is the percentage yield of ammonia if the conditions used to run the process are:
   (i) a temperature of 350 °C and a pressure of 100 atmospheres?
   (ii) a temperature of 550 °C and a pressure of 350 atmospheres?

e The conditions in industry for the production of ammonia are commonly of the order of 200 atmospheres and 450 °C. What is the percentage yield of ammonia using these conditions?

f Why does industry use the conditions stated in (e) if it is possible to obtain a higher yield of ammonia using different conditions?

6 The following results were obtained from a neutralisation reaction between potassium hydroxide and nitric acid. The dilute nitric acid had been found in a cupboard in the school laboratory. The student carrying out the experiment was interested in finding out the concentration of the nitric acid.

The student used 25 cm³ of 0.15 M potassium hydroxide solution in a conical flask to which was added the indicator phenolphthalein. The dilute nitric acid was added from a burette until the indicator just changed colour. The student repeated the experiment four times. Her results are shown below.

	Rough	1	2	3
Final burette reading (cm³)	10.25	13.20	13.90	12.65
Initial burette reading (cm³)	0.00	3.10	3.80	2.50
Volume used (cm³)				

a Copy and complete the table above by calculating the volume of dilute nitric acid used in each titration.

b From the three most accurate results, calculate the average volume of dilute nitric acid required to neutralise the 25 cm³ of potassium hydroxide solution.

c (i) Write a word and balanced chemical equation for the reaction which has taken place.
   (ii) Write down the number of moles of nitric acid and potassium hydroxide shown reacting in the equation.

d (i) Calculate the number of moles of potassium hydroxide present in 25 cm³ of solution.
   (ii) Calculate the number of moles of nitric acid neutralised.

e Calculate the molarity of the dilute nitric acid.

7 Explain the following.
a Dry litmus paper does not change colour when added to dry ammonia gas.
b Ammonia gas cannot be collected over water but can be collected by downward displacement of air.
c Soil often needs artificial fertilisers added to it if it is to continuously support the growth of healthy crops.
d When colourless nitrogen(II) oxide gas is formed, it almost instantaneously turns brown.

8 Nitrogen oxides are emitted from car exhaust pipes. To animals and plants these gases are very harmful.
a Normally, nitrogen gas and oxygen gas do not react. Why, in a car engine, do they react to form these dangerous gases?
b Where does the nitrogen which forms these nitrogen oxides come from?
c How is the production of these nitrogen oxides being minimised in newer cars?

16 SULPHUR

Sulphur—the element

Sulphur is a non-metallic element which has a very important role in the chemical industry. It is a yellow solid which is found in large quantities but in various forms throughout the world (Figure 16.1). It is found in metal ores such as copper pyrites ($CuFeS_2$) and zinc blende (ZnS) and in volcanic regions of the world. Natural gas and oil contain sulphur and its compounds, but the majority of this sulphur is removed as it would cause environmental problems. Sulphur obtained from these sources is known as 'recovered sulphur' and it is an important source of the element. It is also found as elemental sulphur in sulphur beds in Poland, Russia and the US. These sulphur beds are typically 200 m below the ground. Sulphur from these beds is extracted using the **Frasch process**, named after Hermann Frasch.

The Frasch process

Superheated water at 170 °C and hot compressed air are forced underground through pipes, forcing water and molten sulphur to the surface. Sulphur is insoluble in water and so the two substances emerging from the pipes are easily separated. The sulphur is kept molten and sold in this form. The sulphur obtained from this process is about 99.5% pure and can be used directly (Figure 16.2).

Figure 16.1 Sulphur — rhombic (top) and monoclinic

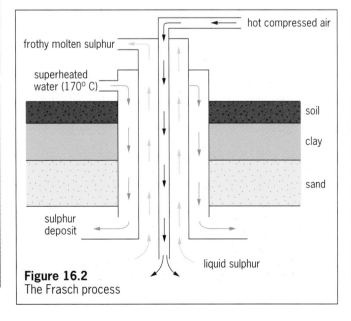

Figure 16.2
The Frasch process

199

Uses of sulphur

The vast majority of sulphur is used to produce perhaps the most important industrial chemical, sulphuric acid. Sulphur is also used to **vulcanise** rubber, a process which makes the rubber harder and increases its elasticity. Relatively small amounts are used in the manufacture of matches, fireworks and fungicides, as a sterilising agent and in medicines.

Allotropes of sulphur

Sulphur is one of the few non-metal elements which exist as allotropes (Chapter 4, p. 52). The main allotropes are called rhombic sulphur and monoclinic sulphur. Both of these solid forms of sulphur are made up of S_8 molecules (Figure 16.3). The fact that there are two different allotropes of sulphur is due to the way in which these S_8 molecules pack together. In rhombic sulphur the molecules are packed more closely than in the monoclinic form (Figure 16.4).

Although sulphur is insoluble in water, it will dissolve in an organic solvent such as methylbenzene. If a solution of sulphur in methylbenzene is heated and allowed to cool then crystals of monoclinic sulphur are produced. When the temperature of the solution falls below 96 °C, rhombic sulphur crystals are produced. Rhombic sulphur is stable below 96 °C and monoclinic sulphur is stable above 96 °C. This temperature is called the **transition temperature**.

When solid sulphur is heated, it melts at 112 °C and forms a runny (mobile) liquid. At this point the S_8 molecules are moving freely around each other, as the weak attractive forces between them have been overcome (Figure 16.5).

However, if the sulphur is heated further the liquid becomes thicker (viscous). This is because the S_8 rings have been broken by the energy given to the

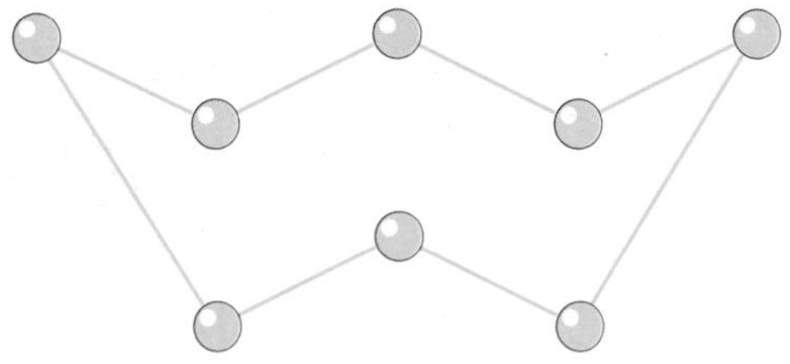

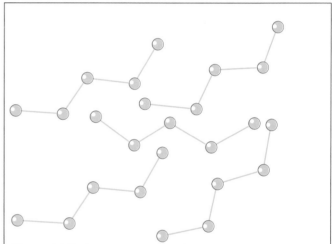

Figure 16.5 At the melting point S_8 rings move freely around one another

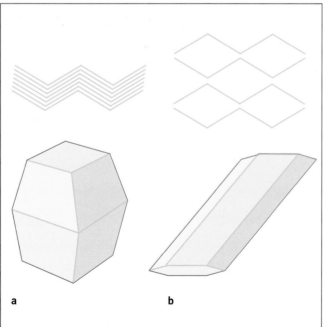

Figure 16.3 S_8 molecule

Figure 16.4 The packing of S_8 molecules in (**a**, left) a rhombic crystal and (**b**, right) a monoclinic crystal of sulphur

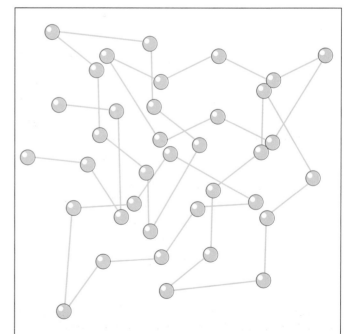

Figure 16.6 A viscous liquid is produced as chains of sulphur atoms are formed and get tangled together

sulphur and they bond together, forming long chains of sulphur atoms which become tangled, making the liquid viscous (Figure 16.6). Continued heating, to 444 °C, makes the liquid more mobile once again as the long chains are broken down into smaller ones which move around one another freely.

If this liquid is poured into a beaker of cold water, a substance called **plastic sulphur** is formed. This is an elastic, rubber-like substance. In plastic sulphur, the sulphur atoms remain bonded together in the form of chains, very similar to chains of carbon atoms in plastics such as polythene. After a few hours, however, the plastic sulphur loses its elasticity and once again becomes solid as the S_8 molecular rings re-form.

Properties of sulphur

Sulphur:

- is a yellow, brittle solid at room temperature
- does not conduct electricity
- is insoluble in water.

Sulphur will react with both metals and non-metals.

- It reacts with magnesium to form magnesium sulphide.

$$\text{magnesium} + \text{sulphur} \rightarrow \text{magnesium sulphide}$$
$$Mg_{(s)} \quad + \quad S_{(s)} \quad \rightarrow \quad MgS_{(s)}$$

- It reacts with oxygen to produce sulphur dioxide gas.

$$\text{sulphur} + \text{oxygen} \rightarrow \text{sulphur dioxide}$$
$$S_{(s)} \quad + \quad O_2{(g)} \quad \rightarrow \quad SO_2{(g)}$$

Sulphur dioxide

Sulphur dioxide is a colourless gas produced when sulphur or substances containing sulphur, for example crude oil or natural gas, are burned in oxygen gas. It has a choking smell and is extremely posionous. The gas dissolves in water to produce an acidic solution of sulphurous acid.

$$\text{sulphur dioxide} + \text{water} \rightleftharpoons \text{sulphurous acid}$$
$$SO_2{(g)} \quad + H_2O_{(l)} \rightleftharpoons \quad H_2SO_3{(aq)}$$

It is one of the major pollutant gases and is the gas principally responsible for **acid rain**. However, it does have some uses: as a bleaching agent, in fumigants and in the preservation of food.

QUESTIONS

a What is meant by the term 'allotrope'?
b 'Sulphur is a non-metallic element.' Discuss this statement, giving physical and chemical reasons to support your answer.

Acid rain

Rainwater is naturally acidic since it dissolves carbon dioxide gas from the atmosphere as it falls. Natural rainwater has a pH of about 5.7. In recent years, especially in central Europe, the pH of rainwater has fallen to between pH 3 and pH 4.8. This increase in acidity has led to extensive damage to forests (Figure 16.7), lakes and marine life.

The amount of sulphur dioxide in the atmosphere has increased dramatically over recent years. There has always been some sulphur dioxide in the atmosphere, from natural processes such as volcanoes and rotting vegetation. Over Europe, however, around 80% of the sulphur dioxide in the atmosphere is formed from the combustion of fuels containing sulphur (Figure 16.8). After dissolving in rainwater to produce sulphurous acid, it further reacts with oxygen to produce sulphuric acid.

Figure 16.7 This forest has been devastated by acid rain

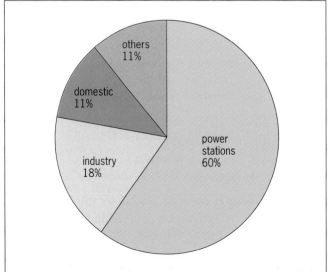

Figure 16.8 Sources of sulphur dioxide

a How could the amount of sulphur dioxide being produced by the above sources be reduced?

b Devise an experiment which you could carry out in the school laboratory to determine the amount of sulphur in two different types of coal.

Industrial manufacture of sulphuric acid—the Contact process

The major use of sulphur is in the production of sulphuric acid. This is probably the most important chemical, and the quantity of it produced by a country has been linked with the economic stability of the country. Many millions of tonnes of sulphuric acid are produced in the UK each year. It is used mainly as the raw material for the production of many substances (Figure 16.9), such as:

- fertilisers
- paints, pigments and dyes
- artificial fibres
- detergents
- chemicals.

The process by which sulphuric acid is produced is known as the **Contact process** (Figure 16.10). The process has the following stages.

- Sulphur dioxide is first produced, primarily by the reaction of sulphur with air.

$$\text{sulphur} + \text{oxygen} \rightarrow \text{sulphur dioxide}$$
$$S(s) \quad + \quad O_2(g) \quad \rightarrow \quad SO_2(g)$$

- Any dust and impurities are removed from the sulphur dioxide produced, as well as any unreacted oxygen. These 'clean' gases are heated to a temperature of approximately 450°C and fed into a reaction vessel, where they are passed over a catalyst of vanadium(v) oxide (V_2O_5). This catalyses the reaction between sulphur dioxide and oxygen to produce sulphur trioxide (sulphur(VI) oxide, SO_3).

$$\text{sulphur} + \text{oxygen} \rightleftharpoons \text{sulphur}$$
$$\text{dioxide} \qquad \text{trioxide}$$
$$2SO_2(g) + O_2(g) \rightleftharpoons 2SO_3(g) \quad \Delta H = -197\,\text{kJ}\,\text{mol}^{-1}$$

This reaction is reversible and so the ideas of Le Chatelier (p. 193) can be used to increase the proportion of sulphur trioxide in the equilibrium mixture. The forward reaction is exothermic and so would be favoured by low temperatures. The temperature of 450°C used is an optimum

Figure 16.9 Products made from sulphuric acid include fertilisers, paints and pigments

Figure 16.10 A Contact process plant used for making sulphuric acid

temperature which produces sufficient sulphur trioxide at an economical rate. Since the reaction from left to right is also accompanied by a decrease in the number of molecules of gas, it will be favoured by a high pressure. The process is actually run at atmospheric pressure. Under these conditions, about 96% of the sulphur dioxide and oxygen are converted into sulphur trioxide. The heat produced by this reaction is used to heat the incoming gases, so saving money.

- If this sulphur trioxide is added directly to water, sulphuric acid is produced. This reaction, however, is very violent and a thick mist is produced.

$$\text{sulphur trioxide} + \text{water} \rightarrow \text{sulphuric acid}$$
$$SO_3(g) + H_2O(l) \rightarrow H_2SO_4(l)$$

This acid mist is very difficult to deal with and so a different route to sulphuric acid is employed. Instead, the sulphur trioxide is dissolved in concentrated sulphuric acid (98%) to give a substance called **oleum**.

$$\text{sulphuric acid} + \text{sulphur trioxide} \rightarrow \text{oleum}$$
$$H_2SO_4(aq) + SO_3(g) \rightarrow H_2S_2O_7(l)$$

The oleum formed is then added to the correct amount of water to produce sulphuric acid of the required concentration.

$$\text{oleum} + \text{water} \rightarrow \text{sulphuric acid}$$
$$H_2S_2O_7(l) + H_2O(l) \rightarrow 2H_2SO_4(l)$$

QUESTION

a Produce a flow diagram to show the different processes which occur during the production of sulphuric acid by the Contact process. Write balanced chemical equations showing the processes which occur at the different stages, giving the essential raw materials and conditions used.

Properties of sulphuric acid

Dilute sulphuric acid

Dilute sulphuric acid is a typical strong **dibasic** acid. A dibasic acid is one with two replaceable hydrogen atoms which may produce two series of salts — normal and acid salts (Chapter 7, p. 85).

It will react with bases such as sodium hydroxide and copper(II) oxide to produce normal salts, called **sulphates**, and water.

- With sodium hydroxide:

$$\text{sodium} + \text{sulphuric} \rightarrow \text{sodium} + \text{water}$$
$$\text{hydroxide} \quad \text{acid} \quad \text{sulphate}$$
$$2NaOH(aq) + H_2SO_4(l) \rightarrow Na_2SO_4(aq) + H_2O(l)$$

- With copper(II) oxide:

$$\text{copper(II)} + \text{sulphuric} \rightarrow \text{copper(II)} + \text{water}$$
$$\text{oxide} \quad \text{acid} \quad \text{sulphate}$$
$$CuO(s) + H_2SO_4(aq) \rightarrow CuSO_4(aq) + H_2O(l)$$

It also reacts with carbonates to give normal salts, carbon dioxide and water, and with reactive metals

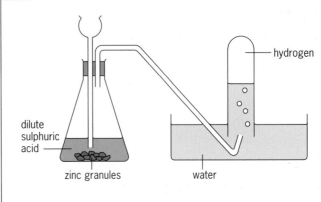

Figure 16.11 The laboratory preparation of hydrogen gas

to give a normal salt and hydrogen gas. The reaction between zinc and sulphuric acid is often used to prepare hydrogen gas in the laboratory (Figure 16.11).

$$\text{zinc} + \text{sulphuric acid} \rightarrow \text{zinc sulphate} + \text{hydrogen}$$
$$Zn(s) + H_2SO_4(aq) \rightarrow ZnSO_4(aq) + H_2(g)$$

The preparation of the acid salt with sodium hydroxide requires twice the volume of acid as that used in the preparation of the normal salt. Therefore, if 25 cm³ of dilute sulphuric acid was required to form the normal salt from a given volume of alkali of the same concentration then 50 cm³ of the same acid solution would be required to produce the acid salt, sodium hydrogensulphate, from the same volume of alkali.

$$\text{sodium} + \text{sulphuric} \rightarrow \text{sodium} + \text{water}$$
$$\text{hydroxide} \quad \text{acid} \quad \text{hydrogensulphate}$$
$$NaOH(aq) + H_2SO_4(aq) \rightarrow NaHSO_4(aq) + H_2O(l)$$

Sulphates

The salts of sulphuric acid, sulphates, can be identified by a simple test tube reaction. To test for a sulphate, add a few drops of dilute hydrochloric acid followed by a few drops of barium chloride. If a sulphate is present, a white precipitate of barium sulphate is produced.

$$\text{barium ions} + \text{sulphate ions} \rightarrow \text{barium sulphate}$$
$$Ba^{2+}(aq) + SO_4^{2-}(aq) \rightarrow BaSO_4(s)$$

Many sulphates have very important uses, as can be seen from Table 16.1

Table 16.1 Uses of some metal sulphates

Salt	Formula	Use
Calcium sulphate	$CaSO_4.\frac{1}{2}H_2O$	'Plaster of Paris' used to set bones
Ammonium sulphate	$(NH_4)_2SO_4$	Fertiliser
Magnesium sulphate	$MgSO_4$	In medicine it is used as a laxative
Barium sulphate	$BaSO_4$	'Barium meal' used in diagnostic medical X-ray studies

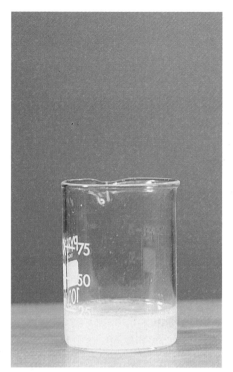

Figure 16.12 The concentrated sulphuric acid has removed the elements of water from the sugar, leaving black carbon

Concentrated sulphuric acid

Concentrated sulphuric acid is a powerful **dehydrating agent**. This means it will take water from a variety of substances. One such substance is cane sugar, or sucrose (Figure 16.12).

$$\text{sucrose(sugar)} \xrightarrow{\text{conc } H_2SO_4} \text{carbon} + \text{water}$$
$$C_{12}H_{22}O_{11}(s) \longrightarrow 12C(s) + 11H_2O(l)$$

Concentrated sulphuric acid will also take water from hydrated copper(II) sulphate crystals, leaving only anhydrous copper(II) sulphate. If a few drops of concentrated sulphuric acid are added to some blue hydrated copper(II) sulphate crystals, they slowly turn white as the water of crystallisation is removed by the acid. Eventually, only a white powder — anhydrous copper(II) sulphate — remains.

$$\begin{array}{l}\text{hydrated} \xrightarrow{\text{conc } H_2SO_4} \text{anhydrous} + \text{water} \\ \text{copper(II)} \qquad\qquad \text{copper(II)} \\ \text{sulphate} \qquad\qquad \text{sulphate} \\ CuSO_4.5H_2O(s) \longrightarrow CuSO_4(s) + 5H_2O(l)\end{array}$$

Concentrated sulphuric acid should be treated very carefully, because it will also remove water from flesh! It is a very corrosive substance and should always be handled with care.

Diluting concentrated sulphuric acid must be done with great care because of its affinity for water. The concentrated sulphuric acid should always be added to the water, not the other way around.

QUESTIONS

a If you were given an unlabelled bottle which was thought to be dilute sulphuric acid, how would you show that the solution contained sulphate ions (SO_4^{2-}(aq)), how would you show that it was an acid and how would you determine the concentration of the acid?

b Write balanced chemical equations for the reactions between dilute sulphuric acid and:
  (i) zinc oxide
  (ii) potassium carbonate
  (iii) aluminium.

Manufacture of a soapless detergent

A more recent use of sulphuric acid is in the production of soapless detergents. These are detergents that can be used more effectively than soap in hard water areas, are fairly cheap to make and are gradually replacing soaps (Chapter 14, p. 184).

The general process involves the reaction of a long, straight-chain alkene, such as dodecene, with benzene.

$$\text{benzene} + \text{dodecene} \longrightarrow \text{dodecylbenzene}$$
$$C_6H_6(l) + CH_3(CH_2)_9CH{=}CH_2(l) \rightarrow C_6H_5(CH_2)_{11}CH_3(l)$$

The molecular formula of dodecylbenzene is $C_{18}H_{30}$. This compound is then reacted with concentrated sulphuric acid to give a compound called a sulphonic acid.

dodecyl-+ sulphuric → dodecylbenzene + water
benzene   acid   sulphonic acid

$C_{18}H_{30}(l)+ H_2SO_4(l) \rightarrow C_{18}H_{29}SO_3H(aq) + H_2O(l)$

Finally, this is reacted with the alkali sodium hydroxide, NaOH.

dodecylbenzene + sodium →   sodium   + water
sulphonic   hydroxide   dodecylbenzene
acid   sulphonate

$C_{18}H_{29}SO_3H(aq) + NaOH(aq) \rightarrow C_{18}H_{29}SO_3{}^-Na^+(aq) + H_2O(l)$
(soapless detergent)

Soapless detergents such as this are to be found in most washing powders.

QUESTION

a State the advantages and disadvantages of using soapless detergents.

Checklist

After studying Chapter 16 you should know and understand the following terms.

Frasch process The process of obtaining sulphur from sulphur beds below the Earth's surface. Superheated water is pumped down a shaft to liquefy the sulphur, which is then brought to the surface.

Allotropes Different structural forms of the same element having the same physical state. For example, carbon exists as the allotropes diamond, graphite and buckminsterfullerene, and sulphur as rhombic and monoclinic sulphur.

Transition temperature The temperature boundary at which one allotropic form of an element is converted into another allotropic form.

Contact process The industrial manufacture of sulphuric acid using the raw materials sulphur and air.

Dibasic acid An acid which contains two replaceable hydrogen atoms per molecule of the acid, for example sulphuric acid, H_2SO_4.

Sulphate A salt of sulphuric acid formed by the reaction of the acid with carbonates, bases and some metals. It is possible to test for the presence of a sulphate by the addition of dilute hydrochloric acid and some barium chloride solution. A white precipitate of barium sulphate is formed if a sulphate is present.

Soapless detergents Substances which are more effective than soap at producing lathers, especially in hard water areas. They are large organic molecules, produced using sulphuric acid.

QUESTIONS

1 Write down what you understand by the following terms.
  a Dehydrating agent.
  b Oxidising agent.
  c Optimum temperature.
  d Acid rain.

2 Explain the following.
  a Chemical plants that produce sulphuric acid are often located on the coast.
  b Even though more sulphuric acid could be produced using high pressures, normal atmospheric pressure is used.
  c Natural rubber cannot be used to produce car tyres but vulcanised rubber can.
  d Sulphur dioxide gas is regarded as a pollutant.
  e Coal-fired and oil-fired power stations produce sulphur dioxide. Some of them are being fitted with flue gas desulphurisation (FGD) units.

3 A type of coal contains 0.5% of sulphur by mass.
  a Write an equation for the formation of sulphur dioxide gas when this coal is burned.
  b If 1500 tonnes of coal was burned, what mass of sulphur would it contain?
  c What mass of sulphur dioxide gas would be formed if 1500 tonnes of coal were burned?
  d What volume would this mass of sulphur dioxide gas occupy, measured at room temperature and pressure (rtp)? (A_r: S = 32; O = 16. One mole of a gas occupies 24 dm^3 at rtp.)

4 Fossil fuels, such as oil, coal and natural gas, all contain some sulphur. When these fuels are burned they produce many different gases. Concern has grown in recent years about the effects of one of these gases, sulphur dioxide. When sulphur dioxide dissolves in rainwater it forms an acidic solution which has become known as acid rain.

QUESTIONS (continued)

Money has been made available to solve the problem of acid rain. Attempts are being made to clean those gases being released from power stations and to look into ways in which the effects of acid rain can be reversed.

The table below and Figure 16.8 (p. 201) give some data about the emission of sulphur dioxide.

	Million tonnes per year
USA	26
Russia/Ukraine	18
Germany	7
UK	5
Canada	5
France	3
Poland	3
Italy	3
Other countries	30

a Using the figures in the table, produce a bar chart to show the amount of sulphur dioxide produced by each of the countries listed.
b What percentage of the world's sulphur dioxide is produced in:
  (i) the UK?
  (ii) North America?
c Using the information above, explain why countries such as the US, Russia and Germany are at the top of the list of sulphur dioxide producers.
d If the total amount of sulphur dioxide produced by the UK is 5 million tonnes per year, what amount is produced by:
  (i) power stations?
  (ii) domestic users?
  (iii) industry?

5 Study the following reaction scheme:

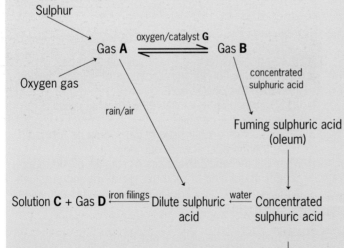

a Identify the substances **A** to **H** by giving their names and formulae.
b Write a balanced chemical equation for the formation of gas **B**.
c (i) Describe a chemical test, and give the positive result of it, to identify gas **D**.
  (ii) Describe a chemical test, and give the positive result of it, to identify gas **F**.
d How would you obtain solid **C** from the solution **C**?
e Which pathway shows the formation of acid rain?
f In which way is the concentrated sulphuric acid acting in its reaction with sucrose?
g Where does the oxygen gas come from to form gas **A**?

6 In a neutralisation experiment, 25 cm³ of dilute sulphuric acid was required to react completely with 40 cm³ of a solution of 0.25 M potassium hydroxide.
a Write a balanced chemical equation for the reaction between dilute sulphuric acid and potassium hydroxide.
b Calculate the number of moles of potassium hydroxide solution used in the reaction.
c How many moles of dilute sulphuric acid would this number of moles of potassium hydroxide react with?
d Calculate the concentration of the dilute sulphuric acid.
e Which indicator could have been used to determine when neutralisation had just occurred?

7 Describe how you would prepare some crystals of hydrated copper(II) sulphate from copper(II) oxide and dilute sulphuric acid. Draw a diagram of the apparatus you would use and write a balanced chemical equation for the reaction.

8 When sulphur is extracted from sulphur beds below the Earth's surface, superheated water is pumped down a shaft into the beds to melt the solid sulphur.
a What is meant by superheated water?
b Why does the water have to be superheated? Why would boiling water not work?
c (i) When the molten sulphur is pumped to the surface it solidifies. Which allotrope of sulphur forms first?
  (ii) What eventually happens to this allotrope as the temperature falls?
Most sulphur obtained from these sulphur beds, in countries such as Poland and France, is exported.
d In which form do you think the sulphur is loaded onto sulphur tankers? Explain your answer.
e What hazards do you think people working in industries that use sulphur face? Give reasons for your answer.

17 RADIOACTIVITY

The photograph in Figure 17.1 shows a nuclear explosion. When an atomic bomb explodes, a huge amount of energy is released as well as a deadly burst of radioactivity. Radioactivity is regularly mentioned on television and in the papers for a variety of reasons.

What is radioactivity?

Certain elements have unstable isotopes. The nuclei of these atoms break up spontaneously with the emission of certain types of radiation. The atoms of these isotopes, with unstable nuclei, are **radioactive** and they are called **radioisotopes**. There are some naturally occurring radioisotopes, for example tritium (an isotope of hydrogen, ^3_1H) and carbon-14 ($^{14}_6C$). However, the vast majority of radioisotopes are artificially produced in nuclear reactors.

When these unstable atoms spontaneously break up they are said to **disintegrate** or **decay**. The result of this process is that energy is released in the form of heat, and **radiation** is emitted from the nucleus.

Radioactive elements can emit one or more of three types of radiation. These types of radiation are

Figure 17.1 When a nuclear explosion takes place a burst of radioactivity is released

Table 17.1 Properties of different types of radiation

Radiation	Nature	Relative charge	Penetration	Effect on nucleus
Alpha particle	2 protons and 2 neutrons (written as $^4_2\alpha$, a helium nucleus)	+2	Stopped by a few sheets of paper	Mass number decreases by 4 Atomic number decreases by 2
Beta particle	Electron	−1	Stopped by a few mm of plastic	Mass number is unchanged but atomic number increases by 1
Gamma rays	Electromagnetic radiation with shorter wavelength than X-rays	0	Stopped by several cm of lead	No change to mass or atomic number but the nucleus does lose energy

207

named after the Greek letters alpha (α), beta (β) and gamma (γ). Some of their properties are shown in Table 17.1 and Figure 17.2.

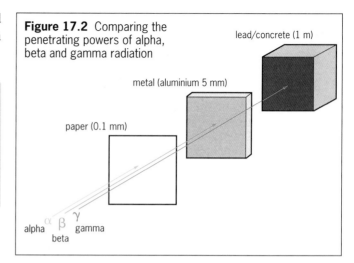

Figure 17.2 Comparing the penetrating powers of alpha, beta and gamma radiation

lead/concrete (1 m)

metal (aluminium 5 mm)

paper (0.1 mm)

alpha α β γ gamma

beta

Radioactive decay

Nuclear scientists have found the following.

- The isotopes whose natural radioactive decay involves emissions of alpha particles usually have a relative atomic mass greater than 210 ($A_r > 210$). They have too much mass to be stable and give out alpha particles to form smaller, more stable atoms. For example, uranium-238 undergoes radioactive decay by losing an alpha particle. This change can be shown by writing a **nuclear equation**.

$$^{238}_{92}U \rightarrow {}^{234}_{90}Th + {}^{4}_{2}\alpha$$

- The isotopes whose radioactive decay involves emission of beta particles often have an $A_r < 210$. Beta particles are usually emitted from heavier nuclei that have too many neutrons compared with the number of protons. One neutron changes to a proton and an electron is emitted from the nucleus as a beta particle.

$$^{1}_{0}n \rightarrow {}^{1}_{1}p + {}^{0}_{-1}\beta$$

The product has, therefore, one more proton. For example, thallium-208 undergoes radioactive decay by losing a beta particle.

$$^{208}_{81}Th \rightarrow {}^{208}_{82}Pb + {}^{0}_{-1}\beta$$

- When alpha particles or beta particles are emitted, gamma rays are often emitted at the same time. When a radioactive isotope emits gamma rays, it becomes more stable because it loses energy.

In both alpha and beta decay the new element formed is called the **daughter isotope**. Different atoms are **not** produced when gamma rays are emitted.

Dangers of radioactivity

Both beta particles and gamma rays can pass quite easily through the skin and can easily damage or even kill cells, causing illness (Figures 17.3 and 17.4). They can cause mutations in a cell's DNA, which means that it cannot reproduce properly, which may lead to diseases such as cancer. Extremely large doses of radiation can be lethal.

Alpha particles cannot pass through the skin and are, therefore, less dangerous. However, they can be extremely dangerous if they get inside your body. This can happen if you inhale radioactive material.

Radiation emitted by a radioactive source can be detected using an instrument called a Geiger–Müller counter (Figure 17.5). This works by counting the electrical signals triggered by the radiation entering the instrument.

Safety and radioactivity

Because of the dangers associated with the radiation emitted by radioactive material, the nuclear power industry has developed strict safety precautions. These precautions are carefully followed to maintain very high safety standards (Figure 17.6). The

Figure 17.3 Anything showing this symbol is radioactive and extreme care should be taken

Figure 17.6 Nuclear power workers have strict safety standards

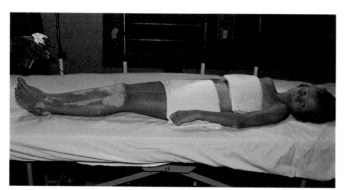

Figure 17.4 A victim of radiation burns. Large doses of radiation can be lethal

precautions taken by workers who deal with radioactive materials include:

- wearing protective suits
- wearing radiation level badges
- checking the radiation levels regularly
- using thick lead-walled containers for transporting radioactive materials
- using remote-control equipment from behind thick glass or lead walls to handle radioactive materials.

> **QUESTION**
>
> **a** Many people object to the use of nuclear power to generate our electricity, even though the safety record of the nuclear industry is extremely good. Make a table or write briefly to state the advantages and disadvantages of using nuclear energy as a source of electricity.

Half-life

Radioactive isotopes decay at their own rate. This is a completely random process and is unaffected by temperature or whether or not the radioactive isotope is present as the element or is combined in a compound. It should be noted that these are very important differences between radioactive decay (involving the nucleus) and ordinary chemical reactions.

The **half-life** ($t_{1/2}$) of a radioactive isotope is the time taken for half of its atoms to decay and is used to indicate the rate of decay of radioactive isotopes. Some radioactive isotopes have a very long half-life

Figure 17.5 A Geiger–Müller counter detects and measures amounts of radiation

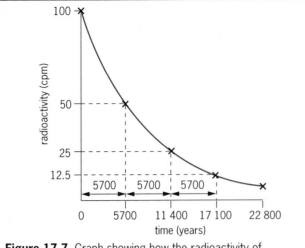

Figure 17.7 Graph showing how the radioactivity of carbon-14 changes with time

while others have an extremely short half-life. For example, the half-life of carbon-14 is 5700 years, while that of unniloctium, one of the recently discovered elements, is 2 milliseconds.

Looking more closely at the half-life of carbon-14, we find that in 5700 years half the atoms will have decayed by beta emission.

$$^{14}_{6}C \rightarrow {}^{14}_{7}N + {}^{0}_{-1}\beta$$

It follows that in another 5700 years half of the remainder of the atoms will have decayed and there will only be a quarter of the original atoms left, and so on. If you plot radioactivity against time for the radioactive decay of carbon-14, a graph like the one shown in Figure 17.7 is obtained.

QUESTIONS

a Use the information below to answer the questions which follow.

Isotope	Half-life
Carbon-14	5700 years
Radium-226	1610 years
Uranium-238	4.5×10^6 years
Uranium-235	7.13×10^8 years
Iodine-135	6.5 hours
Oxygen-14	71 seconds

  (i) Which of the radioisotopes above has the most stable nucleus?
  (ii) Which of the isotope(s) above do you think could be safely used in medical diagnosis? Explain your answer.
  (iii) If a patient was given a dose containing 2 g of iodine-135 to detect a problem with his thyroid gland, how long would it take for the mass of iodine to decay to less than 0.1 g?

Uses of radioactivity

Food preservation

The preservation of food using gamma rays is quite widespread worldwide. Treating food with gamma rays can:

- slow down the ripening of some fruits and the sprouting of potatoes — in both cases this helps storage and increases the shelf-life of the food
- kill highly dangerous micro-organisms such as *Salmonella*
- kill micro-organisms that spoil food.

However, it is still quite a contentious and controversial issue.

Radiocarbon dating

There are three isotopes of carbon, carbon-12, carbon-13 and carbon-14. The isotope carbon-14 is radioactive and has a half-life of 5700 years. Small and constant amounts of this isotope are present in all living things. When an animal or plant dies, no more carbon-14 is taken in and that which is present undergoes radioactive decay. If we measure the amount of carbon-14 left, it is possible to determine the age of the sample (Figure 17.8).

Tracer studies

Tracer techniques can be used to track where substances go to and where leaks may have occurred. Leaks in gas or oil pipes or in ventilating systems can be detected using this technique (Figure 17.9).

Tracer techniques are also important in medicine. Thyroid glands can become underactive or overactive. The activity of the thyroid gland can be

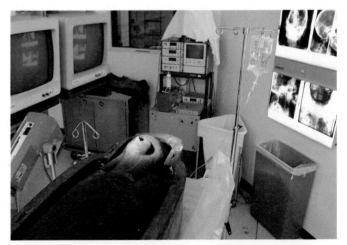

Figure 17.8 The radiocarbon dating of Tutankhamun's remains enabled an accurate date to be obtained for his reign. In addition, X-rays and computed tomography can be used to obtain images of the mummified bodies without damaging the coffins or their contents

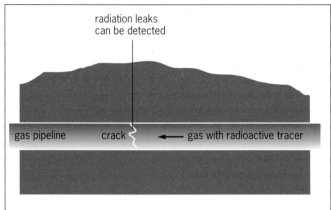

Figure 17.9 Using radioactive tracers to detect a leak in an oil pipeline

monitored by the patient being injected with or asked to drink radioactive iodine. The radioactivity in the vicinity of the thyroid gland is then checked to see how much of the radioactive iodine has settled in the area around the gland.

Nuclear power stations

Nuclear power stations harness the tremendous amounts of energy released when uranium-235 undergoes nuclear fission. The energy released, by this controlled chain reaction, is then used to produce electricity (Chapter 13, p. 163).

In an atomic bomb the nuclear fission of uranium-235 proceeds at an uncontrollable rate, in a fraction of a second, which results in an enormous explosion (see Figure 17.1).

Nuclear fusion

In nuclear fusion the nuclei of elements with a very low atomic number are fused together to make heavier elements. When this takes place it is accompanied by a very large release of energy. In the Sun, hydrogen nuclei are fusing together all the time to make helium nuclei. This is also the process by which a hydrogen bomb works.

A European team of nuclear scientists is currently working, at Culham in Oxfordshire, to attempt to control the fusion process at temperatures in excess of $1 \times 10^8 \,°C$ (Figure 17.10). Their project is called the JET project, which stands for **J**oint **E**uropean **T**orus. If all goes well and they succeed, we will have a major new source of energy.

Industry

Gamma radiation and beta radiation from radio-isotopes can be used to monitor the level of material inside a container (Figure 17.11). If there is a sudden decrease in the amount of radiation reaching the detector, which will happen when the container is

Figure 17.10 The inside of the Joint European Torus, in which deuterium ($^2_1$H) and tritium ($^3_1$H) fuse at temperatures in excess of 100 million °C

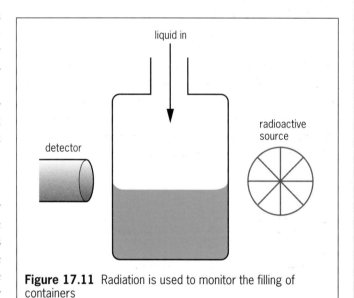

Figure 17.11 Radiation is used to monitor the filling of containers

full, then this can be used as a signal to switch off the flow of substance into the container. A similar method is used to monitor the thickness of sheets of plastic, metal and paper in production.

Medicine

Cancer cells are more easily killed by radiation than healthy cells. Penetrating gamma rays from a cobalt-60 source are used for this purpose. Other cancers such as skin cancer can be treated by less penetrating beta radiation from a strontium-90 source. Surgical instruments can also be sterilised using gamma radiation.

QUESTIONS

a Why do you think it is important for the JET project to succeed?

b Many diabetic patients use disposable plastic syringes to inject their insulin. These syringes must be sterile before they are used and this is done using gamma rays. Glass syringes used to be sterilised by boiling. Why cannot plastic syringes be sterilised in this way?

c (i) Cobalt-60, used for cancer treatment, decays, giving off beta radiation. Write a nuclear equation to show this decay.

(ii) Why is it very important for the instrument which gives the dose of cobalt-60 to be correctly set up before it is used for cancer treatment?

Checklist

After studying Chapter 17 you should know and understand the following terms.

Radioisotope   An atom with an unstable nucleus which disintegrates spontaneously to give off one or more types of radiation.

Alpha (α) radiation   The loss, from the nucleus of a radioisotope, of a particle made up of two protons and two neutrons. The particle is positively charged and is slow moving, being stopped by paper.

Beta (β) radiation   The loss, from the nucleus of a radioisotope, of an electron. The electron is formed from the breakdown of a neutron, forming a proton and an electron. The proton remains in the nucleus, and the electron is emitted at high speed, as a beta particle. The particle is negatively charged and has greater penetrating power than an alpha particle; it can be stopped by sheets of plastic or metal a few millimetres thick.

Gamma (γ) radiation   This type of radiation usually accompanies alpha and beta radiation. Gamma rays are given out, in the form of electro-magnetic radiation, as the result of the nucleus settling down after the decay. This type of radiation travels at the speed of light and can be stopped only by thick blocks of lead or concrete.

Daughter isotope   The isotope produced as the result of radioactive decay.

Half-life ($t_{1/2}$)   The time taken for the amount of a radioactive substance to decay to half of its original amount.

Nuclear fusion   The joining together of two, usually small, nuclei to produce a heavier atom.

QUESTIONS

1 The half-life of a radioisotope is the time taken for half of its atoms to decay, by either alpha or beta radiation. Some radioisotopes have very long half-lives while others have very short ones:

Isotope	Half-life	Decay by
Carbon-14	5700 years	beta radiation
Polonium-216	0.15 seconds	alpha radiation

a Write nuclear equations for the decay of each of these radioisotopes.

b Carbon-14 is a natural isotope of carbon and its presence in living matter is used in radiocarbon dating. If a tree, which died 22 800 years ago, contained 0.1 g of carbon-14, how much would be left today?

c If a sample of polonium had a mass of 128 g, how long would it take for the mass to reduce to 1 g?

2 The table below contains data taken from the measurements made of the radioactivity in counts per minute (cpm) from the radioisotope iodine-128.

Counts per minute	120	102	90	78	69	61	54	47	42
Time (min)	0	5	10	15	20	25	30	35	40

a Plot a graph, using the above data, of cpm against time.

b What is the half-life ($t_{1/2}$) of iodine-128?

c Which instrument could be used to detect the radiation produced by this isotope?

d What is the count rate after:
(i) 8 min?
(ii) 23 min?

e After how many minutes was the count rate:
(i) 55 cpm?
(ii) 95 cpm?

3 The International Union of Pure and Applied Chemistry (IUPAC) is the well established world authority on the names of chemical elements and their compounds. Up to now they have recognised, and have named, the elements up to atomic number 109. This last element has been named unnilennium. Symbols, mass numbers, relative atomic masses and half-lives are given below for elements 104 to 107.

Atomic number (z)	Symbol	Mass number (A)	Relative atomic mass (A_r)	Half-life (s)
104	Unq	261	261.11	65
105	Unp	262	262.114	34
106	Unh	263	263.118	0.9
107	Uns	264	262.12	0.12

a How many electrons, protons and neutrons are in the atoms ^{261}Unq and ^{264}Uns?

b What is the relation between the half-lives shown for the elements in the table and:
  (i) the mass number?
  (ii) the atomic number?

c ^{261}Unq is thought to undergo alpha decay. Write down a nuclear equation to represent this change. Give the symbol of the new element formed.

d Which element is used as a standard against which all the atomic masses of the other elements are compared?

e Why are the relative atomic masses shown in the table not whole numbers?

4 In some areas of the country the naturally occurring, radioactive gas radon leaks through cracks in the Earth to the surface. Some houses in these areas of the country trap this gas in their foundations. The radioisotope of radon in these cases is radon-222.

a How do you think this gas could be removed from the foundations of the affected houses?

b The radon-222 isotope is formed from the alpha decay of radium-226. Write a nuclear equation to show this change.

c Radon-222 also decays by alpha decay.
  (i) Write an equation to show this change.
  (ii) How could the alpha particles produced from this decay be harmful to the occupants of the houses concerned?

d How could the presence of radiation as a result of radon-222 be detected?

5 Design an automatic system, using a radioactive source, which could be used to detect when an industrial machine had correctly counted out 500 sheets of paper for packing. State which type of radiation you would use and the precautions you would take.

6 The counts per minute (cpm) given by a sample of the beta emitter lead-214 are given in the table below.

Counts per minute	150	113	85	64	52	38	28	23	17
Time (min)	0	8	16	24	32	40	48	56	64

a Plot a graph of cpm against time for lead-214.

b Why do the cpm decrease with time?

c What is the half-life for lead-214 according to these data?

d After what time are there:
  (i) 70 cpm?
  (ii) 50 cpm?

e How long does it take for the cpm to drop from:
  (i) 100 to 50 cpm?
  (ii) 140 to 50 cpm?

7 Explain the following:

a Alpha particles and beta particles can be deflected by an electrical field but gamma rays are unaffected.

b Workers in the nuclear industry wear badges which contain a small piece of photographic film.

c Beta radiation is composed of fast-moving electrons emitted from the nucleus of the unstable atom.

d An alpha source would be more dangerous to you than a gamma source if you inhaled it.

e We are always exposed to some radiation.

8 Smoke alarms are common in houses. They contain a very small amount of the radioactive isotope americium-241. In the alarm, a detector picks up the radiation given off by the isotope. If smoke gets between the sample of americium-241 and the detector, the alarm sounds.

a Write a nuclear equation for the decay of americium-241.

b Why would smoke particles stop the alpha particles from reaching the detector?

c Draw a diagram of how you think smoke alarms function. Remember that they all need a battery; what is it for?

d The half-life of americium-241 is 433 years. Why is it important to use a radioisotope with a long half-life?

18 THE PLANET EARTH

Volcanoes

Sleeper Awakens After 100 Years

This was the headline which announced the eruption of Mount St Helens in Washington State, USA, in May 1980 (Figure 18.1). When it exploded into life it sent huge amounts of volcanic ash for over 1000 km and left a crater which was over 3 km wide.

Volcanoes are formed when **magma**, the molten rock material containing dissolved gases and water beneath the Earth's crust, escapes to the surface through cracks (**fissures**) or holes (**vents**) in the crust (Figure 18.2a). The magma appears at the surface as **lava** (Figure 18.2b). Lava flow can engulf vast areas of land around the volcano.

QUESTION

a Explain how a volcano is formed.

Figure 18.1 The eruption of Mount St Helens in May 1980

The structure of the Earth

How do we know that the Earth has the structure shown in Figure 18.3? Scientists have used sound waves as well as the waves sent out by earthquakes in their studies. From the information gathered they have concluded that the Earth consists of the following.

- Core — this is made up of very dense molten rock, which consists mainly of iron and nickel, under great pressure. It is approximately 6930 km in diameter.
- Mantle — this surrounds the core and is made up of cooler, less dense rock which contains a lot of iron. It is approximately 2800 km thick.
- Crust — this is the thin, less dense, solid outer layer. The thickness of the crust under the oceans varies from 5 km to 10 km, while that under the continents varies from 25 km to 50 km.

The core has a temperature of about 4300 °C. This temperature drops as you go into the mantle and the temperature just below the crust is only about 1000 °C! These high temperatures are maintained mainly by:

- the inside of the Earth being insulated by the outer layers
- the radioactive isotopes of the elements potassium, thorium and uranium — the nuclei of these isotopes are unstable (Chapter 17, p. 208) and break up, giving out large amounts of energy as they change into smaller nuclei.

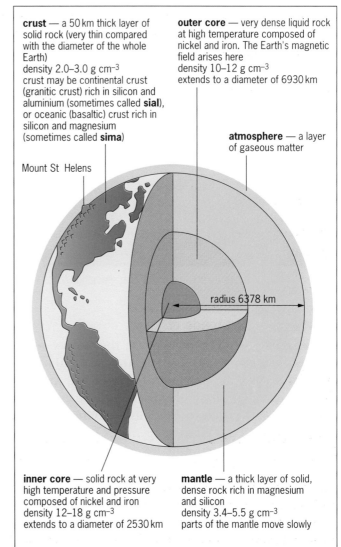

crust — a 50 km thick layer of solid rock (very thin compared with the diameter of the whole Earth)
density 2.0–3.0 g cm⁻³
crust may be continental crust (granitic crust) rich in silicon and aluminium (sometimes called **sial**), or oceanic (basaltic) crust rich in silicon and magnesium (sometimes called **sima**)

Mount St Helens

outer core — very dense liquid rock at high temperature composed of nickel and iron. The Earth's magnetic field arises here
density 10–12 g cm⁻³
extends to a diameter of 6930 km

atmosphere — a layer of gaseous matter

radius 6378 km

inner core — solid rock at very high temperature and pressure composed of nickel and iron
density 12–18 g cm⁻³
extends to a diameter of 2530 km

mantle — a thick layer of solid, dense rock rich in magnesium and silicon
density 3.4–5.5 g cm⁻³
parts of the mantle move slowly

Figure 18.3 The Earth

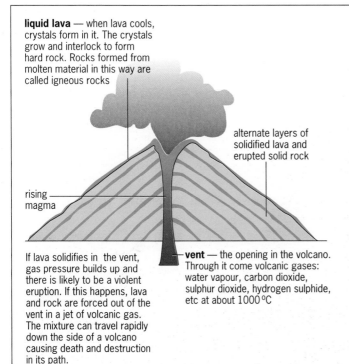

liquid lava — when lava cools, crystals form in it. The crystals grow and interlock to form hard rock. Rocks formed from molten material in this way are called igneous rocks

alternate layers of solidified lava and erupted solid rock

rising magma

If lava solidifies in the vent, gas pressure builds up and there is likely to be a violent eruption. If this happens, lava and rock are forced out of the vent in a jet of volcanic gas. The mixture can travel rapidly down the side of a volcano causing death and destruction in its path.

vent — the opening in the volcano. Through it come volcanic gases: water vapour, carbon dioxide, sulphur dioxide, hydrogen sulphide, etc at about 1000 °C

Figure 18.2 (a, left) A volcano. (**b**, right) Molten lava often flows from a volcano and engulfs land around it

Figure 18.4 A granite erratic moved during glaciation

Figure 18.5 Vesicular lava, otherwise known as basalt

What is the crust made of?

There are many different rocks in the Earth's crust. These different rocks vary in the minerals they contain and in the shape and size of the mineral grains. Geologists have shown that there are three main groups of rocks — igneous, sedimentary and metamorphic.

Igneous rocks are formed when hot magma from the Earth's mantle cools and hardens. Igneous rocks are usually crystalline. There are two main types of igneous rock: intrusive and extrusive.

- Intrusive igneous rocks are formed by crystallisation of the magma underground. Granite is an example of this type of igneous rock (Figure 18.4).
- Extrusive igneous rocks are formed by crystallisation of the magma on the Earth's surface. Basalt is an example of this type of igneous rock (Figure 18.5).

Sedimentary rocks cover approximately 75% of the continents. These are formed when solid particles carried or transported in seas and rivers are deposited. Layers of sediment can pile up for millions of years, and the sediment at the bottom of the pile experiences great pressure; the grains become cemented together, forming the sedimentary rock. Sedimentary rocks have definite layers, or **strata**, associated with them, and you can often see these layers running through the rocks (Figure 18.6a). There is a large variation in their hardness and grain size. Sedimentary rocks often contain fossils (Figure 18.6b).

Limestone is a sedimentary rock which formed beneath the sea. Although it was formed beneath the sea, it is often found well above sea level due to the movement of the Earth's crust. This happens during

Figure 18.6 (**a**, top) Limestone strata. (**b**, bottom) Fossils are often found in limestone

the process of **uplift**. Uplift occurs mainly because of the large-scale lateral forces at work on the Earth's crust, resulting in its crumpling, for example at plate boundaries (Figure 18.12, p. 222).

Limestone is composed mainly of calcium carbonate, which effervesces when it comes into contact with a dilute acid. This property is often used to show the presence of limestone in a rock sample (Chapter 8, p. 98). Sandstone is another example of a sedimentary rock.

Metamorphic rocks are formed when rocks buried deep beneath the Earth's surface are altered by the action of great heat and pressure. Marble is a metamorphic rock and is formed by this type of action on limestone (Figure 18.7, top). Slate is another example of a metamorphic rock (Figure 18.7, bottom), which is formed from mudstone.

QUESTIONS

a Describe the structure of the Earth.
b What is:
  (i) igneous rock?
  (ii) sedimentary rock?
  (iii) metamorphic rock?
Give an example of each type of rock.

Fossils

Fossils are the remains or impressions of living organisms made when animals or plants die in soft silt. When these organisms decay, they leave their impressions in the silt. When the silt becomes rock through the action of temperature and pressure,

Figure 18.7 The Taj Mahal in India and roofing material are made from the metamorphic rocks marble and slate

impressions are left in the rock (Figure 18.6b). In some cases the organism decays and dissolves in the silt, leaving a space. When this happens, certain minerals may seep into the space and take up the shape of the dissolved organism. When the silt turns into rock, through the action of temperature and pressure, a fossil 'mould' is preserved.

As stated earlier in this chapter, fossils are found in sedimentary rocks such as limestone. Sedimentary rocks are layered with the oldest layers (strata) found deeper underground than the younger rock. Geologists have been able to divide time into three **eras**, according to the type of fossil found in the different rock strata (Figure 18.8).

ERAS	PERIODS	EPOCHS	Present	SOME FOSSILS
CENOZOIC The Earth's climate became much colder, resulting in several Ice Ages	QUATERNARY	Holocene 0.01 Pleistocene 2		human skull, mammoth (tooth)
The age of the mammals as well as insects and flowering plants	TERTIARY	Pliocene 7 Miocene 26 Oligocene 38 Eocene 54		snail, bivalve shellfish
Opening of the North Sea MESOZOIC The age of the dinosaurs	CRETACEOUS	Paleocene 65 136		ammonite, lampshell, sea urchin
The great continent of Pangea broke up, forming most of the continents as we know them	JURASSIC	190		lampshell, ammonite, sea urchin, coral
The Earth's climate was generally warm and pleasant	TRIASSIC	225		bivalve shellfish, fish (tooth)
PALEOZOIC Initially, most life was in the sea. Plants appeared on the land in the Silurian era, followed after a few million years by the first amphibians. Towards the end of this period the first reptiles appeared on land	PERMIAN	280		bivalve shellfish, fish (tail), lampshell
	CARBONIFEROUS	355		tree root, coral, amphibian (skull)
	DEVONIAN	395		lampshell, fish
	SILURIAN	440		coral, trilobite, lampshell, graptolite
	ORDOVICAN	500		graptolite, lampshell, trilobite
	CAMBRIAN	570		lampshell, trilobite, trilobite

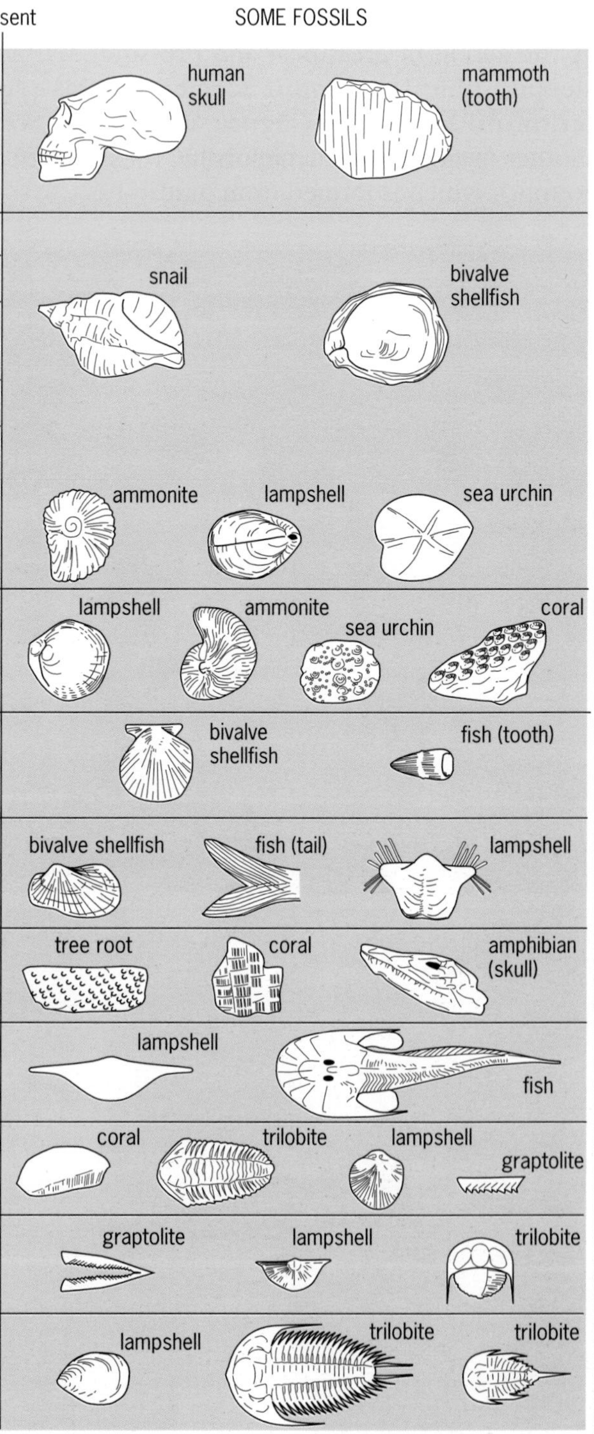

Figure 18.8 Geological time (figures refer to millions of years before present)

- Cenozoic era — this is the most recent era and covers the present to 65 million years ago.
- Mesozoic era — this era covers the time from 65 million years ago to 225 million years ago.
- Palaeozoic — this era covers the time from 225 million years ago to 570 million years ago.

Figure 18.8 shows the way in which the eras have been divided into **periods** and the periods divided further into **epochs**.

Weathering

Weathering is the actual breakdown of exposed rock on the Earth's surface. There are two main ways that rock can be broken down: by chemical means and by physical means.

Chemical means

Rainwater contains dissolved carbon dioxide (Chapter 8, p. 101) as well as other gases such as sulphur dioxide (Chapter 16, p. 201) and nitrogen dioxide (Chapter 11, p. 145). The effect of these substances is to reduce the pH to quite a low value. This means that this now 'acid rain' can dissolve rocks such as limestone quite easily.

$$\text{calcium carbonate} + \text{carbon dioxide} + \text{water} \rightarrow \text{calcium hydrogencarbonate}$$

$$CaCO_3(s) + CO_2(aq) + H_2O(l) \rightarrow Ca(HCO_3)_2(aq)$$

Minerals may also be oxidised. Oxygen from the air can combine with iron silicates to form iron(III) oxide. This leads to a brown stain on the surface of rocks containing this mineral.

Some minerals combine with water molecules and take them into the crystal structure. They become hydrated, which causes expansion, leading to stresses within the rock structure. This causes the rocks to break up. An example of this type of weathering takes place with haematite (Fe_2O_3). With water it forms limonite ($Fe_2O_3.H_2O$).

Physical means

These include the actions of wind, water and temperature. Rainwater enters cracks in rocks. When water freezes, its volume expands and it forces the rock apart. Stresses can also be built up in a rock formation by temperature changes. Minerals within the rock will expand and contract with change in temperature at different rates. In the temperate areas of the world, such as the UK, where alternate freezing and thawing happens a lot, you find that the pieces of rock which break off fall down mountain sides, forming **scree** (Figure 18.9).

Figure 18.9 This scree slope in Gorsdale Scar, Yorkshire was formed by physical weathering

Erosion

Erosion involves the wearing away of rock and its transportation to another place. The photographs in Figure 18.10 show the four main ways by which erosion takes place.

Taking into account the four main methods of erosion, it has been found that the erosion rate for the land area is between 8 cm and 9 cm of depth per 1000 years.

Figure 18.10 (**a**, top left) This deep gorge was formed by the eroding action of the river. (**b**, top right) This stack was formed by wave action. (**c**, bottom left) Glaciers erode the mountain to which they are attached. (**d**, bottom right) Wind erosion caused these formations

Soil

The smaller pieces of rock produced by the different weathering processes are transported by the different methods shown in the photographs in Figure 18.10 and are deposited (deposition) to cover the surface of the Earth.

Humus, which is decayed (or partly decayed) organic material from plants and animals, mixes with the different types of rock material. This mixture of humus and rock particles is called soil. It takes 400 years for 1 cm of soil to form.

The soil provides nutrients and minerals as well as water for the plants that grow in it. Plants remove nitrates (Chapter 15, p. 192) as well as elements such as sodium, potassium, magnesium and copper from the soil.

The pH of a soil depends on the type of rock from which the soil was formed as well as the amount of humus present. For example, soil formed in limestone areas tends to be alkaline, while soil formed in granite areas tends to be acidic. Also, the more humus present, the lower of the pH of the soil.

There are different types of soil.

- **Loams**. These are the ideal soils for agriculture. This type of soil has sufficient clay (20%) to retain nutrients and hold moisture, sufficient sand (40%) to ensure the soil is well aerated and to prevent waterlogging, and silt (40%) which acts like an adhesive, holding the clay and sand together.
- **Sandy soils**. These are drained and aerated. They are, therefore, easy to cultivate as they allow crop roots to penetrate them. However, because they lack humus they are vulnerable to drought. Also, they need large amounts of fertiliser as nutrients drain away quite quickly after rain.
- **Peaty soils**. These grow excellent crops but unfortunately the 'peaty' material can oxidise to carbon dioxide. It also gets blown away quite easily, which means that the level of this type of soil tends to fall by many metres a century.

The rock cycle

The pattern of change we have been discussing, which takes place on the Earth's surface, is known as the **rock cycle**. This concept was first developed by James Hutton. Figure 18.11 shows the full rock cycle. In this cycle the rocks on the upland areas are weathered and the particles are carried away by erosion to form sediments which eventually become sedimentary rocks. These are then brought to the surface by the Earth's movements (uplifting) or they may be heated and compressed to form metamorphic rocks. If these metamorphic rocks are pushed deep below the surface, they will melt to form magma in the mantle. The magma may then be squeezed upwards towards the surface, and igneous rocks are formed. So one type of rock may be recycled to form another type of rock.

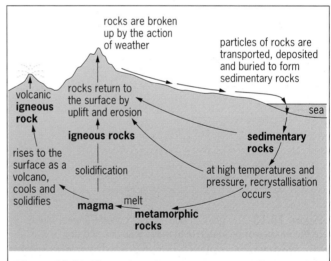

Figure 18.11 The rock cycle

Plate tectonics

Evidence from geologists shows that the Earth's crust is not a continuous structure but is divided into sections called **plates** (Figure 18.12). The majority of these plates have continents sitting on top of them. These plates are actually moving very slowly about the Earth's surface. The driving force behind this movement is thought to be convection currents within the mantle.

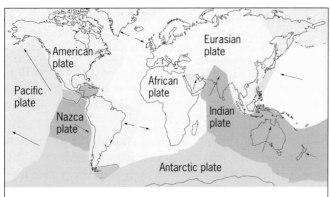

Figure 18.12 The world's plate boundaries. The arrows show the directions in which the plates are moving

Figure 18.13 The giant continent was called Pangea. Much of the interior of this huge land mass was hot and dry

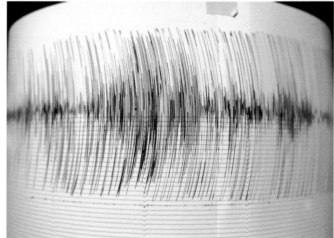

Figure 18.14 (**a**, top) The 1994 earthquake in Los Angeles caused buildings and bridges to collapse. (**b**, bottom) The drum seismogram reading of the Los Angeles earthquake

The study of the way these plates behave is called **plate tectonics**. Geologists think that the continents were originally joined together into one giant continent, which they called Pangea (Figure 18.13). It is thought that it has taken approximately 200 million years for the continents, as we know them today, to drift to the positions they are now in through their plate movement. Evidence for this has been obtained from fossils found in the US. These fossils show that animals and plants found in the US 200 000 million years ago also lived in Europe about that time.

Where plates join, enormous forces are generated. This can, and does, create earthquakes (Figure 18.14) as well as volcanoes and mountains. Earthquakes may be caused in one of two ways:

- where plates are scraping past each other on a 'fault'
- at margins where one plate is descending into the mantle.

Geologists monitor the earthquake activity around the Earth through hundreds of **seismic stations**. These stations are able to detect earthquake waves using an apparatus called a **seismometer** (Figure 18.15). Geologists hope to be able to predict earthquake activity through gathering information in this manner.

There are three types of waves produced by earthquakes.

- **Primary (P) waves**. These are longitudinal waves, produced by pushing and pulling forces which cause the rock to shake backwards and forwards. This type of wave travels at speeds of up to 6 km per second and travels through both solids and liquids.
- **Secondary (S) waves**. These are transverse waves and cause the rock to shake at right angles to the direction of movement of the waves. This type of wave travels at speeds of up to 3 km per second and travels only through solids.

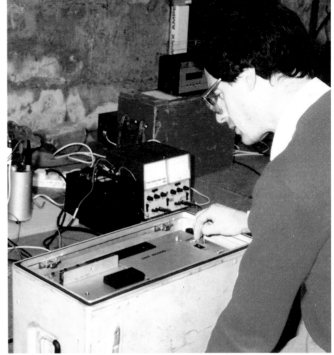

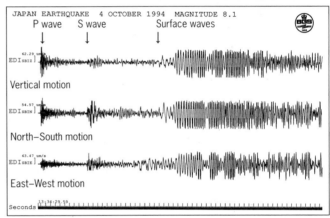

JAPAN EARTHQUAKE   4 OCTOBER 1994   MAGNITUDE 8.1

P wave   S wave   Surface waves

EDI SBIZ   Vertical motion

EDI SBIN   North–South motion

EDI SBIE   East–West motion

Seconds

Figure 18.15 (**a**, top left) A seismometer site. (**b**, top right) Taking a reading from the seismometer. (**c**, bottom left) Analysing the data from the seismometer. (**d**, bottom right) A seismogram showing the three different waves

- **Surface waves**. These waves have long wavelengths and are responsible for the largest land movements and, therefore, cause the most damage. This type of wave travels more slowly than either the P or the S wave.

The actual power of an earthquake is measured using the **Richter scale**. Each unit on this scale represents a ten-fold increase on the previous unit. In theory, the

Table 18.1   The Richter scale

Richter scale unit	Destruction level
2–3	Hardly noticed
3–4	Slightly noticed
4–5	Minor
5–6	Damaging
6–7	Destructive
7–8	Major destruction
8+	Enormously destructive

scale does not have a limit; however, earthquakes are rarely encountered above 8 (Table 18.1).

The recent (January 1995) earthquake in Kobe, Japan, measured 7.2 on the Richter scale. Almost 5300 people died and approximately 27 000 were injured. In terms of devastation, around 110 000 buildings were damaged or destroyed. The cost of rebuilding has been estimated at £65 billion.

The meeting of plates also leads to the formation of high mountain ranges, for example the Himalayas. This is where the crust is at its weakest and most unstable.

> **QUESTIONS**
>
> **a** Explain why earthquakes usually happen at the boundaries between the world's tectonic plates.
> **b** Use your research skills to find the earthquake rating, on the Richter scale, of the Los Angeles earthquake of 1994.

Checklist

After studying Chapter 18 you should know and understand the following terms.

Volcano A hole (vent) or crack (fissure) in the Earth's crust through which molten rock (magma) and hot gases escape to the surface during an eruption.

Magma Molten rock which includes dissolved water and gases.

Lava Molten rock material that surges from a volcanic vent or fissure.

Core The central part of the Earth, composed of iron and nickel. It is found 2900 km below the Earth's surface.

Mantle The part of the Earth between the core and the crust, 40-2900 km below the Earth's surface.

Crust The outermost layer of the Earth.

Igneous rocks Rocks formed when magma cools and solidifies. Igneous rocks are usually crystalline. There are two main types of igneous rocks: intrusive (for example granite) and extrusive (for example basalt).

Sedimentary rocks These rocks cover approximately 75% of the continents. They are formed when solid particles carried or transported in seas and rivers are deposited. Layers of sediment can pile up for millions of years, and the sediment at the bottom of the pile experiences great pressure, causing the grains to become cemented together. Sedimentary rocks, for example limestone, have definite layers or strata associated with them, and you can often see these layers running through the rocks.

Metamorphic rocks These are formed when rocks buried deep beneath the Earth's surface are altered by the action of great heat and pressure. For example, marble is a metamorphic rock and is formed by this type of action on limestone.

Fossils These are traces of prehistoric life which have been preserved by natural processes in rocks.

Geological time The geological division of time into three eras: Cenozoic, Mesozoic and Palaeozoic. Each era is subdivided into periods and these are further subdivided into epochs.

Weathering The action of wind, rain and frost on rock.

Erosion The removal and transportation of material.

Soil A mixture of mineral particles and organic matter or humus.

Rock cycle The cycle of natural rock change in which rocks are uplifted, eroded, transported, deposited and possibly changed into another type of rock and then uplifted to start a new cycle.

Plate tectonics The Earth's crust is not a continuous structure but is divided into sections called plates. The majority of these plates have continents sitting on top of them. These plates are actually moving very slowly about the Earth's surface. The driving force behind this movement is thought to be convection currents within the mantle. Plate tectonics is the study of the way these plates behave.

Earthquake The movement of the Earth's surface caused by plates scraping past each other on a 'fault', or at margins where one plate is descending into the mantle.

Seismometer An instrument used to monitor the magnitude of earthquake waves.

Richter scale A scale used to measure the power of an earthquake.

QUESTIONS

1 If mountains are created where plates are joined, are the plates moving away from each other or towards each other? Explain your answer.

2 Dilute hydrochloric acid can be used to distinguish between pieces of limestone and granite.
  a Which of the two rock samples would give a reaction with the dilute acid?
  b Write a word and balanced chemical equation to represent the reaction taking place in (a).
  c Name two other types of rock which would give a similar reaction with dilute hydrochloric acid.

3 The average size of the crystals in a sample of granite is larger than that found in a sample of basalt.
  a What type of rock are granite and basalt?
  b Explain the difference in the average size of the crystals found in the two samples.
  c Granite is an *intrusive* rock and basalt is an *extrusive* rock. Explain the meaning of the terms in italics.

4 The *core* of the Earth is maintained at a temperature of about 4300 °C. This temperature drops as you go into the *mantle* and the temperature just below the *crust* is only about 1000 °C! One of the reasons for these high temperatures is the *radioactive decay* of the *isotopes* of *elements* such as potassium, thorium and uranium.
  a Explain the meaning of the terms in italics.
  b Why does the radioactive decay of the isotopes mentioned help to maintain the high temperatures?
  c One of the isotopes of uranium present is uranium-238. This undergoes alpha decay.
    (i) Write a nuclear equation for the decay of uranium-238.
    (ii) The product of this decay undergoes further decay by beta particle emission. Write a further nuclear equation to represent this change.

5 a With reference to the rock cycle (Figure 18.11), describe how magma undergoes a change to a metamorphic rock.
  b Describe the differences in the driving forces behind the water cycle and the rock cycle.

6 Plate tectonics is the study of the movement of the Earth's plates.
  a What are tectonic plates?
  b Why are these plates able to move?
  c Explain, as fully as you can, one way the movement of these plates can cause:
    (i) earthquakes to happen
    (ii) mountains to be formed.

7 A seismometer is an instrument used to monitor the magnitude of earthquake waves.
  a Describe how a seismometer works.
  b There are three types of waves produced by earthquakes.
    (i) Name the three types of waves.
    (ii) Which of these waves would be able to travel through the Earth's mantle? Explain your answer.
  c The waves produced are either *longitudinal* or *transverse*. Describe the meaning of these terms as applied to waves.

8 The diagram below shows a volcano.

Copy the diagram into your notes and label the following:
  a (i) a vent
    (ii) lava
    (iii) magma flow
    (iv) volcanic ash.
  b Also label the place where:
    (i) basalt is likely to be found
    (ii) granite is likely to be found
    (iii) weathering and erosion will take place
    (iv) uplift occurs.
  c Explain how a volcano is formed.

GCSE Exam
Questions

GCSE EXAM QUESTIONS

1 Everyday experiences can often be explained using scientific ideas about the way particles are arranged in solids, liquids and gases.

a Use boxes like the ones below to show how the particles are arranged in a typical solid, liquid and gas (in each case, use the symbol x to represent each particle):

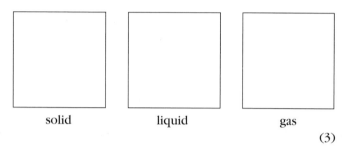

| solid | liquid | gas |

(3)

b When a jar of coffee is opened, people in all parts of the room soon notice the smell. Use ideas about particles to explain how this happens. (2)

c Salt soon disappears when stirred in water, but all parts of the water then taste salty. (2)

d When a saucepan of water is heated, the temperature rises steadily at first, then remains constant while water boils away. If the pan is left boiling for some time, droplets of water form on the windows of the room. Use ideas about particles and about energy to explain these effects.

  (i) The temperature remains constant while the water boils. (2)

  (ii) Droplets of water form on the windows. (2)

(MEG, 1995)

2 The diagram shows some changes of state which are possible.

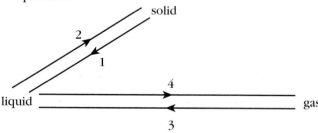

a Copy and finish the table by writing the name of each change of state. The first one has been done for you. (3)

	Change of state
1	melting
2	
3	
4	

b The diagram shows the arrangement of particles in a solid, a liquid and a gas.

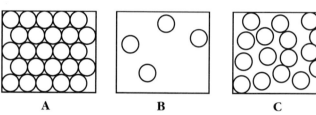

Identify **A**, **B** and **C** as either a solid, a liquid or a gas. (3)

(MEG, 1995)

3 **a** Use the processes below to answer the questions that follow.

chromatography   distillation
condensation   evaporation
crystallization   filtration

For these questions no process should be used more than once.

Which process:

  (i) can be used to remove leaves from swimming pools? (1)

  (ii) makes windows steam up in cold weather? (1)

  (iii) is used to separate the different colours in black ink? (1)

  (iv) causes loss of water from lakes during hot weather? (1)

  (v) is used to separate the fractions from crude oil? (1)

b Sodium chloride can be separated from rock salt by first adding water to dissolve the salt. Give **two** different ways by which you could make the salt dissolve as fast as possible.

(1 mark each)
(SEG, 1991)

4 a Here is a list of materials:

metals   glass   ceramics   fibres   plastics

Use this list to classify each of the materials **A**, **B** and **C**.

(i) **A** is a good conductor of heat and electricity. It can be beaten into shape and has a high melting point.

(ii) **B** is transparent, hard and brittle. It does not conduct electricity. **B** melts, without decomposing, at a high temperature.

(iii) **C** is very hard and opaque. It cannot be melted by simple means. (3)

b Copper is often used for piping in central heating systems.
Suggest **two** properties of copper which make it very suitable for this purpose. (2)

c Nylon and other plastics are gradually replacing copper for central heating pipes.

(i) Suggest **one** advantage of using plastics rather than copper for central heating pipes.

Central heating systems normally include radiators for warming the rooms. The hot water in the system passes through the radiators. Plastics are not suitable for making radiators.

(ii) Suggest **one** reason why plastics are not suitable for this use. (2)
(ULEAC, 1995)

5 Cher wants to separate the liquid from the dye in black ink. She uses this apparatus to distil the liquid.

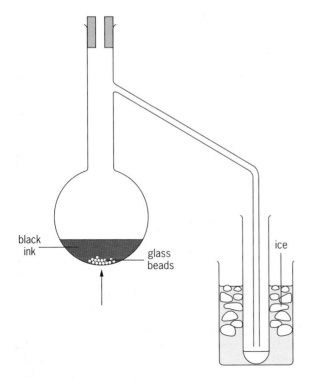

black ink
glass beads
ice

a Copy out the diagram and add to the diagram a thermometer, in the correct position, to measure the boiling point of the liquid which distilled over first. (1)

b What is the reason for putting ice and water in the beaker around the test-tube? (1)

c The black dye in the ink was thought to have a boiling point from this list.

58°C   93°C   150°C

Cher collected the first liquid in the test-tube and found that its boiling point was 100°C. The liquid was colourless.
State, with a reason, which of the three temperatures could be the boiling point of the dye in the ink. (2)

d Cher tested another black ink to see if more than one colour was present in it.
She made chromatograms by placing drops of the liquids to be tested on a pencil line drawn on filter (chromatography) paper.
She stood the paper vertically in a suitable solvent for some hours.
Figure 1 shows the chromatograms of some known dyes.
Figure 2 shows the chromatogram of the black ink.

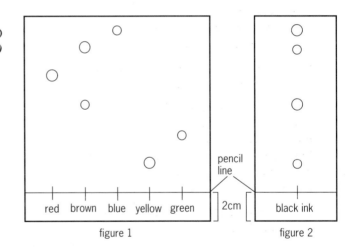

red   brown   blue   yellow   green   2cm   black ink

pencil line

figure 1   figure 2

(i) Which dyes are present in the black ink? (2)
(ii) Suggest why the starting line is drawn in pencil. (2)
(MEG, 1995)

6 Copy and complete each sentence by using a word from the following list.
(Any word may be used more than once.)

convecting   evaporating
condensing   filtering
conducting   melting
dissolving   precipitating

a Glass is made by _____ a mixture of silicon dioxide and metal oxides and then cooling it. (1)

b Sand and gravel beds remove fine particles of sewage by _____ them from water. (1)

c The Dead Sea has a high salt content which is increased by the _____ water from the salt solution. (1)

d The first step in separating salt from a mixture of sand and salt is carried out by _____ the salt in water. (1)

e Clouds are formed by water vapour in the air _____ to form droplets of water. (1)

(MEG, 1995)

7

I	II		GROUPS		III	IV	V	VI	VII	O

(Periodic table grid showing: Period 1 with L in group O; Period 2: Q in group I, E G J Z M in groups III–VII; Period 3: R D; Period 4: T; Period 5 empty)

This question is about the periodic table of the elements. The letters shown in the table are **not** the symbols of the elements. You will need to use these in some of your answers.

Each letter may be used once, more than once or not at all.

a How many protons does an atom of element L contain? (1)

b Which element shown forms ions with a single negative charge? (1)

c Which metallic element is more reactive than R? (1)

d Give the formula of the compound formed between elements D and M. (1)

e Which element has its electrons arranged in four shells? (1)

f How many neutrons are there in an atom of element M (mass number 19)? (1)

g State one *chemical* property that the **oxides** of elements Q, R and T have in common. (1)

h The diagram below shows a molecule with one atom unlabelled.

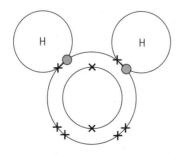

Using a letter from the periodic table above, identify the unlabelled atom. (1)

i The table below gives data about three additional elements from one group of the periodic table.

Physical state at room temperature	Colour
Gas	Yellow/green
Liquid	Red/brown
Solid	Grey/black

What group are these three elements in? (1)

j What is the electron arrangement of the atom J? (1)

(ULEAC, 1991)

8 a The table shows the electronic structure of the atoms of sodium and chlorine.

Element	Number of electrons in one atom	Electronic structure
Sodium (Na)	11	2, 8, 1
Chlorine (Cl)	17	2, 8, 7

(i) Copy and complete the diagram below to show what happens to the electrons when these elements combine to form sodium chloride. (2)

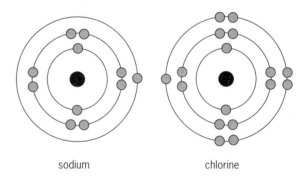

sodium   chlorine

(ii) How many protons does the nucleus of chlorine contain? (1)

b An element X forms an ion X^{2+}. This ion contains 54 electrons. For this element give:

(i) its name. (1)
(ii) its group in the periodic table. (1)
Give a reason for your answer. (1)
(iii) the formula of the compound it forms with chlorine. (1)

c Sodium chloride is an ionic compound. Methane (CH_4) is a covalent compound. Copy and complete the following table to compare the properties of the two compounds. (3)

Property	Sodium chloride	Methane
Melting point	High	
Electrical conductivity when solid	Does not conduct	
Electrical conductivity when liquid		Does not conduct

(MEG, 1992)

9 a Atoms are the basic building blocks of elements. The diagrams below show a variety of ways in which two different atoms can combine.

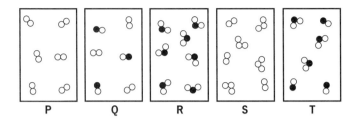

P   Q   R   S   T

Copy and complete the table below by choosing the code letter for the diagram which best matches the description given in the first column. (2)

The atoms are arranged to make:	Code letter
one compound only	
a mixture of an element and a compound	

b The table below gives some of the properties of four elements. The letters are **not** the chemical symbols of the elements.

Element	A	B	C	D
Boiling point in °C	−246	59	1317	1322
Electrical conductivity when solid	None	None	Good	Good
when molten	None	None	Good	Good
when put in water	None	Fair	Reacts	Reacts

Which **two** are most likely to be metals? Give **two** reasons for your choice. (3)

c The diagram below shows part of the periodic table with information about two elements enlarged.

(i) Sodium is in group 1 of the table. What can you say about the rest of the elements in this group? (1)

(ii) [A] How many protons does an atom of sodium have? (1)
[B] How many electrons does an atom of chlorine have? (1)
[C] As well as electrons and protons, most atoms contain a third kind of particle. What is its name? (1)

(iii) When sodium forms compounds it usually does so as a positive ion with one unit of charge.
[A] What change in electron structure occurs when a sodium atom becomes a sodium ion? (1)
[B] What change, if any, occurs in the nucleus when the ion is formed? (1)

(iv) When chlorine forms an ionic compound it gains one electron. What symbol is used to represent the chloride ion formed in this way? (1)

(v) Explain why the formula for the compound formed when sodium and chlorine react is $NaCl$ and **not** $NaCl_2$. (2)

(vi) What does the information given below tell you about the **structure** of each substance?

Chlorine melts at −101 °C.
Sodium chloride melts at 801 °C. (4)
(SEG, 1995)

10 Copy and complete the table by writing the missing numbers in the empty boxes. (6)

Symbol of atom or ion	Atomic number	Mass number	Number of protons	Number of neutrons	Electron arrangement
C	6	12	.....	6	2, 4
Al	13	27	13	14	.....
H	1	.....	1	0	1
Na⁺	.....	23	11	12	2, 8
F⁻	9	19	9	.....	.....

(NEAB, 1994)

1 H 1.00																	1 (H) 1.00	2 He 4.00
3 Li 6.94	4 Be 9.01											5 B 10.8	6 C 12.0	7 N 14.0	8 O 16.0	9 F 19.0		10 Ne 20.2
	12 Mg 24.3											13 Al 27.0	14 Si 28.1	15 P 31.0	16 Si 32.0		18 Ar 39.9	
19 K 39.1	20 Ca 40.1	21 Sc 45.0	22 Ti 47.9	23 V 51.0	24 Cr 52.0	25 Mn 54.9	26 Fe 55.9	27 Co 58.9	28 Ni 58.7	29 Cu 63.5	30 Zn 65.4	31 Ga 69.7	32 Ge 72.6	33 As 74.9	34 Se 79.0	35 Br 79.9	36 Kr 83.8	
37 Rb 87.5	38 Sr 87.6	39 Y 88.9	40 Zr 91.2	41 Ho 92.9	42 Mo 96.0	43 Tc 99	44 Ru 101	45 Rh 103	46 Pd 107	47 Ag 108	48 Cd 112	49 In 115	50 Sn 119	51 Sb 122	52 Te 128	53 I 127	54 Xe 131	
55 Cs 133	56 Ba 137	57-71 La 139	72 Hf 178	73 Ts 181	74 W 184	75 Re 186	76 Os 190	77 Ir 192	78 Pt 195	79 Au 197	80 Hg 201	81 Tl 204	82 Po 207	83 Bi 209	84 Po 210	85 Al 211	86 Rn 222	

11
Na
Sodium
23

17
Cl
Chlorine
35·5

11 **a** Copy and complete the table by correctly filling in the blank spaces. (7)

Substance	Formula	Bonding diagram	Type of bonding
Sodium chloride		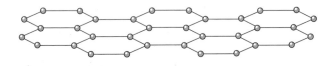	
Methane	CH_4		Covalent
Carbon dioxide	CO_2		
	MgO		

b The diagrams **A** and **B** below represent two different forms of carbon.

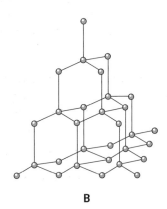

A

B

(i) Name these two forms of carbon. (2)

(ii) What type of bonding and structure is shown by both forms of carbon? (2)

(iii) Copy and complete the following table. Sodium is given as an example. (3)

Substance	Electrical conductivity	Melting point
Sodium	Good	Low
A		
B		

(SEG, 1990)

12 Sodium is in group 1 of period 3 of the periodic table.

↓ group 1

```
          H                    He
          Li Be      B C N O F Ne
period 3→ Na Mg      Al Si P S Cl Ar
          K Ca                 Br Kr
```

a **(i)** Show, by means of diagrams or otherwise, the arrangement of electrons in an atom of sodium.

(ii) Sodium can be burned in chlorine to form the compound sodium chloride. This is a white solid with very high melting and boiling points. Relate these physical properties of sodium chloride to its structure. (3)

(ULEAC, 1995)

13 The table below gives some information about the elements of the third period of the Periodic Table and their chlorides.

Element	Melting Point (°C)	Boiling Point (°C)	Chloride Formula	Chloride Melting Point (°C)
Sodium	98	883	NaCl	801
Magnesium	649	1107	$MgCl_2$	714
Aluminium	660	2467	$AlCl_3$	190
Silicon	1410	2355	$SiCl_4$	−70
Phosphorus	44	280	PCl_3	−161
Sulphur	119	443	SCl_2	−78
Chlorine	−101	−38	—	—
Argon	−189	−186	No Compound	—

You will find the above data useful in answering the following questions.

a **(i)** Which element has the highest melting point? (1)

(ii) Which element has the lowest boiling point? (1)

b Which element is liquid over
   (i) the smallest temperature range? (1)
   (ii) the largest temperature range? (1)

c **(i)** Which element in this period has allotropes? (1)
   (ii) What is an allotrope? (3)

d **(i)** Explain fully why argon does not form a chloride. (2)
   (ii) Give one use of argon. (1)

e Suggest how the type of bonding present in the chlorides changes across the period. Give a reason for your suggestion. (4)

f Describe how you would
   (i) chemically confirm the presence of chloride in aluminium chloride. Give the observation you would make. (3)
   (ii) distinguish between aluminium chloride and magnesium chloride using sodium hydroxide solution. Give the observations you would make. (7)

(NISEAC, 1993)

14 a Fluorine, chlorine, bromine and iodine are members of Group VII of the Periodic Table.

(i) For these four elements, construct a table using the following headings:

name; relative atomic mass, A_r; molecular formula;

physical state at r.t.p.: number of outer shell electrons.

(ii) Use the ideas of electron sharing and electron transfer to explain why sodium chloride exists as the ions Na^+Cl^- rather than the covalent molecule Na-Cl. (6)

b Why is chlorine used to sterilise drinking water? (1)

c An element X is in Group V of the Periodic Table. Use this information to state

(i) the number of electrons in the outer shell of an atom of X,

(ii) whether the element X is metallic, or non-metallic,

(iii) the formula of a chloride of X. (3)

(ULEAC, 1993)

15 The element carbon is in group IV of the periodic table. Carbon atoms contain 6 protons, 6 neutrons and 6 electrons.

a (i) What is the atomic number of carbon? (1)

(ii) To the nearest whole number, what is its relative atomic mass? (1)

(iii) How many electrons are in the outer layer of a carbon atom? (1)

b Parts of the structures of two different forms of solid carbon are shown below:

diamond   graphite

3-D giant molecular   flat layer giant molecular
lattice   lattice
4 bonds per carbon atom   3 bonds per carbon atom

Use the information about these structures to explain why

(i) diamond is one of the hardest substances known (2)

(ii) graphite is a good lubricating agent (2)

(iii) graphite conducts electricity. (2)

When carbon burns, carbon dioxide is formed. This has a simple molecular structure.

c (i) Draw a diagram to illustrate the bonding in a carbon dioxide molecule. (1)

(ii) Explain how the physical properties of carbon dioxide are related to its structure. (3)

(MEG, 1995)

16 A student carried out a reaction using 0.095g of magnesium powder with 20cm³ (an excess) of dilute hydrochloric acid and 10cm³ of water. She obtained the following results:

Time (s)	5	10	20	30	40	50
Volume of gas collected (cm³)	32	52	78	93	95	95

A second student repeated the experiment but forgot to add the 10cm³ of water.

A third student then repeated the experiment. He remembered to add the water but used 0.095g zinc powder instead of magnesium.

$$Zn + 2HCl \rightarrow ZnCl_2 + H_2$$

Plot a graph of the results obtained by the first student. Label this graph **A**.

Using the same axes, sketch a graph of the sort of results which the second student would have obtained. Label this graph **B**. Explain why graph **B** differs from graph **A**.

Calculate the total volume of hydrogen which the third student would have obtained. Sketch the third student's results on the same axes and label the graph **C**. Explain why graph **C** differs from graph **A**. (Zn=65. One mole of gas occupies 24dm³ at room temperature and atmospheric pressure.) (10)

(ULEAC, 1992)

17 Washing soda crystals (hydrated sodium carbonate) can be represented by the formula $Na_2CO_3.xH_2O$. When heated the crystals break down and the water is lost leaving anhydrous sodium carbonate.

To find the value of x the following steps were taken.

Step 1 A clean **dry** evaporating dish was weighed.

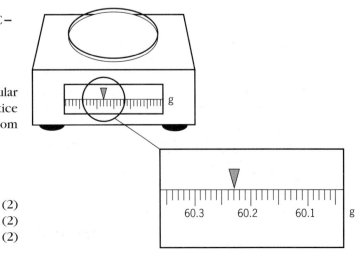

Step 2 Some washing soda was added to the dish. The dish and contents were then reweighed.

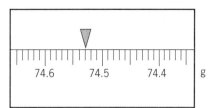

Step 3 The dish was heated with the contents being stirred from time to time. It was then allowed to cool before being reweighed.

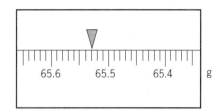

Step 4 The dish and contents were reheated for a few minutes, then allowed to cool before being reweighed.

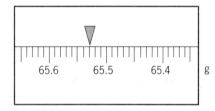

a Copy and complete the results table using the readings from **steps 1**, **2**, **3** and **4** above. (6)

mass of evaporating dish + washing soda	= ....g
mass of dish	= ....g
Therefore mass of washing soda	= ....g
mass of dish + anhydrous sodium carbonate	= ....g
Therefore mass of anhydrous sodium carbonate	= ....g
Therefore mass of water lost	= ....g

b (i) Explain why it is important that the dish in **step 1** is dry. (1)

(ii) Explain why the contents of the dish were stirred in **step 3**. (2)

(iii) Explain why the dish and contents were reheated in **step 4**. (2)

c The value of x can be calculated as follows:

(i) Work out the ratio of:

$$\frac{\text{mass of anhydrous sodium carbonate}}{106} : \frac{\text{mass of water}}{18}$$

(ii) Work out the simplest ratio by dividing each answer in (i) by the smallest of the two.

(iii) Give the value of x. (4)

(SEG, 1992)

18 a Copper(II) oxide was heated in a stream of hydrogen in a test tube. The word equation for the reaction is given below:

copper(II) oxide + hydrogen → copper + water

What chemical process has taken place to convert the copper(II) oxide into copper? (1)

b The following results were obtained in the experiment.

Mass of test tube + copper(II) oxide	=	21.86 g
Mass of the empty test tube	=	20.60 g
Mass of the copper(II) oxide	=	____ g
Mass of test tube + copper produced	=	21.72 g
Mass of the empty test tube	=	20.60 g
Mass of copper produced	=	____ g

Copy and complete the table of results by calculating

(i) the mass of copper(II) oxide (1)

(ii) the mass of copper (1)

c How much oxygen was combined with copper in the copper(II) oxide? (1)

d A sample of copper(II) oxide was found to contain 1.92 g of copper and 0.48 g of oxygen.

(i) Calculate the number of moles of copper in 1.92 g. (1)

(ii) Calculate the number of moles of oxygen in 0.48 g. (1)

(iii) By considering your answers to (i) and (ii), give the simplest formula of the copper(II) oxide. (1)

(MEG, 1992)

19 A packet of washing soda crystals has been left open to the atmosphere for some time. The crystals have formed a white powder which may be represented by the formula $Na_2Co_3.xH_2O$. The value of x may be found by dissolving a known quantity of the powder in water and titrating the solution with acid of known concentration.

In one experiment a student weighed out 1.59 kg of the powder and dissolved it in water. The volume of the solution was made up to exactly 250 cm³. 25 cm³ of this solution was titrated with hydrochloric acid of concentration 0.10 mol/dm³ (mol per litre). 25.6 cm³ of the acid was required for complete reaction. The equation for this reaction is:

$$Na_2CO_3 + 2HCl \rightarrow 2NaCl + H_2O + CO_2$$

a The burette which you are to use for the titration has been left with an unknown liquid in it from a previous experiment. Describe in detail the steps you would take to prepare the burette containing hydrochloric acid for the titration. The hydrochloric acid is supplied in a 2.5 dm³ (litre) bottle. (5)

b (i) Calculate the number of moles of hydrochloric acid used in the titration. (2)

(ii) Calculate the number of moles of sodium carbonate in the $25\,cm^3$ sample of solution which reacted with the hydrochloric acid. (2)

(iii) How many moles of sodium carbonate were in the $250\,cm^3$ of solution? (1)

(iv) Calculate the mass of sodium carbonate in the $250\,cm^3$ of solution. (Relative atomic masses: C = 12, O = 16, Na = 23.) (2)

(v) Using your answer to part **(iv)**, what mass of water was in the original powder? (1)

(vi) How many moles of water were present in the original powder? (Relative atomic masses: H = 1, O = 16.) (2)

(vii) Using your answers to parts **(iii)** and **(iv)**, calculate the value of x. (2)

c Describe an experiment which you would perform to show that water was present in a sample of the white powder from the packet. (5)

(NISEAC, 1995)

---

20 Silicon carbide, SiC, has a diamond-like structure and a very high melting point. It is made by heating pure sand, SiO_2, with carbon at a very high temperature. Carbon monoxide is the other product.

a (i) Construct a balanced equation for this reaction.

(ii) Calculate the mass of sand required to produce 3.00 kg of silicon carbide. (4)

---

b (i) Draw a possible structure for silicon carbide and explain why this compound has a high melting point.

(ii) Suggest two other physical properties that silicon carbide is likely to have. (6)

(ULEAC, 1994)

---

21 The diagram shows three electrolysis cells in series.

Cell **1** contains silver electrodes dipping into aqueous silver nitrate solution.
Cell **2** contains copper electrodes dipping into aqueous copper(II) sulphate solution.
Cell **3** contains chromium electrodes dipping into an aqueous chromium salt solution.

a Aqueous copper(II) sulphate solution contains Cu^{2+}, SO_4^{2-}, H^+ and OH^- ions.
Which ions are present in aqueous silver nitrate? (1)

b One product formed in cell **2** is copper. Write ionic equations for the reactions which take place at the electrodes in cell **2**. (2)

c A current was passed through the circuit. If 0.540 g of silver was deposited at the cathode in cell **1** and 0.130 g of chromium at the cathode in cell **3**,

(i) calculate the mass of copper deposited at the cathode in cell **2**. (2)

(ii) calculate the charge on the chromium ion in cell **3**. (3)

(MEG, 1995)

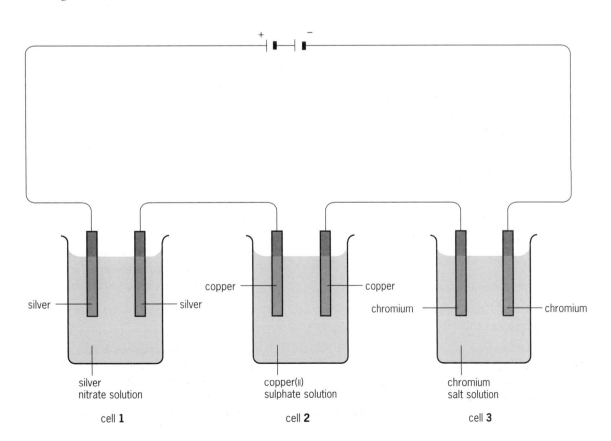

22 Read the following passage before answering the questions which follow.

Aluminium Ore

Aluminium is produced from bauxite ore which contains 50–70% aluminium oxide. Jamaica is the world's third largest producer of bauxite ore which accounts for about 70% of that country's foreign trade. The American and Canadian companies which own the bauxite factories are the largest single employers.

Aluminium is obtained from purified bauxite ore by electrolysis. The main impurities in bauxite ore are iron(III) oxide (20%), silica (5%) and titanium(IV) oxide (2%). Purification involves treatment with concentrated sodium hydroxide at 120°C which dissolves aluminium oxide but leaves the other oxides undissolved. The aluminium oxide is converted in this process into soluble aluminate, $NaAlO_2$. The equation for the reaction is:

$$Al_2O_3 + 2NaOH \rightarrow 2NaAlO_2 + H_2O$$

The aluminate solution produced by the above process is pumped into large precipitator tanks. The sodium aluminate reacts with water and on cooling pure aluminium oxide trihydrate precipitates leaving sodium hydroxide in solution. The aluminium oxide trihydrate is removed by filtration and dehydrated in rotary kilns at about 1000°C.

The aluminium oxide, which melts at 2013°C, is dissolved in molten cryolite and electrolysed to obtain pure aluminium.

The waste from the purification process is in the form of a slurry containing about 20% solids. It is dumped in ponds and called 'red mud' by the local residents. Each year enough mud is produced in Jamaica to cover the equivalent of 700 football pitches to the depth of their goalposts. The mud has a bad smell and a pH of 12–13. Ground water has been polluted, plants and trees near the ponds have died and the sodium hydroxide from the mud reacts with metals such as aluminium in nearby buildings. The 'red mud' also reacts with air to form sodium carbonate and sodium hydrogencarbonate. These solids are blown away by the winds and local people blame this dust for causing illness.

a Give one advantage and one disadvantage to Jamaica of bauxite production.

Advantage

Disadvantage (2)

b Why is the ore purified before export? (2)

c Give the balanced symbol equation for the reaction which takes place in the rotary kilns. (2)

d Why is the mud red? (2)

e Describe an experiment which you would perform in the laboratory to show that the mud contained iron(III) oxide. (Iron(III) oxide is an insoluble base.) (5)

f Give the word equation and the balanced symbol equation for the reaction between aluminium metal and sodium hydroxide solution.

Word equation (1)
Balanced symbol equation (2)

g Name the gas present in air with which the 'red mud' reacts to give sodium carbonate. (1)

h What type of bonding is present in aluminium oxide? Give an explanation for your answer.

Type of bonding (1)
Explanation (2)

(NISEAC, 1994)

23 Zinc has been used for over two thousand years. The major ore is zinc blende, ZnS, which is mined in Canada, the USA and Australia.

a (i) Describe how zinc is extracted from zinc blende. (3)

(ii) Sulphur dioxide is formed during this extraction. Sulphur dioxide is used to preserve food.
Why does it slow down the rate at which food goes 'bad'? (1)

b Pure zinc can be obtained by the electrolysis of aqueous zinc sulphate. This solution is made from impure zinc oxide. Impurities in the solution include iron(II) sulphate, copper(II) sulphate and cobalt(II) sulphate.

(i) Describe how a solution of zinc sulphate could be made from the insoluble compound zinc oxide. (3)

(ii) Manganese(IV) oxide, with a base, changes the aqueous iron(II) sulphate into iron(III) hydroxide.
What is the colour of iron(III) hydroxide and how could it be separated from the solution?
Colour of iron(III) hydroxide
Method of separation (2)

(iii) Powdered zinc is now added to remove the other metals, leaving only zinc sulphate in solution.
Explain how zinc removes the other metals from solution. (2)

(iv) During the electrolysis, zinc and oxygen are formed at the electrodes.
What remains in the solution? (1)

c An important use of zinc is in the manufacture of batteries.

(i) Why are batteries needed? (1)

(ii) Describe how to find the order of reactivity of copper, silver and zinc by measuring the voltage of suitable cells. (3)

(ULEAC, 1993)

24 Minerals, like salt, are very useful as raw materials for making chemicals we need. Salt can be extracted from the ground by pumping water down boreholes into the salt-bearing rock. The salt is then carried to the surface as brine.

Electrolysis is used to turn salt into many useful products. The diagram shows how this is done.

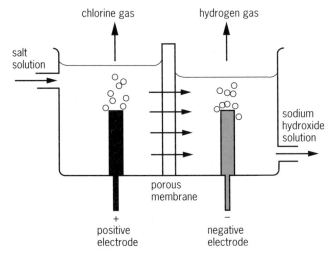

a Explain why hydrogen gas is released at the negative electrode (cathode) and **not** sodium metal. (2)

b Write a balanced equation to show what happens at the positive electrode. (2)

c The electrolysis process uses large amounts of electricity. 1 g of chlorine gas (Cl_2) is released by a charge of 2704 C.
How much electric charge is needed to release a mole of chlorine gas? (2)

(MEG, 1995)

25 Magnesium is made by the electrolysis of a mixture of magnesium chloride ($MgCl_2$) and sodium chloride (NaCl). The diagram shows a cell used to make magnesium.

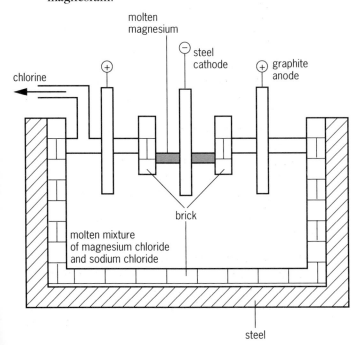

a Write down the formula of **one** ion present in molten magnesium chloride. (1)

b Why does the electrolysis mixture need to be molten? (1)

c Suggest a reason why:

   (i) the anodes are made of graphite rather than steel

   (ii) the magnesium and chlorine, produced during electrolysis, are kept separate (2)

d The table gives the melting points of the materials involved in the electrolysis.

Substance	Melting point (°C)
Magnesium	650
Magnesium chloride	708
Sodium chloride	801
Sodium chloride and magnesium chloride mixture for electrolysis	730

   (i) From the information given in the table suggest why it might **appear** that the electrolysis would be better carried out on pure magnesium chloride.

   (ii) Suggest why a mixture of sodium chloride and magnesium chloride is used. (4)

e Here are three metals in order of their rates of reaction with water/steam.

Quickest   sodium   magnesium   iron   **Slowest**

This table gives some details of the reactions of these metals with water/steam.

	Steam	Hot water	Cold water
Sodium	✔	✔	✔
Magnesium	✔	✔	✗
Iron	✔	✗	✗

Key ✔ = readily reacts
   ✗ = doesn't readily react

Explain how you would find, by experiment, the position of the metal calcium in the order of rate of reaction. Your answer should give brief details of the experiments you would carry out and how you would interpret your results. (8)

(ULEAC, 1995)

26 a The following table shows how salts can be prepared. Copy and complete the table. (3)

Suitable chemicals	Salt formed
Magnesium+	Magnesium sulphate
+ sulphuric acid	Potassium sulphate
+ sulphuric acid	Copper(II) sulphate

b Use the information about solubility of substances in your Data Booklet to help you answer this question. Describe how you would separate a solid mixture of potassium sulphate and lead sulphate. (5)

c What would you see when barium chloride solution is added to magnesium sulphate solution? (2)

(SEG, 1992)

27 A student uses these instructions to make some crystals of zinc sulphate.

Step 1 Pour 50 cm³ of dilute sulphuric acid into a beaker.

Step 2 Add excess zinc powder to react with the 50 cm³ of acid.

Step 3 When all the acid has reacted, filter the solution.

Step 4 Evaporate some of the water from the filtered solution.

Step 5 Allow the solution to crystallise.

Apparatus available for this experiment are shown below.

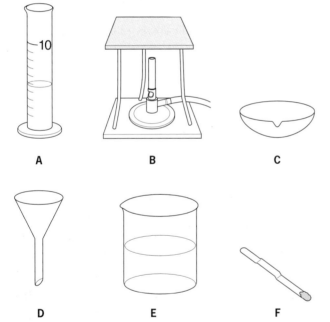

a Choose the apparatus (**A** to **F**) needed for each step and state how the apparatus is used for that step. (You may need more than one piece of apparatus for each of the steps.)

(**i**) Step 1 (3)
(**ii**) Step 2 (4)
(**iii**) Step 3 (3)
(**iv**) Step 4 (3)
(**v**) Step 5 (2)

b Give **two** safety precautions you would take while doing this experiment. (2)

(SEG, 1991)

28 a Sulphuric acid (H_2SO_4) is used at some stage in the production of many of the goods that we buy. In its concentrated form sulphuric acid does not have typical acid properties but it is a useful dehydrating agent. In its dilute form sulphuric acid behaves as a typical acid.

(**i**) Choose and describe **one** reaction in which sulphuric acid acts as a dehydrating agent. (3)

(**ii**) Explain why dilute sulphuric acid behaves as a typical acidic solution but the concentrated acid does **not**. (2)

b Sulphuric acid is a component of acid rain. Discuss the causes and seriousness of the acid rain problem and suggest **two** possible ways to reduce the problem. (4)

(SEG, 1995)

29 a The amount of calcium carbonate in a sample of limestone can be found from the volume of carbon dioxide formed when it reacts with excess hydrochloric acid.

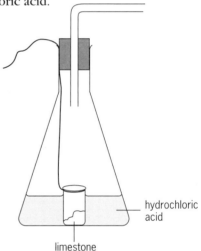

(**i**) Copy and complete the diagram above to show how you could collect and measure the volume of carbon dioxide formed. (3)

(**ii**) Why is the **excess** of hydrochloric acid used? (1)

(**iii**) Write equations for the reaction between calcium carbonate and hydrochloric acid. (2)

b The volumes of carbon dioxide formed from 1 g limestone samples reacting with excess hydrochloric acid are shown below.

% calcium carbonate in limestone sample	Volume carbon dioxide in cm³ formed from limestone sample
100	240
80	192
60	144
40	96
20	48
0	0

(i) Plot the data above and draw the graph line on a grid like that below. (3)

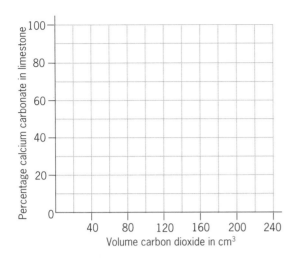

(ii) From your graph what is the percentage of calcium carbonate in a 1 g sample of limestone that gives 220 cm³ of carbon dioxide when it reacts with dilute hydrochloric acid? (1)

c When calcium carbonate (limestone) is heated to a high temperature it decomposes to form calcium oxide (quicklime).
Relative atomic masses: Ca 40, C 12, O 16.

$$CaCO_3 \rightarrow CaO + CO_2$$

(i) Work out the formula mass of
calcium carbonate (1)
calcium oxide (1)

(ii) Calculate how much calcium oxide (quicklime) would be formed from 500 g of calcium carbonate (limestone). (2)

(iii) Why would the amount of quicklime formed by heating 500 g of limestone be less than the mass found by calculation? (1)

(iv) Look at the flow chart below and then deduce the chemical names of **D**, **E**, **F** and **G**. (4)

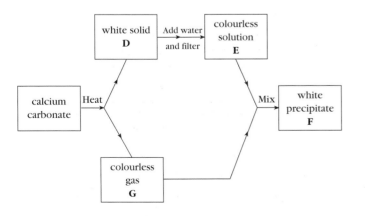

(SEG, 1991)

---

30 Marble is a naturally occurring form of calcium carbonate. When heated, marble breaks down in an endothermic chemical reaction to form quicklime.

a (i) Name substance **P**. (1)
(ii) What is added to **P** to make **R**? (1)
(iii) Write down the formula of **R**. (1)

b What happens in an **endothermic** reaction? (1)

c Apart from marble, name another naturally occurring form of calcium carbonate. (1)

d (i) Calculate the formula mass of calcium carbonate ($CaCO_3$) and calcium oxide (CaO). (Relative atomic masses Ca = 40, C = 12, O = 16) (12)

(ii) Calculate the mass of calcium carbonate needed to make 2.8 g of calcium oxide. (2)

e Give **one** use of calcium oxide (1)

f (i) What substance in the air will react with the calcium hydroxide to form calcium carbonate? (1)

(ii) Explain why these two substances react together. (2)

(SEG, 1993)

---

31 The table below is taken from a magazine called 'Gardening Which'.

| pH of original soil | Application rate (grams per square metre) | | | | | |
| | Limestone | | | Hydrated lime | | |
	Clay	Loam	Sandy	Clay	Loam	Sandy
6.0	420	270	140	315	200	100
5.5	800	540	270	600	380	200
5.0	1200	800	420	900	600	315
4.5	1200	1080	540	900	800	380

Do not apply more than these amounts at one time, little and often is best. Check pH next year and add more if necessary.

It compares the amounts of limestone and hydrated lime needed to give soil a pH value of 6.5.

a Use the information in the table to answer the following questions.

(i) Which of the soil types, clay, loam or sandy, is most acidic? Give a reason for your answer. (2)

(ii) Why might a gardener prefer hydrated lime to limestone for changing the pH value of the soil? (1)

239

(iii) How much limestone per square metre is needed to increase the pH value of a sandy soil from 4.5 to 6.0? (1)

(iv) Suggest a reason why it is recommended that no more than the amounts of hydrated lime given in the table should be used on the soils. (1)

b Another part of the 'Gardening Which' article is given below.

Ground limestone
This is ground or pulverised rock. The more finely it is ground the more quickly it acts.

Suggest a reason why the finer the limestone is ground the more quickly it acts. (1)

(MEG, 1993)

32 Below is information from the British Steel Corporation about the percentage composition of ores, limestones and blast furnace slags.
Use the information to help you answer the questions.

	Iron % (by mass)	Calcium oxide % (by mass)	Silicon oxide % (by mass)	Phosphorus oxide % (by mass)
Local ore	23.0	19.0	14.0	0.7
Foreign ore	64.5	0.4	5.4	0.1
Limestone	0.0	54.0	2.5	0.0
Blast furnace slag	0.3	41.0	36.0	0.0

a (i) What is the advantage of using foreign rather than local ore to make iron? (2)

(ii) Which **compounds** are more plentiful in local ore than in foreign ore? (1)

(iii) Name a compound which must be present in the slag. (1)

b Limestone is mainly calcium carbonate, $CaCO_3$. When it is heated in the furnace the following reaction takes place

$$CaCO_3 \rightarrow CaO + CO_2$$

(i) Calculate the mass of one mole of calcium carbonate. (1)

(ii) Calculate the mass of one mole of calcium oxide. (1)

(iii) How many tonnes of calcium oxide are produced from 50 tonnes of calcium carbonate? (2)

c Give a word equation and a symbol equation for the burning of coke in the blast furnace. (2)

(MEG, 1993)

33 a

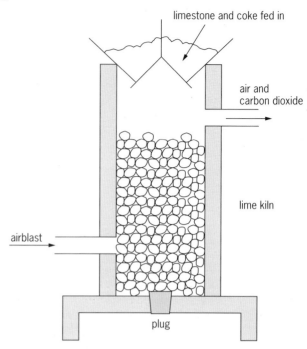

The diagram above shows a lime kiln used for the industrial production of quicklime (calcium oxide, CaO) from limestone (calcium carbonate, $CaCO_3$) in accordance with the equation

$$CaCO_3(s) \rightarrow CaO(s) + CO_2(g)$$

(i) Calculate the relative molecular masses of calcium carbonate and calcium oxide.

$(A_r(C) = 12; A_r(O) = 16; A_r(Ca) = 40.)$ (2)

(ii) Using the above molecular masses and the equation, calculate the mass of limestone required to produce 28 tonnes of calcium oxide.

(1 tonne = 1000 kg) (2)

b What is the purpose of the coke in the charge for the furnace? (1)

c Limestone is obtained from natural deposits by quarrying. If a firm were making out a case for re-opening a limestone quarry in North Wales instead of opening a new quarry elsewhere, a public enquiry would probably be held. Give **two** arguments that could be put forward in favour of such a proposal and **two** arguments against. (4)

d Quicklime (calcium oxide) is used to make slaked lime (calcium hydroxide) by the addition of water. Calcium carbonate **or** calcium hydroxide are sometimes added to certain soils to improve the yield of crops in a process known as *liming* of the soil.

(i) Give a formula equation for the formation of calcium hydroxide from calcium oxide. (1)

(ii) What effect does *liming* have on the soil? (1)

e Oxygen is produced from carbon dioxide in the leaves of plants and trees by a process called photosynthesis.

(i) Give a brief description of the way in which photosynthesis takes place. (2)

(ii) Considerable worry has been expressed in recent times about the increase in the burning of fossil fuels and the large scale deforestation in certain areas of the world such as Brazil. Explain the reason for this worry. (2)

(WJEC, 1993)

34 Over seventy of the known elements are metals. Only about half of them are of any great economic importance. The rest are either found in very small quantities or have no practical use at present. This position may change as further scientific development occurs. For example, before the second world war the only use of uranium was in making yellow glass. Uranium is now in world-wide demand.

Only a few metal elements occur free in nature. Most metals occur in compounds. Metals like gold, silver and copper that occur free have been used for a very long time. Other metals, like tin, known in early times, were those that could be easily extracted from their compounds. Bronze, an alloy of tin and copper, was used so much that it gave its name to the Bronze Age. Iron was more difficult to extract from iron ore so that weapons and tools were not made from iron until later.

a (i) What was the main use of uranium before the second world war?

(ii) Why is uranium in considerable demand today?

(iii) Suggest **two** reasons why about half of the known metal elements are not of economic importance.

(iv) Why are metals like gold and copper found free in nature?

(v) Suggest **two** reasons why iron replaced bronze as the major tool making material. (7)

Metals high in the reactivity series, like aluminium, are obtained from their ores by electrolysis. Aluminium is a very useful element. It is the most abundant metal in the earth's crust. The main ore is bauxite which is found in Africa, Australia and America. Bauxite (Al_2O_3) melts at 2045 °C. In 1886 it was discovered that molten cryolite would dissolve bauxite. The melting point of this mixture is 900 °C. This discovery enabled aluminium to be extracted on a commercial basis.

b (i) Why is it **not** possible to extract reactive metals like aluminium from their ores by simple chemical means?

(ii) What discovery made it possible to produce aluminium on an economic basis?

(iii) Suggest reasons why aluminium extraction plants are usually found near power stations or coalfields and deep water ports.

(iv) Suggest **two** reasons why aluminium is preferred to copper for use in overhead power cables even though copper is the better electrical conductor.

(v) Suggest why aluminium is not obtained by the electrolysis of a solution of an aluminium compound in water. (9)

c Many drink cans are now made from aluminium. There are collective sites for used cans so that the metal can be recycled. What factors must be considered when deciding whether aluminium cans should be recycled? (5)

d Magnesium is extracted from molten magnesium chloride by electrolysis. The process is similar to the extraction of aluminium from bauxite.

In the extraction of magnesium:

(i) At which electrode will magnesium be produced?

(ii) Which gas will be produced in the extraction process?

(iii) Why are special precautions required to remove this gas?

It is dangerous to allow molten magnesium to come into contact with water.

(iv) What gas is produced when magnesium reacts with water?

(v) Why is this gas dangerous? (5)

(ULEAC, 1991)

35 This question is about tin (a group IV element) and some of its compounds.

a Tin occurs naturally as the ore cassiterite, SnO_2. Tin is obtained by reducing this ore with carbon.

Copy and complete the following equation to show the reaction occurring when tin is produced.

$$SnO_2 + \text{.......} \rightarrow \text{.......} + \text{......} CO$$

(2)

b One of the chief uses of tin is for coating steel to make tin-plate. Name an article which might be made from tin-plate and suggest the reason for the plating. (2)

c A mixture of tin and lead is known as solder.

(i) What name is given to a mixture of two or more metals? (1)

(ii) What property of solder makes it useful for joining meals together? (1)

d Tin can exist in three allotropic forms with their transition temperatures shown below.

grey tin $\underset{}{\overset{13°C}{\rightleftharpoons}}$ white tin $\underset{}{\overset{162°C}{\rightleftharpoons}}$ rhombic tin

241

(i) Name the allotrope which is stable at room temperature, 18 °C. (1)

(ii) Grey tin has a structure similar to that of diamond. Suggest how the atoms of tin might be arranged in grey tin. Your answer can be in the form of a diagram if you wish. (2)

e Some information is given in the table below about the two chlorides of tin.

Name	Formula	Melting point (°C)	Boiling point (°C)
Tin(II) chloride	SnCl₂	247	603
Tin(IV) chloride	SnCl₄	−36	114

(i) What is the physical state at room temperature (20 °C) of
tin(II) chloride?
tin(IV) chloride? (2)

(ii) What type of structure would you expect for tin(IV) chloride? (1)

(iii) Tin(II) chloride, $SnCl_2$, can be prepared by treating tin with hydrochloric acid. Write a symbol equation for this reaction. (2)

f Tin(IV) chloride, $SnCl_4$, can be prepared by reacting tin with chlorine.

$$Sn + 2Cl_2 \rightarrow SnCl_4$$

This reaction can be used to recover tin from scrap tin-plate.
When 1000 g of scrap tin-plate was treated with chlorine, 26.1 g of tin(IV) chloride was obtained.
(Relative atomic masses: Cl = 35.5, Sn = 119.)

(i) Calculate the mass of 1 mole of tin(IV) chloride. (1)

(ii) What mass of tin is present in 26.1 g of tin(IV) chloride? (2)

(iii) From your answer to **(ii)**, calculate the percentage of tin in the scrap tin-plate. (1)

(NEAB, 1991)

36 a When magnesium is heated in steam it reacts rapidly forming a white solid and hydrogen gas.

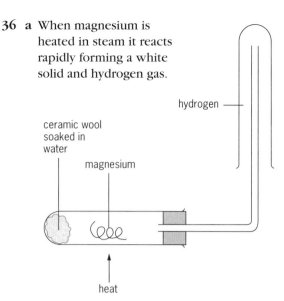

(i) Why can hydrogen be collected as shown in the diagram above? (1)

(ii) How would you show that the gas collected in the test-tube is hydrogen? (2)

(iii) Write a word equation and a balanced chemical equation for the reaction of magnesium with steam.
Word equation (1)
Balanced chemical equation (2)

b Use the reactivity series in the Data Booklet to help you to answer these questions.

(i) Choose **one** metal, from the reactivity series, that will **not** react with steam. (1)

(ii) Choose **one** metal, from the reactivity series, that will **safely** react with dilute sulphuric acid. (1)

(iii) Name the salt formed when your chosen metal in **(ii)** reacts with sulphuric acid. (1)

(iv) In each of the experiments below a piece of metal is placed in a metal salt solution.

Copy and complete the table of observations below the diagrams. (6)

		zinc	zinc	tin	silver	copper
Diagram		tin(II) chloride solution	copper(II) sulphate solution	copper(II) sulphate solution	copper(II) sulphate solution	silver nitrate solution
At start	Colour of metal	grey		silver-coloured	silver-coloured	
	Colour of solution	colourless		blue	blue	colourless
At end	Colour of metal	coated with silver-coloured crystals		coated with a brown solid	silver-coloured	coated with silver-coloured crystals
	Colour of solution	colourless		colourless	blue	

(v) Use these results to place the metals copper, silver, tin and zinc in order of reactivity. (3)

c (i) Give **one** reason why saucepans are sometimes made from aluminium rather than steel, even though steel is a cheaper and a less reactive metal. (1)

(ii) Steel is an alloy of iron. Explain what you understand by this statement. (2)

(iii) Describe a different type of experiment than that in **b (iv)**, by which you could show that aluminium is more reactive than iron. (4)

(SEG, 1990)

37 The table below shows the solubility in water and the poisonous nature of certain barium (Ba) compounds.

Compound	Solubility in water	Poisonous
Barium chloride ($BaCl_2$)	Soluble	Yes
Barium nitrate(......)	Soluble	Yes
Barium carbonate ($BaCO_3$)	Insoluble	Yes
Barium sulphate (......)	Insoluble	No

a Copy and complete the table by writing in the formulae of barium nitrate and barium sulphate. (2)

b (i) Which barium compound would you give to a patient who was to have a stomach and intestine X-ray? (1)

(ii) Give a reason for your answer. (1)

c An industrial worker swallowed a small quantity of aqueous barium hydroxide which is poisonous.

(i) Which soluble magnesium salt would the worker be given as first aid treatment? (1)

(ii) Explain your answer in **(i)**. (2)

d The table below gives information about elements in group II of the periodic table.

Element	Atomic number	Reactivity with water	Reactivity with air
Beryllium	4	No reaction	Beryllium turnings burn when strongly heated
Magnesium	12	Reacts very slowly with cold water	Burns with a blinding bright flame on heating
Calcium	20	Reacts slowly with cold water	Reacts slowly without heating
Strontium	38	Reacts quickly with cold water	Reacts quickly without heating
Barium	56	Reacts very quickly with cold water	Can catch fire in air without heating
Radium	88		

Use the information in the table to help you answer the questions.

d (i) Suggest how radium
  1 reacts with water (1)
  2 reacts with air (1)

(ii) As a result of your answers to **(i)** suggest how radium should be stored. (1)

(iii) Suggest a connection between the atomic number and reactivity of the elements with air. (1)

(iv) Use the information in the table to suggest a use for magnesium. (1)

(MEG, 1992)

38 Aluminium and iron are important metals used in the construction industry. In both cases, considerable amounts of energy are required for the extraction of the metals from their ores, but that required for aluminium is much greater. The methods used for the extraction of the metals from their ores are related to their positions in the reactivity series. It was known over 3000 years ago that iron oxides when heated in a charcoal fire gave a spongy mass of iron that could be beaten into shape and used for tools and weapons, and iron was produced in crude furnaces in the Middle Ages. Aluminium, however, was only discovered about 150 years ago and has been produced commercially for about 70 years. Both aluminium and iron are **recycled** and this is much cheaper than direct extraction of the metal from the ore.

a Give **two** properties of aluminium which make it a more attractive construction material than iron and state **one** industrial use based on these properties. (3)

b Why was aluminium discovered and produced commercially so many years after iron? (2)

c (i) State the name of the aluminium compound contained in the main ore of aluminium. (1)

(ii) Give a **brief** description of the extraction of aluminium from its purified ore. Diagrams and chemical equations are **not** required. (3)

(iii) Why does the process require so much energy? (1)

(iv) What is done to decrease the amount of energy used in this process? (1)

d (i) State what is meant by **recycling** of metals and give **one** example in **each** case of the types of material that are recycled in order to obtain aluminium and iron. (2)

(ii) Why is recycling of aluminium cheaper than the extraction of aluminium from its ore? (1)

(iii) Apart from economic reasons, what other advantage is there in the recycling of metals? (1)

(WJEC, 1990)

243

39 a The water in rivers, lakes and canals may be polluted in three main ways:

(i) domestic waste
(ii) industrial waste
(iii) intensive farming.

For **each** of these three ways name a pollutant and outline how it causes pollution. (Choose a different pollutant in each case). (5)

b Outline the various processes that take place at a water treatment works to ensure that water is made fit to drink. (3)

c A sample of water is tested in the laboratory at a treatment works and found to contain calcium ions and sulphate ions.

(i) Outline the tests, with observations, that you could perform to indicate the presence of each of these ions in the water.
(ii) Is this water fit to drink? Explain your answer.
(5)
(MEG, 1992)

40 a A group of students looked at the reactions of metal sulphate solutions with different metals. They used the same concentration and always added an excess of the metal. Their observations are given in the table below.

Metal sulphate	Colour of sulphate solution	Metals			
		Chromium	Cobalt	Copper	Magnesium
Chromium	Green		No reaction	No reaction	Colourless solution Grey solid
Cobalt	Pink	Green solution Grey solid		No reaction	Colourless solution Grey solid
Copper	Blue	Green solution Grey solid	Pink solution Grey solid		Colourless solution Grey solid
Magnesium	Colourless	No reaction	No reaction	No reaction	

(i) Which metal reacted with all the other metal sulphate solutions? (1)
(ii) Which metal did not react with any of the other metal sulphate solutions? (1)
(iii) Put the metals in their order of reactivity, with the most reactive first. (2)
(iv) Iron is slightly more reactive than cobalt. With which other metal sulphate solution will it react? (1)

b Magnesium was formerly used for making racing cars.

(i) Suggest **one** reason for its use. (1)

For safety reasons, magnesium has now been succeeded in some cases by carbon fibre.

(ii) Why is magnesium considered to be dangerous? (1)

c The properties of some materials are given below.

Material	Property
A	high tensile strength, corrodes easily
B	high tensile strength, brittle
C	tough, low density, good electrical insulator
D	good electrical conductor, low density
E	good electrical conductor, brown colour

Complete the table below by choosing the **code letter** which best describes each substance's properties and give **one** use for that substance.

Substance	Code letter	Use
Aluminium		
Copper		
Glass		
Iron		
Plastic		

(10)
(NISEAC, 1995)

41 Catalytic converters are used in car exhausts to reduce pollution. The diagram below shows the main features of a converter. Use the diagram to answer the questions below.

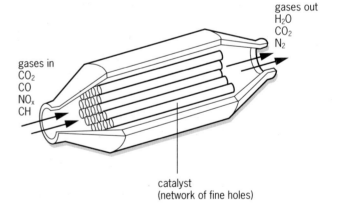

gases out
H_2O
CO_2
N_2

gases in
CO_2
CO
NO_x
CH

catalyst
(network of fine holes)

a Which substance in car exhausts is **not** affected by the catalyst? (1)

b (i) Explain what is meant by a *catalyst*. (2)
(ii) Suggest why the catalyst is a network of fine holes. (2)

c (i) What type of substance is represented by the formula CH? (1)
(ii) Is CH a correct molecular formula? Explain your answer. (2)

d Nitrogen oxides, which are pollutants, are represented by the formula NO_x. How are nitrogen oxides produced in a car exhaust? (2)

e Suggest why lead-free fuel must be used by cars fitted with a catalytic converter. (1)

f Does a catalytic converter stop **all** air pollution from cars? Explain your answer. (2)

(SEG, 1992)

42 This question is mainly concerned with air pollution.

a Increased amounts of carbon dioxide in the atmosphere are thought to be causing an increase in the 'greenhouse effect' which will cause global warming. In a class discussion, a student suggested that carbon dioxide from power stations could be removed from the gases produced when coal or fuel oil burns. The gases would be passed up towers down which a mixture of calcium hydroxide and water falls.

$$Ca(OH)_2(aq)+CO_2(g) \rightarrow CaCO_3(s)+H_2O(l)$$

Calcium hydroxide is made by roasting limestone:

$$CaCO_3(s) \rightarrow CaO(s)+CO_2(g)$$

and then adding water to the residue.

$$CaO(s)+H_2O(l) \rightarrow Ca(OH)_2(s)$$

(i) How are large amounts of carbon dioxide produced by power stations?

(ii) Does the suggested method of absorbing carbon dioxide seem promising? Explain your answer.

(iii) The first of the three equations represents a reaction commonly used in school laboratories. For what is the reaction used? (5)

b Car exhausts contain additional pollutant gases including carbon monoxide, CO, and nitrogen monoxide, NO.

(i) How is carbon monoxide formed in the engine?

(ii) What is the source of the nitrogen in the nitrogen monoxide?

(iii) Which of the gases coming from the exhaust contributes to 'acid rain'?

(iv) What is particularly dangerous about carbon monoxide? (5)

c Some vehicles are fitted with a catalytic converter which changes the exhaust gases into a more acceptable form of emission.

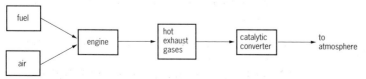

One design of catalytic converter is shown below.

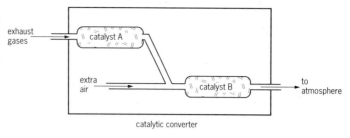

catalytic converter

Each of the catalysts, which only work at a fairly high temperature, brings about one of the following reactions:

$$2CO(g)+O_2(g) \rightarrow 2CO_2(g)$$
$$2NO(g) \rightarrow N_2(g)+O_2(g)$$

(i) What is a catalyst?

(ii) Which of the gases is removed (converted) by catalyst **B**? Give a reason for your answer.

(iii) If the volume of the nitrogen monoxide, NO, was much greater than the volume of carbon monoxide, CO, one feature of the design could be removed. Which part of the converter could be removed? Explain why.

(iv) One catalyst is platinum which is coated on aluminium beads. State **two** reasons for coating beads rather than using platinum on its own.

(v) If a large excess of air were drawn into the engine with the fuel, any carbon monoxide which was formed in the hot engine would burn away. This would avoid the problem of carbon monoxide. Suggest why this is not done. (Hint: many engine parts in contact with the very hot gases are made of iron.) (10)

(ULEAC, 1993)

43 A new method has been proposed to remove the sulphur dioxide contained in the gases produced in power stations. Current methods of removal usually involve the use of large quantities of limestone as a raw material but a major disadvantage is the large amount of low-grade gypsum produced.

The new method uses large quantities of sea water to neutralise the acidic gases. Sulphur dioxide is converted into harmless sulphate salts in the process which is summarised below.

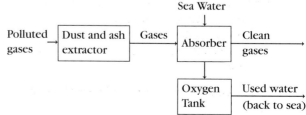

In the absorber sea water removes sulphur dioxide from the polluted gases with 90 per cent efficiency. The used sea water which is now acidic then passes into a tank where oxygen is bubbled through it. This converts potentially harmful sulphite ions into sulphate ions and neutralises the acidity of the effluent water before it is discharged back to the sea. The oxygen bubbles also displace carbon dioxide gas from the water and oxygenate the water.

The advantages of the process are that it requires no bulk chemicals and has low capital and operating costs. However environmentalists are worried that heavy metals from the coal, particularly those with atomic numbers 34 and 80, may be picked up by the water and discharged into the sea.

a Explain why sulphur dioxide is produced as a by-product of power stations. (2)

b Explain why it is important to remove sulphur dioxide from gases produced by power stations. (3)

c Give TWO disadvantages of using limestone to remove sulphur dioxide from the gases produced by power stations. (2)

d Give the chemical formula of

   (i) limestone (1)
   (ii) gypsum (2)

e Give TWO uses of gypsum. (2)

f Give a balanced symbol equation for the reaction of sodium sulphite with oxygen. (2)

g How will the pH of the water be affected by removal of carbon dioxide? Explain your answer. (2)

h Why is it important to oxygenate the water before it is discharged to the sea? (1)

i Name the two heavy metals which environmentalists fear may be picked up from the coal by the sea water. (2)

(NISEAC, 1992)

44 Norway is the main producer of magnesium in Europe. First of all, magnesium oxide is obtained from sea-water by the following scheme.

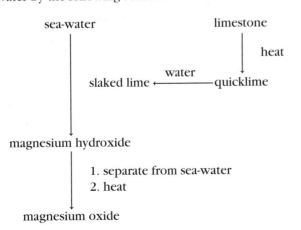

a What are the chemical names for limestone, quicklime and slaked lime?

limestone
quicklime
slaked lime (3)

b Suggest how the precipitate of magnesium hydroxide would be removed from sea-water. (1)

c How many tonnes of magnesium could be obtained from 1000 tonnes of sea-water, given that sea-water contains 0.13% of magnesium, by mass.

. . . . . . tonnes (2)

d When magnesium hydroxide is heated, it decomposes to form magnesium oxide and water. How would you show in the laboratory that water is produced when dry magnesium hydroxide is heated? (2)

e The magnesium oxide obtained in the process is sold to various industries. One of the products made from it is *Milk of Magnesia*, which reacts with hydrochloric acid in the stomach and cures acid indigestion.

   (i) Name the **two** products formed from the reaction of hydrochloric acid with magnesium oxide. (2)
   (ii) What is the name given to the reaction of an acid with a base? (1)

f The magnesium oxide to be used for magnesium manufacture is made to react with chlorine to form magnesium chloride which is then electrolysed.

   (i) Why is *molten* magnesium chloride used in the electrolysis? (1)
   (ii) At which electrode is the magnesium formed? (1)
   (iii) Suggest why Norway is one of the few countries that is able to use electrolysis to produce magnesium cheaply. (1)

(ULEAC, 1993).

45 a Clean, dry, air contains about 79% by volume of nitrogen.

   (i) Name the gases which make up the remaining 21% by volume of clean air.

A sample of polluted air was found to contain both carbon monoxide and sulphur dioxide.

   (ii) Suggest where each of these two pollutants may have come from.
   (iii) Describe a test which you could carry out to confirm the presence of sulphur dioxide in polluted air. (4)

b Respiration and rusting are chemical processes which involve oxygen.

(i) Show clearly how oxygen takes part in each of these reactions.

(ii) One method of rust prevention is known as 'sacrificial protection'.

Describe how this method of rust prevention could be used on an oil pipeline and explain why it works. (6)

(ULEAC, 1993)

46 Ozone, O_3, is a form of oxygen found naturally in the ozone layer of the upper atmosphere. It is formed as part of a natural cycle similar to the nitrogen and carbon cycles. The ozone in the upper atmosphere is essential because it absorbs some of the dangerous ultraviolet radiation from the sun that would be harmful to life on earth. The first step in the formation of ozone is the decomposition of oxygen molecules into atoms by low energy ultraviolet (UV) light:

$$O_2 + \text{low energy UV light} \rightarrow O + O \quad \text{step 1}$$

This is followed by the reaction between oxygen molecules and oxygen atoms which forms ozone molecules:

$$O_2 + O \rightarrow O_3 \quad \text{step 2}$$

Ozone molecules, however, absorb UV light of higher energy and are decomposed in the process:

$$O_3 + \text{high energy UV light} \rightarrow O_2 + O \quad \text{step 3}$$

a Use the information above to copy and complete the diagram below showing the 'ozone cycle' by filling in the boxes. (3)

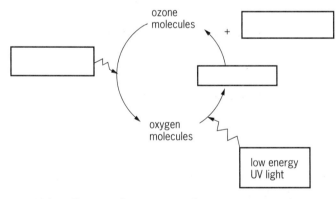

Chlorofluorocarbons, CFCs, destroy ozone in the upper atmosphere. One example of a CFC is the compound CFC-113 with the formula $C_2F_3Cl_3$.

CFC-113 is an inert, synthetic compound used in the electronics industry.

b Copy the table below and fill in the spaces.

Compound	m.p. in °C	b.p. in °C	Physical state at room temperature	Electrical conductivity at room temperature
$C_2F_3Cl_2$	−35	48		

c What characteristic of this CFC allows its vapour to reach the upper atmosphere chemically unchanged? (1)

Once the CFC vapour reaches the upper atmosphere the following reactions occur in order:

Reaction 1: CFC + UV light → CFC fragment + Cl
Reaction 2: $Cl + O_3 \rightarrow ClO + O_2$
Reaction 3: $ClO + O \rightarrow Cl + O_2$

d Describe and explain what has happened in reaction 2 in terms of the processes of oxidation and reduction. (2)

e It has been suggested that one chlorine atom destroys many ozone molecules. Suggest, by reference to reactions 2 and 3, how this could happen. (3)

Some data about CFC-113 are given below.

Bond	Energy needed to break the bond in kJ per mole
C—Cl	+330
C—C	+346
C—F	+450

The structure of the CFC-113 molecule is

The energy of UV light is equivalent to 400 kJ per mole.
Use these data in your answers to the following questions.

f (i) Suggest reasons why the CFC is decomposed by UV light to form chlorine atoms rather than fluorine atoms.

(ii) Suggest a further bond change you would expect to occur in the CFC molecule. (3)

g The formation of ozone in the upper atmosphere can be represented by a single equation.

$$3O_2(g) \rightarrow 2O_3(g) \quad \Delta H = +286 \, \text{kJ mol}^{-1}$$

Draw an energy level diagram for this reaction. (3)

(ULEAC, 1991)

47 This passage is based on a newspaper article. Read the passage and use it to answer the questions.

Toxic Metal Leak Poisons Water Supply

Millions of gallons of rust-coloured water have been pouring out of an old tin mine in Cornwall. The "acid water" could cause the worst pollution problem that Britain has ever known.

Scientists believe that there could be very high levels of heavy metals in the polluted water. For example, arsenic, cadmium, iron, zinc and copper. Therefore shellfish caught on the coast are being tested for metal poisoning. Tests have shown that the levels of cadmium are 100 times higher than the European Union allows.

Some houses get their water from bore-holes or wells. People are being told not to drink this "acid water" but to use bottled water.

There are many underground passages in the mines. These passages go in different directions, so if the flow is stopped in one part it will come out somewhere else. It will cost millions of pounds to solve the problem.

a The water has a pH less than 7. How do we know this from information in the passage? (1)

b (i) The passage says that arsenic, As, is a metal. We do classify it as a metal. [Use pages 6 and 7 of the Data Book to] suggest why you would not classify arsenic as a metal. (1)

(ii) The passage says that there were metals in the water. It is more likely that they were metal ions. Explain why. (1)

(iii) Name two heavy metals which the passage says could cause metal poisoning in shellfish. (1)

c Although water is supplied to households from bore-holes or wells, they have been told to use bottled water. Suggest why. (1)

d Why will it cost millions of pounds to stop the flow of water? (1)

e The level of cadmium set by the European Union should be no higher than 0.005 mg per litre. What level of cadmium is present in the polluted water according to the passage? (1)

f One way of removing harmful metal ions from solution is to add a more reactive, but harmless metal. Use a Data Book to help you to name a metal which could be used to react with the zinc and copper ions in the polluted water. (1)

(NEAB, 1994)

48 The table below shows USSR's foreign trade in chemicals in 1989.

Product	Export (%)	Import (%)
Basic chemicals	22.1	5.0
Agrochemicals	40.4	5.6
Polymers	15.6	24.6
Paints	0.3	6.8
Dyes	1.9	1.8
Photochemicals	0.5	2.0
Detergents	0.9	3.0
Pharmaceuticals	4.4	27.0
Cosmetics	0.3	8.9
Other	14.0	15.2

a (i) Some cosmetics contain soft waxes. Suggest the name of the raw material from which these waxes are obtained. (1)

(ii) The pharmaceutical aspirin could be made from ethanoic acid and a compound which contains a C–OH group. What type of compound is aspirin? (1)

b 'Basic chemicals' are used to manufacture other chemicals. A typical 'basic chemical' is ammonia. The reaction used in the Haber Process to manufacture ammonia is reversible.

$$N_2 + 3H_2 \rightleftharpoons 2NH_3$$

(i) How is the nitrogen obtained for the reaction? (2)

(ii) What are **two** of the conditions of the reaction? (2)

(iii) Name a chemical that is manufactured from ammonia. (1)

(iv) Give another example of a reversible reaction. Explain how changing a reaction condition, such as temperature or pressure, would affect it.

Example
Change in reaction condition (2)

c Photochemicals are those used in photography. A film, which is coated with silver bromide particles, is exposed to light and the image 'captured' on the film. When this film is developed, the developer changes only those silver ions that were exposed to light into silver atoms. Finally, the unreacted silver ions are removed.

(i) Complete the equation for the changing of silver ions into silver atoms.

$$Ag^+ + \ldots \ldots \rightarrow \ldots \ldots$$ (1)

(ii) What type of reagent is the developer? (1)

(iii) Suggest a reason why the developed film should not be exposed to light until the unreacted silver ions have been removed. (1)

(iv) Suggest two factors that would increase the rate of reaction between the developer and the silver ions in the solid silver bromide. (2)

(v) The diagram shows the image on the film and the same film after developing.

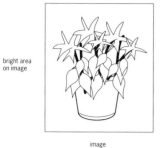

bright area on image

same area dark when film developed

image

image on film after developing

Explain why some areas of the developed film are darker than others. (3)

(ULEAC, 1993)

49 In an experiment to investigate the rate of reaction, a piece of clean magnesium ribbon was allowed to react with excess dilute hydrochloric acid at room temperature (20 °C). The experiment was then repeated with the same mass of magnesium and the same volume of hydrochloric acid but of different concentration as shown in the table below.

Experiment	Mass of Mg	Volume of acid	Concentration
1	0.1 g	10 cm³	1 mol dm⁻³
2	0.1 g	10 cm³	2 mol dm⁻³

The results of the experiment are shown in the graph below.

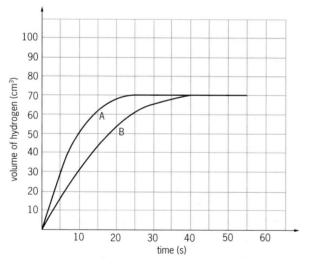

a Hydrogen gas was given off during the reaction and the volume obtained at different times is shown for each experiment by the lines **A** and **B** on the graph.

(i) Explain why the last part of both lines **A** and **B** is horizontal. (1)

(ii) Which line, **A** or **B**, shows the faster reaction? (1)

(iii) Which experiment, 1 or 2, gave line **A**? (1)
Give a reason for your answer. (1)

b Why is it important that the same mass of magnesium and same volume of acid are used in both experiments? (1)

c The equation for the reaction is shown below.

$$Mg(s) + 2HCl(aq) \rightarrow MgCl_2(aq) + H_2(g)$$

(i) How many moles of hydrogen gas would be produced if 0.5 mole of magnesium was used in the reaction? (1)

(ii) What is the mass of 1 mole of hydrogen chloride (HCl) molecules?

$$(A_r(H) = 1; A_r(Cl) = 35.5)$$
(1)

d Explain the following:

(i) Powdered magnesium is used in flares rather than magnesium ribbon. (2)

(ii) Many foodstuffs will keep longer if they are kept in a refrigerator. (2)

(WJEC, 1993)

50 a Describe an experiment you have seen, or carried out, to show how the rate of a reaction varies with temperature. Your account should include references to:

what you would do
what measurements you would make to ensure a fair test
how you would use the results to reach a conclusion. (5)

b The Haber process for the manufacture of ammonia from nitrogen and hydrogen is designed to produce a reasonable yield of ammonia in a short period of time. The reaction is exothermic.

(i) State the pressure, temperature and catalyst used.

(ii) Explain how increasing the temperature can increase the rate of this reaction.

(iii) Suggest why higher pressures and temperatures are not used. (5)

(MEG, 1991)

51 Magnesium reacts with hydrochloric acid to produce a salt and hydrogen.

a (i) You are to design an experiment to determine how the rate of the reaction varies with temperature. Your account should include

a clear account of what you would do
a clear statement of the measurements you would make (6)

(ii) State **two** ways, other than a change of temperature, by which the rate of the reaction could be decreased. (2)

b The equation represents the exothermic reaction between sulphur dioxide and oxygen:

$$2SO_2(g) + O_2(g) \rightleftharpoons 2SO_3(g) \quad \Delta H = -196 \, kJ$$

In industry this reaction is carried out at about 450 °C. Suggest why this temperature is used rather than 900 °C. Your answer should include advantages and disadvantages of both temperatures. (3)

(MEG, 1992)

52 The rate of reaction between dilute hydrochloric acid and marble chips, calcium carbonate, was investigated using the following apparatus:

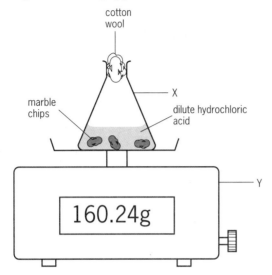

cotton wool

marble chips

X

dilute hydrochloric acid

Y

160.24g

a (i) Name the pieces of apparatus labelled X and Y. (2)
(ii) Why was the cotton wool placed in the neck of apparatus X? (1)
(iii) Name the compound left in the solution at the end of the experiment. (1)
(iv) Why does the mass decrease as the reaction proceeds? (1)

b A student added 50 cm³ of a given hydrochloric acid solution to a large excess of marble chips. The total loss in mass of the apparatus and chemicals was recorded at various intervals of time. The student's results are shown in the table below.

Time (s)	0	10	25	50	100	150	200	250	300
Total loss in mass (g)	0	0.32	0.60	0.83	1.04	1.13	1.19	1.20	1.20

(i) Plot the results on a graph and draw a smooth curve through the points.
(ii) Another student did the experiment again. The student used 50 cm³ of hydrochloric acid of half the concentration used by the first student. On the same graph, sketch another curve to show how the total loss in mass would change with time. (2)

c Why does the reaction eventually stop? (1)

d State two ways the reaction could be made to go faster **without** using a catalyst or changing the concentration of the reagents. (2)

(NEAB, 1994)

53 The hydrocarbons methane, ethane and butane are members of the same homologous series.

a (i) What is meant by hydrocarbons? (1)
(ii) Name a raw material rich in hydrocarbons. (1)

b State **two** general characteristics of any homologous series. (2)

c Write the formula for:
(i) methane (1)
(ii) butane (1)

d Butane exhibits isomerism.
(i) What is meant by isomerism? (1)
(ii) Butane has two isomers. Draw the structural formula of **each**. (2)

e (i) Describe how you would *distinguish* between ethane and ethene. Name the reagent used and give the observations. (3)
(ii) Give **one** large scale use of ethene. (1)

(WJEC, 1990)

54 Many foods contain a lot of starch. A number of chemicals can be produced from starch as shown in the flow chart below.

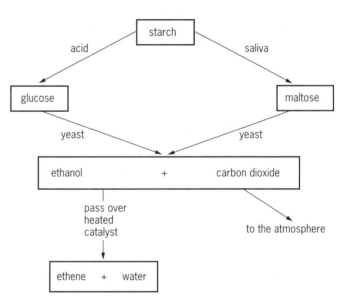

starch

acid

saliva

glucose

maltose

yeast

yeast

ethanol   +   carbon dioxide

pass over heated catalyst

to the atmosphere

ethene   +   water

a (i) Name **two** foods which contain a lot of starch.
(ii) Maltose is a carbohydrate. Which three elements does it contain?
(iii) Give a chemical change shown in the flow chart that is known as fermentation.
(iv) Give **two** major uses of ethanol. (8)

Ethene, C_2H_4, is an important material used in the plastics industry.

b (i) Draw the structure of ethene.
(ii) Ethene is an unsaturated hydrocarbon. What does the term unsaturated mean?

(iii) Starch is not used as a raw material by industry for making ethene.
Which raw material is used to produce ethene?

(iv) Name **two** examples of everyday objects where plastics have replaced traditional materials like wood, metal or ceramics in an everyday object, e.g. a chair. In each case give a reason why plastics are now used. (5)

Part of the carbon cycle is shown in the diagram below.

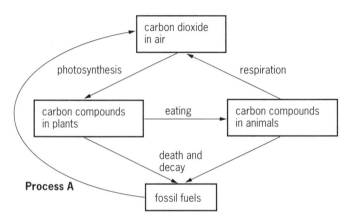

A large amount of carbon dioxide is given off into the atmosphere by process **A**.

c (i) What is the chemical name for process **A**?

(ii) Name two fossil fuels.

(iii) Until recently the percentage of carbon dioxide in the atmosphere was fairly constant. The percentage of carbon dioxide in the atmosphere is now rising. This is one of the causes of the greenhouse effect. Use the carbon cycle diagram to explain how the following changes will affect the amount of carbon dioxide in the atmosphere.
Destruction of the rain forests
Increase in the amount of coal and oil being burnt
Increase in the world population (6)
(ULEAC, 1991)

55 a Ethene, C_2H_4, is the starting material for making plastic carrier bags.

(i) Name the type of chemical change shown above. (1)

(ii) Name the product formed by this reaction. (1)

(iii) Ethene is made by cracking alkanes. Describe a simple test that could show that ethene is present. (2)

b Most plastic carrier bags are difficult to get rid of.

(i) Explain why plastic carrier bags should **not** be thrown away. (2)

(ii) Explain why most plastic bags are **not** recycled.

(iii) Give **one** advantage that a plastic carrier bag has compared to a paper carrier bag. (1)

c This label is found on some plastic carrier bags.

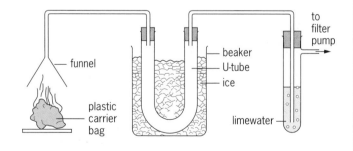

This carrier bag is made from only the elements hydrogen and carbon. When it burns it produces carbon dioxide and water. These are natural substances and will not harm the environment.

(i) What is the meaning of the word *element*? (1)

(ii) What is the name given to compounds that contain only the elements hydrogen and carbon? (1)

(iii) When the plastic carrier bag burns it gives off heat energy. What name is used to describe reactions that give off heat energy? (1)

(iv) The plastic carrier bag could give off a toxic gas when it burns. What is the name of this toxic gas? (1)

d The apparatus below was used to investigate the products formed when a plastic carrier bag burns in air.

(i) First the experiment was done with no plastic carrier bag burning under the funnel. What is the purpose of doing an experiment without the burning carrier bag? (2)

(ii) When the plastic carrier bag burns, some black solid collects on the inside of the funnel. What is the chemical name or symbol of this black solid? (1)

(iii) Water is one of the products formed. Explain how the water is collected by this apparatus. (2)

(iv) Give a simple chemical test to show the presence of water. (2)

(v) Carbon dioxide is the other product formed. Explain how the carbon dioxide is detected in this apparatus. (2)

(SEG, 1991)

56 a 'Cracking' is an important process in the petrochemical industry. In this process, large molecules are split into smaller more useful molecules. This can be shown in the laboratory by using the apparatus below.

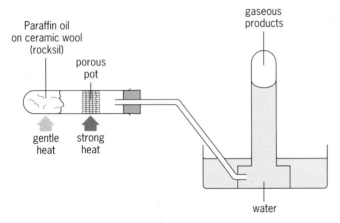

Paraffin oil on ceramic wool (rocksil)

porous pot

gaseous products

gentle heat

strong heat

water

(i) What is the purpose of the porous pot in this experiment? (1)

(ii) The gas formed is found to decolourise bromine water. Give an explanation for this. (2)

b The equation below illustrates the type of reaction that occurs in cracking.

$$C_{10}H_{22} \rightarrow C_8H_{18} + C_2H_4$$
decane   octane   ethene

State and explain **one** use of **either** octane **or** ethene. (2)

c Hydrocarbons such as decane can be shown to contain carbon and hydrogen by heating them with copper(II) oxide and testing for the production of carbon dioxide and water respectively.

(i) What type of reagent is copper(II) oxide in this reaction? (1)

(ii) How could you show by a chemical test that water was produced in this reaction? (2)

d In the Gulf War of 1990–1991, Iraq set fire to all the oil wells in Kuwait. Pictures on television showed these burning fiercely with a red smoky flame. In order to stop an oil well burning, it is necessary (A) to blow the flame out and (B) to 'cap' the well by pouring in mud or concrete to stop further oil escaping. This procedure is in accordance with the *fire triangle* principle used in fire fighting.

(i) State what is meant by the *fire triangle* and explain how the above procedure succeeds in putting out an oil fire. (3)

(ii) What is the cause of the smoke in an oil fire? (1)

e Styrene ($C_6H_5CH = CH_2$) is a product of the petrochemical industry. Styrene can be polymerised in two different ways to make either (A) polystyrenes or (B) cross-linked polystyrenes, which are thermoplastics and thermosets respectively.

(i) State the *difference* between a thermoplastic and a thermoset. (1)

(WJEC, 1993)

57 Petroleum (crude oil) is a mixture of several compounds which are separated in a refinery by means of an apparatus as shown below.

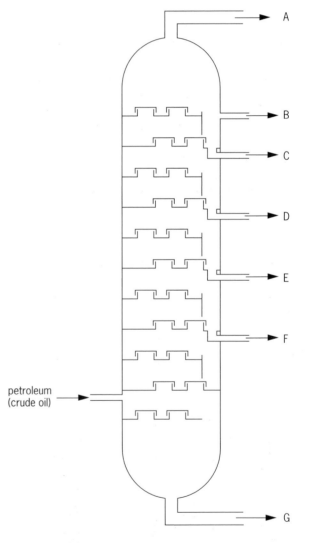

A

B

C

D

E

F

petroleum (crude oil)

G

a What is the name of the apparatus? (1)

b What is the name of the process which is used in separating crude oil? (1)

c On what physical property of the compounds in the mixture does the separation depend? (1)

d Use the letters **A** to **G** to describe where the following could be found.

   (i)   The fraction that represents **gases**.   (1)

   (ii) The fraction with the **largest molecules**.   (1)

   (iii) The fraction that represents **liquids** with the lowest boiling point.   (1)

   (iv) The fraction which is often broken down into lighter fractions using heat and/or a catalyst to give smaller and more useful molecules.   (1)

(WJEC, 1993)

58 a Ethene is an important industrial chemical produced from petroleum (crude oil) fractions by '**cracking**'.

   (i)   Give the structural formula of ethene.   (1)

   (ii) Explain what is meant by '**cracking**', and show how ethene can be formed from another hydrocarbon by this method.   (2)

b Ethene is used in making poly(ethene) (polythene) by a process of **addition polymerisation**. Poly(ethene) is a **thermoplastic** material.

   (i)   State what is meant by **addition polymerisation** and give an equation for the formation of poly(ethene) from ethene.   (2)

   (ii) Describe the difference in properties between thermoplastics and thermosets and give a named example of a thermoset.   (2)

   (iii) Explain, by means of a diagram, how the differences in properties between thermoplastics and thermosets are related to their structures.   (3)

c Ethanol is manufactured from ethene but can also be made by fermentation. Bottles of wine containing ethanol have the stated percentage by volume of ethanol on a label as shown in the picture of a label on a bottle of claret which has 12% by volume of ethanol.

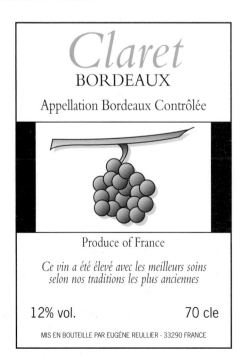

In quality control of wine, the percentage of alcohol can be measured by a technique called gas/liquid chromatography. The amount of ethanol in a given sample is measured by the area under the peak on a graph.

In one experiment, the following results were obtained:

Percentage by volume of ethanol	1	3	5	7	9	12	15
Area under the peak	200	550	920	1280	1535	1905	2210

Note: No special units are used for the peak area.

   (i)   Plot these results on graph paper and draw a smooth curve through them.   (3)

   (ii) What would be the area recorded for a white wine known to contain 10 per cent (10%) by volume of ethanol?   (1)

   (iii) A home-made elderberry wine showed a peak area of 1100. What is the percentage by volume of ethanol in the wine?   (1)

(WJEC, 1990)

59 *Polymers and plastics*

Polymers and plastics are made by joining together hundreds of small molecules called monomers into long chains. Polyethene (polythene), which has an average molecular mass of 140,000, is made from the alkene, ethene, obtained from the thermal cracking of petroleum fractions. Polythene is a very long chain alkane and its many uses depend on its lack of reactivity. Polypropene (polypropylene) and PVC are two other common polymers made from petroleum fractions.

Their lack of reactivity makes polymers and plastics very difficult to recycle and also to separate from other materials such as metals. In the assembly of the average car 150 kg of plastics are used. This helps to reduce the overall mass of the car and improves fuel consumption but makes recycling of scrap cars more difficult.

The chemical industry is developing new methods of recycling plastics. For example, it is now possible to separate PVC from other plastics using a machine which includes an X-ray source to detect the chlorine present. However, most used plastics are still dumped in rubbish tips although some are burned in large incinerators where the energy produced is used for local heating schemes. The burning of PVC produces hydrogen chloride gas which dissolves in rain water.

a Why are polymers and plastics difficult to recycle?   (1)

b Why is it important to find new methods for recycling polymers and plastics?   (1)

253

c Give one advantage and one disadvantage of burning polymers and plastics in incinerators.

Advantage:
Disadvantage: (2)

d Give the advantage to the motorist of using plastics in cars. (1)

e Give one use of

(i) polythene (1)
(ii) PVC (1)

f Explain the meaning of the term 'thermal cracking'. (2)

g (i) Draw the structural formula of ethene. (2)
(ii) Calculate the relative molecular mass of ethene. (Relative atomic masses: H=1, C=12.) (1)
(iii) Calculate the number of ethene molecules, on average, which join together to form a polythene chain. (2)

h (i) What do the letters PVC stand for? (1)
(ii) Draw the structure of the monomer used to make PVC. (2)
(iii) Draw the structure of the polymer PVC. (2)

(NISEAC, 1993)

60 Alcohol (ethanol) is a product of many fermentation reactions. The molecular formula of ethanol is C_2H_5OH.

a Draw the graphical formula for ethanol. (1)

b When ethanol is heated with an excess of acidified potassium dichromate, it is converted to ethanoic acid.

$$C_2H_5OH \rightarrow CH_3COOH$$
ethanol   ethanoic acid

What type of chemical reaction is this? (1)

c Some synthetic flavourings are made by reacting an alcohol with a carboxylic acid.

alcohol + carboxylic acid $\rightarrow$ ester + water

What other substance and conditions are needed to carry out this reaction? (2)

d The structure of the ester propyl-butanoate is shown here.

```
    H   H   H   O       H   H   H
    |   |   |   ||      |   |   |
H - C - C - C - C - O - C - C - C - H
    |   |   |           |   |   |
    H   H   H           H   H   H
```

Draw the graphical formula of the carboxylic acid from which this ester is made. (1)

e Monosodium glutamate is often used as a food additive. State **one** advantage and **one** disadvantage of adding monosodium glutamate to food. (2)

(MEG, 1995)

61 In 1986, several people died in Italy as a result of drinking wine which contained methanol. The methanol had been added to the wine to increase its alcohol content. Samples of the wine were reported to contain as much as 3% methanol, by volume. A level of 0.03% is the maximum allowed by law.

a (i) What name is given to the process by which grape-juice is turned into wine? (1)
(ii) At what temperature is this process best carried out? (1)
(iii) What is added to grape-juice for the process to occur? (1)

b Calculate how much methanol would be legally allowed in a litre bottle of wine.

. . . . . . cm³ (2)

c The chemicals present in the wine, and some of their properties, are listed below.

substance	melting point/°C	boiling point/°C	solubility in water
methanol	–94	65	mixes with water
ethanol	–117	79	mixes with water
water	0	100	—

(i) Draw a labelled diagram of the apparatus you would use, in the laboratory, to obtain pure methanol from the wine. (4)
(ii) How would you know whether the methanol obtained was pure? (2)

d Methanol and ethanol are members of a homologous series.

(i) Draw the molecular structures of methanol and ethanol. (2)
(ii) Explain what the term *homologous series* means. (2)

e Both methanol and ethanol burn in an excess of air to form the same products.

(i) Name these products. (2)
(ii) Suggest why it is not possible to set fire to the contents of a bottle of wine. (1)

62 A European Economic Community (EEC) statement in 1985 laid down that no drinking water should contain more than 50 mg per litre of nitrate. High levels of nitrate can cause medical problems. Nitrate ions are found in water largely because of the use of fertilisers which contain nitrates.

a Name the **two** elements present in a nitrate ion. (2)

b Drinking water is treated before it is used. However, the treatment does not remove nitrates.

(i) State *briefly* how water is treated to make it fit to drink. (2)

(ii) Suggest a means by which pure water may be obtained from drinking water which contains nitrate. (1)

c In the laboratory, aluminium can be used to test for the presence of nitrates. Explain how this test may be carried out. (3)

d Ammonium nitrate is a commonly used fertiliser.

(i) What element in ammonium nitrate causes it to be used as a fertiliser? (1)

(ii) Name **two** other elements which are commonly present in fertilisers but which are not present in ammonium nitrate. (2)

(ULEAC, 1993)

63 a Biogas can be made from rotting plant and animal substances. Biogas contains about 60% methane, CH_4, and the rest is carbon dioxide.

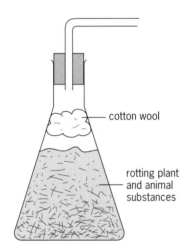

— cotton wool

— rotting plant and animal substances

(i) Copy and complete the diagram of the apparatus above to show how the biogas can be **collected over water**. (3)

(ii) What is the purpose of the cotton wool? (2)

(iii) Why should the apparatus be kept in a warm place? (1)

(iv) **Suggest** one use for the waste substances left in the flask at the end. (1)

(v) Copy and complete the word equation for the burning of biogas in a plentiful supply of air.

biogas + air →

(2)

b Natural gas is a mixture of about 95% methane, 4% other hydrocarbons and the rest is carbon dioxide.

(i) About what percentage of carbon dioxide is in natural gas? (1)

(ii) Name and give the chemical formula of any other hydrocarbon, except methane. (2)

(iii) Give a balanced chemical equation for the burning of methane in air. (2)

(iv) Explain why the reaction in **(iii)** causes atmospheric pollution. (2)

c Both biogas and natural gas contain carbon dioxide.

(i) How would you do a simple chemical test to show that these gases contain carbon dioxide? (2)

(ii) Give **one** advantage and **one** disadvantage of using biogas rather than natural gas as a fuel. (2)

(SEG, 1990)

64 a (Calor gas mainly consists of the **fourth** member of the alkanes called butane.) Methane (CH_4) is the first member of the alkanes and butane is the fourth member.

(i) Give the formula of the **third** member of the alkanes. (1)

(ii) The structural formula of butane is

$$H-\overset{\overset{\displaystyle H}{|}}{\underset{\underset{\displaystyle H}{|}}{C}}-\overset{\overset{\displaystyle H}{|}}{\underset{\underset{\displaystyle H}{|}}{C}}-\overset{\overset{\displaystyle H}{|}}{\underset{\underset{\displaystyle H}{|}}{C}}-\overset{\overset{\displaystyle H}{|}}{\underset{\underset{\displaystyle H}{|}}{C}}-H$$

Draw the structural formula of the isomer of butane. (2)

b Two small cylinders of lighter fuel, one containing propane, the other butane, were used to heat two $500\,cm^3$ samples of water.

The results of the two experiments are given below.

	Using propane	Using butane
Temperature of the water before heating	20°C	20°C
Temperature of the water after heating	66°C	60°C
Mass of the gas cylinder before use	64.5g	64.5g
Mass of the gas cylinder after use	60.0g	62.5g

(i) Which gas produced the greatest rise in temperature? (1)

(ii) 4.5g of propane was used. How much butane was used? (1)

(iii) Which gas produced the greatest rise in temperature per gram of fuel burned? (1)

c When butane is burned the following reaction takes place:

$$2C_4H_{10} + 13O_2 \rightarrow 8CO_2 + 10H_2O$$

(i) How many moles of butane have to be burned to produce 4 moles of carbon dioxide? (1)

(ii) Calculate the mass of one mole of butane (C_4H_{10}). (1)

(iii) Calculate the mass of one mole of carbon dioxide. (1)

(iv) How many grams of carbon dioxide would be produced if 5.8g of butane were burned? (3)

(MEG, 1992)

65 *Chemistry and food*

Natural foodstuffs differ from laboratory chemicals mainly in the complex mixture of substances that they contain; for example 42 chemicals have been identified in orange oil extracted from orange peel, including 12 alcohols, 2 esters and 14 hydrocarbons.

Preservatives are often added to food to make it last longer. There are some 47 preservatives which are permitted to be added to foods. They are mostly simple chemicals, often weak acids, for example vinegar (ethanoic acid). Sometimes metal nitrates are used for preserving cured meat products. This has given some cause for concern since a possible route exists by which nitrates can be converted by bacteria to chemicals which are known to cause cancers in animals.

One of the most widely used preservatives is sulphur dioxide. As a gas, this is very toxic in high concentrations, causing breathing difficulties. When used in foodstuffs, the gas is in solution and is present only in very small quantities.

If food doesn't look right, it doesn't taste right. There are only 4 basic tastes – sweet, sour, salt and bitter – and our appreciation of flavour is mostly due to smell and to a lesser degree visual appearance. Thus, colours are added to enhance natural colours and, in the case of canned foods, such as strawberries and peas, to replace colour lost in the processing. Without added colour, canned strawberries would be dull brown and peas khaki.

There are 46 colours which can be used by the food industry, of which 23 are plant pigments or simple derivatives of them. Artificial colours have great advantages over natural colours and are widely used. They are very rich in hue, which means that very small quantities can be used; they are also more stable than many natural colours. All permitted colours have been thoroughly tested and shown to be safe.

a Is orange oil an element, a mixture or a compound? Give a reason for your choice. (2)

b What **TWO** elements are present in all hydrocarbons? (2)

c Name **ONE** other natural source of esters. (1)

d Why do you think only weak acids such as ethanoic acid, and not strong acids, are used as preservatives? (1)

e Why is the use of nitrates described as giving cause for concern? (1)

f (i) Potassium nitrate is often used in preserving meats. Write a WORD equation to show how potassium nitrate may be formed from potassium carbonate. (1)

(ii) Write a balanced symbol equation for this reaction. (2)

g Write a balanced symbol equation to show what happens when sulphur dioxide dissolves in water. (2)

h Give **ONE** other use for sulphur dioxide not related to food. (1)

i (i) Describe briefly how the colours in a sample of peas may be identified to make sure that only permitted colours are present. (5)

(ii) What name is given to describe such an experiment? (1)

(NISEAC, 1994)

66 **HYDROGEN**

Fuel of the Future

It has been suggested that hydrogen could be used as a fuel instead of the fossil fuels that are used at present. The equation below shows how hydrogen burns in air.

$$2H_2 + O_2 \rightarrow 2H_2O + heat$$

The hydrogen would be made from water using energy obtained from renewable sources such as wind or solar power. The water splitting reaction requires a lot of energy.

a Hydrogen was successfully used as a fuel for a Soviet airliner in 1988.
Why would hydrogen be a good fuel for use in an aeroplane? (2)

b The water splitting reaction is shown in the equation below.

$$2H_2O \longrightarrow 2H_2 + O_2$$

(i) Calculate the energy needed to split the water molecules in the equation into H and O atoms.

$$2H_2O \rightarrow 4H + 2O$$
(2)

(ii) Calculate the energy change when the H and O atoms join to form H_2 and O_2 molecules.

$$4H + 2O \rightarrow 2H_2 + O_2$$
(2)

(iii) Is the overall reaction:

$$2H_2O \rightarrow 2H_2 + O_2$$

exothermic or endothermic?

Use your answers to **(i)** and **(ii)** to explain your choice. (4)

c In the periodic table, hydrogen is placed on its own at the top and in the middle. It is difficult to position because it has properties of metals and non-metals.

(i) Where would you expect hydrogen to be placed in the periodic table on the basis of the arrangement of electrons in hydrogen atoms? (1)

Explain your answer. (1)

(ii) Give **one** way in which hydrogen behaves like a metal. (1)

(iii) Give **one** way in which hydrogen behaves like a non-metal. (1)

(NEAB, 1994)

67 a The rocket engines of the space shuttle shown below use hydrogen as their fuel. The engines are fitted wih two sets of tanks, one filled with liquid hydrogen, the other filled with liquid oxygen.

When the shuttle is launched its engines produce large quantities of water. The engines are very hot and the water is in the form of steam.

(i) Apart from allowing the crew to breathe, suggest and explain **one** reason why the shuttle has to carry its own oxygen. (2)

(ii) Explain why water is produced by the engines. (2)

b Some engineers believe that in the future hydrogen will replace petrol as the fuel for road vehicles.

Hydrogen has already been tested successfully in car engines. However, liquid hydrogen is difficult to store because it boils at –252 degrees Celsius (°C) and has to be kept under high pressure.

(i) Copy and complete the sentence below.

When a substance boils it changes from a

.... to a .... (1)

(ii) Suggest **two** design features for any tank that is to be used to store liquid hydrogen. (2)

(iii) When a car is involved in an accident the fuel may leak out. If this happens with a hydrogen-powered car there is **less** risk of a fire starting than with a petrol-powered car. Suggest an explanation for this. (2)

(SEG, 1995)

68

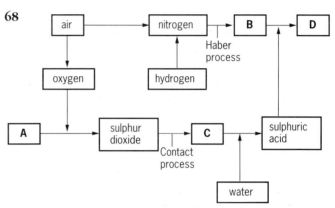

Two industrial processes (**Haber** process and **Contact** process) are shown in the above diagram. **Each** process uses **air** for **one** of the starting materials.

a (i) Give the name of the process by which air is separated into oxygen and nitrogen. (1)

(ii) Give the name of **one** other important product of this process. (1)

b Name the substances which are represented by the letters **A**, **B**, **C** and **D**. (4)

c Name the catalyst used in
(i) the Haber process (1)
(ii) the Contact process (1)

d Give **one** other important use for nitrogen and oxygen and in **each** case explain why air is **not** a suitable substance for that particular use. (4)

(WJEC, 1990)

69 Ammonia is manufactured from nitrogen and hydrogen by the Haber process. The gases are passed under pressure over a hot catalyst.

A simplified flow diagram is given below.

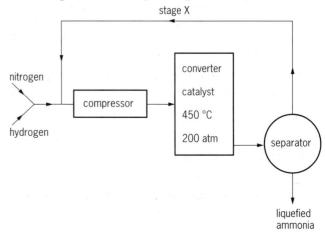

a (i) Copy and complete the equation for the reaction of the nitrogen with the hydrogen.

$$N_2 + \text{........} H_2 \rightleftharpoons \text{........} NH_3$$
(1)

(ii) What is meant by the symbol $\rightleftharpoons$ in the equation? (1)

b Name the source of the nitrogen used in this process. (1)

c Explain why stage X in the flow diagram is important to the economics of the Haber process. (2)

d The graph below shows the percentage yield of ammonia under differing conditions of temperature and pressure.

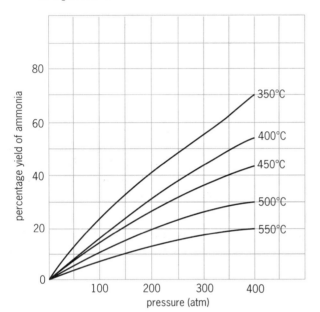

(i) Under the conditions of 200 atmospheres and 450 °C, what is the percentage yield of ammonia? (1)

(ii) What happens to the percentage yield of ammonia if the temperature is increased at constant pressure? (1)

(iii) What happens to the percentage yield of ammonia if the pressure is increased at constant temperature? (1)

(iv) From your answer to **(ii)** suggest why a temperature of 450 °C is used. (2)

e Ammonia can be converted into two compounds which are used as fertilisers; urea, $CO(NH_2)_2$, and ammonium nitrate, NH_4NO_3. Show by a calculation that urea contains more nitrogen per gram than ammonium nitrate.
(Relative atomic masses: H=1, C=12, N=14, O=16)
(3)
(NEAB, 1992)

70 Sulphuric acid is produced in the United Kingdom from sulphur. Sulphur for the process is imported from several countries and is transported in the liquid state.

a Suggest why it is more economical to transport sulphur in the liquid state, rather than as a solid. (1)

b The three main reactions in the production of sulphuric acid may be represented by the equations:

1. $S + O_2 \rightarrow SO_2$
2. $SO_2 + \frac{1}{2}O_2 \rightarrow SO_3$
3. $SO_3 + H_2O \rightarrow H_2SO_4$

Which one of these reactions
(i) requires a catalyst? (1)
(ii) produces a flame? (1)
(iii) is **not** an oxidation reaction? (1)

c Reaction 3, if carried out exactly as shown in the equation, would cause misty fumes to be produced. This would be a problem in the factory. Explain how this reaction is carried out to avoid the problem. (2)

d What mass of sulphuric acid would be produced from 96 tonnes of sulphur?
(Relative atomic masses: H=1, O=16, S=32) (2)

e Name two items of protective clothing a tanker driver should wear when loading and unloading concentrated sulphuric acid. (2)

f In the 19th century the amount of sulphuric acid produced by a country was used as an indication of the economic wealth of that country.
(i) Suggest why it was a useful indicator. (1)
(ii) Suggest another chemical which could be used to give an indication of economic wealth today. (1)
(NEAB, 1993)

71 The Haber process is used to make ammonia (NH_3) which is an important substance.

The equation below shows the reaction in which ammonia is formed.

$$N_2(g) + 3H_2(g) \rightleftharpoons 2NH_3(g) + \text{Heat}$$

The graph opposite shows how temperature and pressure affects how much ammonia is produced in the reaction.

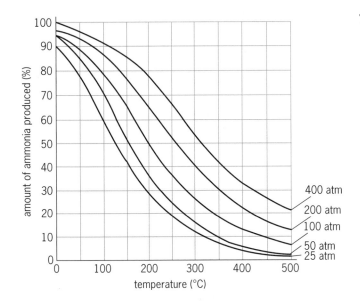

In the industrial process a mixture of nitrogen and hydrogen is passed over iron at a temperature of about 450 °C and 200 atmospheres pressure.

a Use the graph to find the percentage of ammonia present when the temperature and pressure are 450 °C and 200 atmospheres. (2)

b Explain why the nitrogen and hydrogen mixture is passed over iron. (2)

c Explain, as fully as you can, using the graph and your knowledge of the Haber process why 450 °C and 200 atmospheres were chosen as the conditions for this process. (8)

(NEAB, 1994)

72 a A garden centre sells a simple soil testing kit. It consists of a tube containing a small amount of an off-white solid and an instruction card.

Below are the instructions given:

1. Take a sample of soil from 10 cm below the surface of the soil.
2. Quarter fill the tube with the soil sample.
3. Three quarters fill the tube with water.
4. Shake the tube for half a minute and then allow to settle.
5. Compare the colour of the liquid with the chart. (On the instruction card.)

Chart colour	Soil
Bright orange	Very acid
Pale orange	Acid
Pale green	Neutral
Dark green	Alkaline

(i) How was the colour produced in the tube? (3)
(ii) What scale do scientists use to measure acidity and alkalinity? (1)
(iii) Give the name of one substance that a gardener would use to remove the acidity from the soil. (1)
(iv) Gardeners sometimes treat an alkaline soil by adding peat. What does this tell you about the chemical nature of peat? (1)

b The acidity/alkalinity of the soil is important for the growth of plants. Other substances which are called nutrients must also be present in the soil.

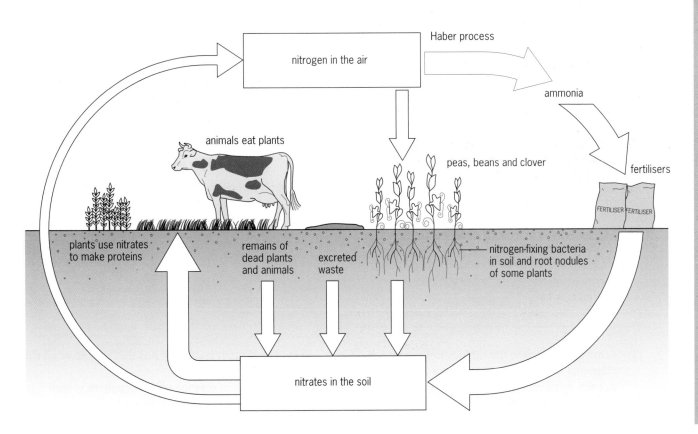

Nitrogen is the most important nutrient for plant growth. Use the diagram of the nitrogen cycle where appropriate to answer the following questions.

(i) Write the word and balanced symbol equations for the Haber Process.
Word equation (1)
Balanced symbol equation (2)

(ii) Give the name and formula of a common fertiliser made from the product of the Haber Process. (2)

(iii) Many farmers use the excreted waste from their animals as manure or slurry to spread on their fields. Give TWO reasons for this practice. (2)

(iv) If the soil of a field is low in nitrogen a farmer will often plant peas, beans or clover. Suggest a reason for this practice. (1)

c Another main nutrient for plant growth is phosphorus which is absorbed as the phosphate ion (PO_4^{3-}). It is applied to the soil either as ammonium phosphate or as calcium phosphate.

(i) Write the chemical formula for calcium phosphate. (1)

(ii) Name the acid used to produce phosphoric acid from phosphate rock. (2)

(iii) Would you expect ammonium phosphate to be soluble or insoluble? Give a reason for your answer. (2)

(NISEAC, 1992)

73 a The major elements needed for plant growth are nitrogen, N, phosphorus, P, and potassium, K. The table below shows some of the different types of fertiliser.

Type of fertiliser	Substances present in the fertiliser	Main elements needed for plant growth present in the fertiliser
Nitram	Ammonium nitrate, NH_4NO_3	N
Nitro-chalk	Ammonium nitrate, NH_4NO_3, chalk, $CaCO_3$	....
Compound fertiliser	Ammonium nitrate, NH_4NO_3, ammonium dihydrogenphosphate, $NH_4H_2PO_4$, potassium chloride, KCl	....

(i) Copy and complete the table, showing which elements needed for plant growth are present in each fertiliser. (2)

(ii) Which one of the above type of fertilisers should be used on an acidic soil? (1)

b Ammonia is needed to make fertilisers. Ammonia is made by the Haber process:

$$N_2 + 3H_2 \rightleftharpoons 2NH_3$$

Sulphuric acid is needed to make some compound fertilisers. It is made using the Contact process:

$$S + O_2 \rightarrow SO_2$$
$$2SO_2 + O_2 \rightleftharpoons 2SO_3$$
$$SO_3 + H_2O \rightarrow H_2SO_4$$

What is meant by the symbol $\rightleftharpoons$ in the equations above? (1)

c The raw materials used to make fertilisers, together with their sources, are shown in the table below. Copy and complete the table. (3)

Name of raw material	Source	Used to make
Air	The atmosphere	Ammonia and sulphuric acid
Water	Reservoirs in Teesdale	Ammonia and sulphuric acid
Natural gas	....	Ammonia
Potassium chloride	Mines in North Yorkshire	Compound fertilisers
Phosphate rock	Imported from overseas	....
....	Imported from overseas	Sulphuric acid

d Large quantities of fertilisers are made on a single site at Billingham. The map shows the location of Billingham. Use the information above and also the map to help you to answer the questions which follow.

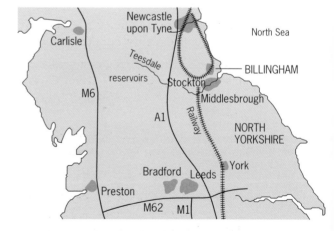

(i) Give one reason why it is useful to have all the factories on one site. (1)

(ii) Give three reasons why Billingham is well situated for making fertilisers. (3)

(NEAB, 1994)

74 Sulphur dioxide is one of the chief sources of acid rain.

a How is acid rain formed from sulphur dioxide in the atmosphere? (3)

b Why does acid rain attack carbonate rocks faster than normal rain of pH 6, even if the concentration of acid in both the normal and acid rain are the same? (1)

c Nitrogen oxides also contribute to acid rain by being converted into nitric acid in the atmosphere. Nitric acid reacts with calcium carbonate to form calcium nitrate, carbon dioxide and water.

$$2HNO_3(aq) + CaCO_3(s) \rightarrow Ca(NO_3)_2(aq) + CO_2(g) + H_2O(l)$$

$$Ca = 40, C = 12, O = 16$$

(i) How many moles of calcium carbonate are there in 25g of calcium carbonate? (2)

(ii) Calculate the mass of carbon dioxide formed when this amount of calcium carbonate reacts with excess nitric acid. (3)

(MEG, 1995)

75 Isotopes are helpful in many areas of chemical research. Some isotopes are radioactive, some are stable. Both types can be used.

Two frequently used stable isotopes are ^2_1H and $^{18}_8O$ which can replace the common isotopes ^1_1H and $^{16}_8O$. A useful radioactive isotope is $^{14}_6C$, which decays by beta-emission. This is used instead of the common isotope $^{12}_6C$.

a (i) Why is there a difference in mass between ^{16}O and ^{18}O atoms? (1)

(ii) What property of isotopes of a particular element makes them suitable for studying chemical reactions? (1)

(iii) What instrument can be used to detect the presence of a radioactive isotope? (1)

(iv) What is a beta particle? (1)

b When making any measurement of radioactivity the background radiation must be taken into account.

(i) State one source of background radiation. (1)

(ii) How is this background radiation allowed for when measuring the radioactivity of a particular source? (1)

c One particular chemical reaction studied by use of radio-isotopes is shown by the equation:

$$CH_3CO_2^- + BrCN \rightarrow CH_3CN + CO_2 + Br^-$$

It was suggested that the CN from BrCN pushed out CO_2 from the $CH_3CO_2^-$ and took its place. Chemists investigated the reaction starting with $CH_3CO_2^-$ containing ^{14}C, and found the following results:

$$CH_3{}^{14}CO_2^- + BrCN \rightarrow CH_3{}^{14}CN + CO_2 + Br^-$$

Explain, with reasons, whether the suggestion was correct or not. (2)

(NEAB, 1992)

76 a The element radium is radioactive. When radium decays it emits three types of radiation: alpha-particles, beta-particles and gamma-rays.

(i) Why are gamma-rays not called gamma-particles? (1)

(ii) When an electric field is placed in the path of the radiation coming from a sample of radium, the gamma-rays are unaffected and continue in a straight line. Copy the diagram below and draw in the path taken by the alpha-particles and the beta-particles, showing clearly which particle is which.

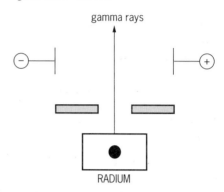

(2)

b In an experiment to measure the half-life of the iodine-128 isotope, the following results were obtained. Background radiation was 12 counts/second and the results have been corrected for this.

Time (min)	0	10	20	30	40	50	60	70	80	90
Count rate (counts per second)	120	90	69	54	42	33	25	19	15	13

(i) What is an isotope? (2)

(ii) Give one source of background radiation. (1)

(iii) State the measured (uncorrected) count rate after 10 minutes. (1)

(iv) Plot a graph of count rate against time. (3)

(v) Use your graph to determine the half-life of the isotope. (1)

c Give two uses of radio-isotopes. (2)

(NEAB, 1993)

77 In an experiment to find the half-life of a radioactive isotope ($^{234}_{91}Pa$) the teacher used a special detector to measure the radioactivity. The detector produced an electrical pulse when a radioactive particle entered it. The pulses were counted by an electrical counter.

a Name a suitable radioactivity detector. (1)

b How many protons, neutrons and electrons does an atom of ($^{234}_{91}Pa$) contain? (3)

The teacher first used the detector to measure the background radiation level. The background count rate was found to be 24 counts per minute.

c Suggest **two** causes of background radioactivity. (2)

d A bottle containing a solution of the radioactive isotope was then brought close to the detector as shown in the diagram below.

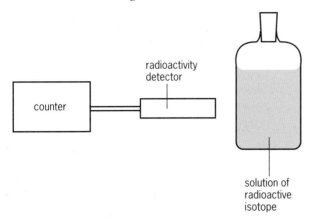

radioactivity detector

counter

solution of radioactive isotope

The teacher was very careful when using the radioactive solution and said that it was very important that none of the solution should get onto the hands.

Explain, as fully as you can, how your hand could be damaged if the radioactive solution was spilt onto it. (4)

e The graph below shows how the amount of radioactivity reaching the detector changes with time. The amount of radioactivity is measured as a count rate in counts per minute. The count rates have been corrected for the background count rate.

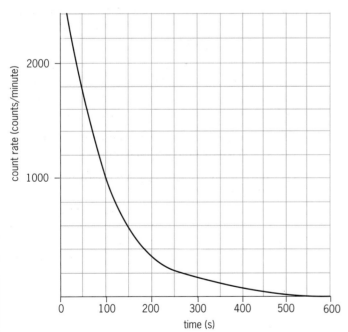

(i) Use the graph to estimate the half-life of the isotope. (*Show your working*) (3)

(ii) Estimate what the actual count rate (including background) would have been after 2 minutes. (2)

f Radioactive substances can be used to find out if there are cancerous growths inside the body without the need for open surgery. The radioactive substance is injected into the patient's body and its movement is followed using a suitable detector. The cancerous growth can be detected because it takes up more of the radioactivity than the healthy tissues around it. The radioactive substance used must emit radiation which can pass through the body to reach the detector and must not make the patient permanently radioactive.

Choose an isotope which might be suitable for this application. Explain, in detail, why you have chosen this isotope. (5)

(NEAB, 1994)

78 Americium-241 is a radioactive isotope. The atomic number of americium-241 is 95.

a (i) Apart from neutrons, what other particles are present in an atom of americium-241?

(ii) How many neutrons are present in one atom of americium-241? (2)

b Americium-242 is another isotope of this element. Compared to an atom of americium-241, how is the structure of an atom of americium-242:

(i) similar?

(ii) different? (3)

(ULEAC, 1995)

79 The diagrams and information below were taken from a brochure explaining how the cliffs at Charmouth in South West England were formed.

The rocks at Charmouth are of great interest because they contain many well preserved fossils.

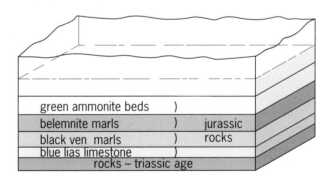

green ammonite beds)
belemnite marls) jurassic
black ven marls) rocks
blue lias limestone)
rocks – triassic age

JURASSIC

During the Jurassic period a warm tropical sea covered large parts of Britain and in it, life flourished. Hundreds (possibly thousands) of feet of rock were laid down over about 60 million years.

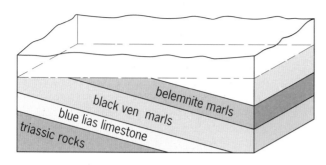

CRETACEOUS/JURASSIC BORDER

At this time the Jurassic rocks became tilted down to the east and may have formed land. The rocks also became cracked in places, forming faults. The rocks were gradually eroded leaving a flat surface.

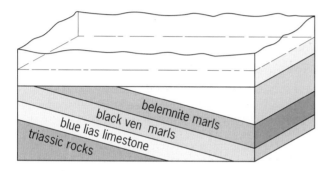

CRETACEOUS

In the Cretaceous period the sea had deepened and advanced over the land and sands were laid down. As the sea continued to deepen, calmer waters developed in which the chalk was formed.

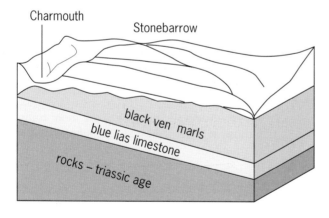

PRESENT DAY CHARMOUTH

In the Tertiary period, the rocks were once again above sea level and became eroded. The chalk was completely removed leaving only a flinty soil while the river Char exploited weakness in rocks, a fault, to cut its valley. Eventually, the landscape we see today evolved.

a (i) The rocks of the Charmouth cliffs are all sedimentary. Give **one** piece of information from the text which would strongly suggest that the rocks are sedimentary. (1)

(ii) Explain your answer. (1)

b (i) Place the rocks formed in the Jurassic period in order of their age with the oldest rock type first. (1)

(ii) Explain why you have placed the rocks in this order. (1)

c (i) Give **two** pieces of evidence from the text that movements in the Earth's crust have taken place during the formation of the Charmouth cliffs. (2)

(ii) About how long ago did this movement take place? (1)

d The rocks at Whitby on the Yorkshire coast are also famous for the fossils they contain and were formed during the same period of time as the rocks at Charmouth. Suggest **one** way in which you could prove that samples of rock from Charmouth and Whitby were roughly the same age. (1)

e The theory of plate tectonics is believed to explain why mountains are formed.

(i) What are tectonic plates? (2)

(ii) Explain, as fully as you can, **one** way in which tectonic plates cause mountains to be formed. (3)

f A scientist wanted to find out the age of a sample of rock which was much older than the Jurassic rocks of Charmouth.

The scientist used a method which involved measuring the amounts of certain radioactive isotopes in the rock.

The isotope potassium-40 ($^{40}_{19}K$) which is present in many rocks slowly undergoes radioactive decay to form argon-40 ($^{40}_{18}Ar$).

When the rock was formed the potassium-40 isotope was trapped in the rock and there was no argon-40. All the argon-40 formed by decay of the potassium-40 also stayed trapped in the rock.

The scientist found that the rock contained three times more argon-40 than potassium-40.

(i) What fraction of the original potassium-40 is left in the rock? (1)

(ii) What further information from the Data Book do you need to find the age of the rock? (1)

(iii) Calculate the age of the rock. (2)

(NEAB, 1994)

80 The diagram shows a section through some strata of rocks found in a cliff. The cliff is located many miles from the sea at about 300m above sea level.

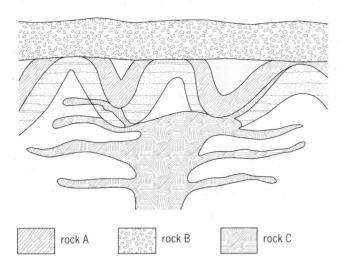

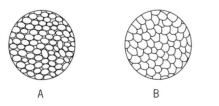

(Rocks **A** and **B** are sedimentary rocks. Rock **C** is igneous.)

a How can you account for the shape of the layer of rock **A**? (2)

b A geologist deduces from the diagram that rock **A** is older than rock **B**. Explain how he reaches this conclusion. (2)

c Some igneous rocks contain the radioactive element thorium. Explain what will happen to the thorium atoms over a long period of time.

(2)

d Limestone is deposited in tropical seas. Explain how Britain once had a climate suitable for limestone deposition. (2)

(MEG, 1995)

81 This sketch was made by a geologist on a field trip to Scotland.

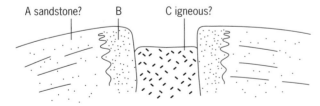

These drawings show what the three rocks look like under a microscope.

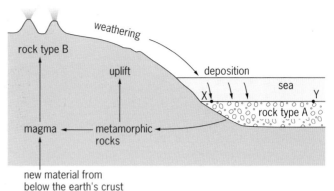

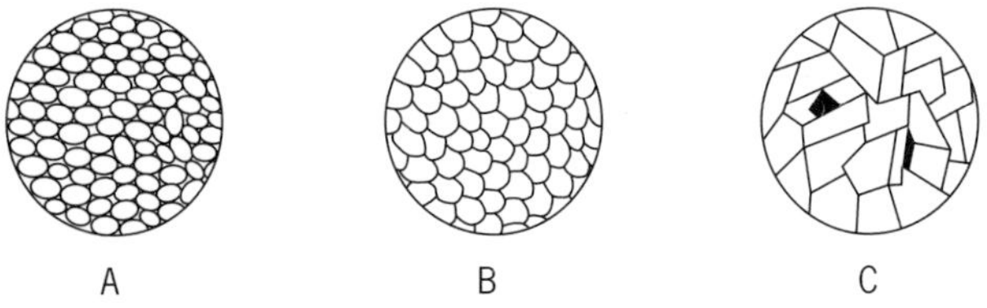

Use the above evidence to explain the formation and history of these rocks. (9)

(MEG, 1995)

82 The diagram below shows part of the rock cycle.

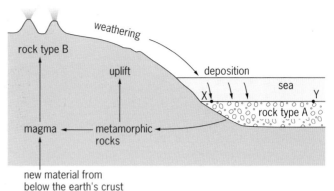

a (i) What is rock type **A**? (1)
  (ii) What changes take place to the magma to form rock type **B**? (1)
  (iii) Suggest **one** way that rain can act as a weathering agent on exposed rocks. (1)
  (iv) What happens to rock type **A** while it is changing to metamorphic rock? (2)
  (v) Suggest **one** difference between sediments found at **X** and **Y** on the diagram. (1)

b The diagram is a sketch map of a coastal area. The distance between **Y** and **Z** is about 300 metres.

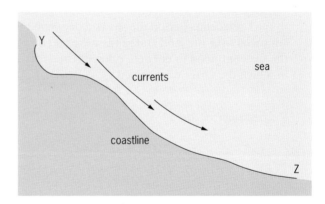

Pebbles found on the beach at **Z** are more rounded than those at **Y**. Explain why the pebbles at **Z** are more rounded. (3)

(SEG, 1995)

Answers to Numerical Questions

ANSWERS TO NUMERICAL QUESTIONS

Chapter 1

Gas laws

c $54.27 \, cm^3$
d $23.81 \, cm^3$
e $14.35 \, cm^3$

Questions

8 $3.45 \times 10^5 \, Pa$
9 a $76.18 \, m^3$

Chapter 5

Calculating moles

a (i) $0.10 \, mol$
  (ii) $0.16 \, mol$
  (iii) $2 \, mol$
b (i) $3.20 \, g$
  (ii) $160 \, g$
  (iii) $5.75 \, g$
c (i) $0.10 \, mol$
  (ii) $1 \, mol$
  (iii) $10 \, mol$
d (i) $162 \, g$
  (ii) $8.50 \, g$
  (iii) $55.83 \, g$
e (i) $0.08 \, mol$
  (ii) $10 \, mol$
  (iii) $0.0008 \, mol$
f (i) $7.20 \, dm^3$
  (ii) $2.40 \, dm^3$
  (iii) $48 \, dm^3$
g (i) $2 \, mol \, dm^{-3}$
  (ii) $0.20 \, mol \, dm^{-3}$
h (i) $7.98 \, g$
  (ii) $40.40 \, g$

Calculating formulae

a CaO
b Empirical formula = CH_3
  Molecular formula = C_2H_6

Moles and chemical equations

a $32 \, g$
b $320 \, g$
c $127 \, g$
d $24 \, dm^3$
e $500 \, cm^3$
f $0.27 \, M$

Questions

1 a (i) $28 \, g$
   (ii) $40 \, g$
  b (i) $20 \, g$
   (ii) $9 \, g$
2 a $300 \, dm^3$
  b $3.60 \, dm^3$
3 a $0.001 \, mole$
  b $5 \, moles$
4 SiO_2
5 a C_2H_6O
  b $C_4H_{12}O_2$
6 d $224 \, tonnes$
7 a $StCl$
8 c $0.001 \, mole$
  d $24 \, cm^3$
  e $20 \, cm^3$

Chapter 6

Calculations in electrolysis

b (i) $2 \, Faradays$
  (ii) $1 \, Faraday$
  (iii) $3 \, Faradays$
  (iv) $2 \, Faradays$
c (i) $193\,000 \, coulombs$
  (ii) $96\,500 \, coulombs$
  (iii) $193\,000 \, coulombs$
d $0.2 \, Faradays$

Questions

4 $5594.2 \, s$ (or $93.24 \, min$)
5 $72 \, dm^3$
6 $0.15 \, Faradays$

Chapter 7

Crystal hydrates

a (i) 36.07%
  (ii) 62.94%
  (iii) 36.29%

Solubility of salts in water

a (i) $135 \, g$ per $100 \, g$ of water
  (ii) $46 \, g$ per $100 \, g$ of water
  (iii) $50 \, g$ per $100 \, g$ of water

c **(i)** 35 g per 100 g of water
   (ii) 44 g per 100 g of water
   (iii) 58 g per 100 g of water
d **(i)** 14 g (± 0.50 g)
   (ii) 9 g (± 0.50 g)
   (iii) 23 g (± 0.50 g)
e 3.30 g (± 0.50 g)

Titration
a 0.224 M
b 0.160 M

Questions
4 c 3 M
6 b **(i)** 21.50 g per 100 g of water
      38.70 g per 100 g of water
      72.00 g per 100 g of water
      118.00 g per 100 g of water
   (ii) 79.30 g (± 0.50 g)
   (iii) 15.20 g (± 0.50 g)
8 e **(i)** 36%
   (ii) 0.05 mol of water
      0.01 mol of copper(II) sulphate
   (iii) $CuSO_4.5H_2O$

Chapter 8

Questions
5 e **(iii)** 2.40 dm³
8 a 61.60 tonnes

Chapter 9

Questions
8 c 623.08 tonnes

Chapter 10

Questions
1 b **(i)** 19.30 cm³
   (ii) 21.44%
5 c 574.40 s (or 9.57 min)
 d **(ii)** 107 500 tonnes
6 a 216 000 tonnes
 b 3.09×10^{12} dm³
7 a 4×10^{10} dm³

Chapter 11

Surface area
b **(iii)** 26 cm³ (± 0.5 cm³)
   (iv) 1 min 51 s (± 3 s)

Questions
2 f 46 s (± 1 s)
 g 43 s (± 1 s)
4 g 12 min
6 c 0.309 g
7 e **(i)** 6.16 g
   (ii) 3.36 dm³
   (iii) 13.44 dm³

Chapter 12

Alkanes
a **(i)** 126 °C
  (ii) 216 °C
e **(i)** 4 moles CH_4 : 1 mole Cl_2
  (ii) 1 mole CH_4 : 4 moles Cl_2

Alkenes
a 63 °C

Questions
7 a C_2H_5
 b C_4H_{10}

Chapter 13

Chemical energy
a **(i)** −1299.50 kJ mol⁻¹
b **(i)** 364 kJ
  (ii) 3640 kJ
  (iii) 182 kJ
c **(i)** 114 kJ
  (ii) 14.25 kJ
  (iii) 114 kJ

Questions
1 b **(i)** −1461 kJ mol⁻¹
4 b **(iii)** −114 kJ
 c **(iii)** 143.75 kJ
5 a 0.50 g
 b 10 080 J
 c 10.08 kJ
 d 20.16 kJ
 e 46 g
 f 927.36 kJ
6 b 100 g
 c 2100 J (2.10 kJ)
 d 0.05 mol
 e 42 000 J (42 kJ)
8 d **(i)** 110 g
   (ii) 1110 kJ
   (iii) 555 kJ
   (iv) 185 000 kJ

Chapter 14

Biotechnology
b **(i)** whisky/brandy: 400 cm³
  (ii) 35%
     350 cm³
  (iii) 16 litres (dm³)

Questions
5 d **(i)** 363 kJ
     2904 kJ
     90.75 kJ
   (ii) 12 dm³

Chapter 15

Artificial fertilisers

a Ammonium nitrate: 35.00%
Ammonium phosphate: 28.19%
Ammonium sulphate: 21.21%
Urea: 46.67%

Questions

4 a Sodium nitrate: 16.47%
Potassium nitrate: 13.86%

  e 39.38 kg

5 d (i) 32%
    (ii) 19%

  e 25%

6 a 10.25   10.1   10.1   10.15

  b 10.12 cm^3

  c (ii) KOH: 1 mol
        HNO$_3$: 1 mol

  d (i) 0.0038 mol
    (ii) 0.0038 mol

  e 0.37 M

Chapter 16

Questions

3 b 7.5 tonnes

  c 15 tonnes

  d 5 625 000 dm^3

4 b (i) 5%
    (ii) 31%

  d (i) 3×10^6 tonnes
    (ii) 550 000 tonnes
    (iii) 900 000 tonnes

6 b 0.01 mol

  c 0.005 mol

  d 0.2 mol dm^{-3} (or 0.2 M)

Chapter 17

Half-life

a (iii) 32.5 h

Questions

1 b 0.0625 g

  c 1.05 s

2 b 25.50 min (± 0.50 min)

  d (i) 94 cpm (± 1)
    (ii) 64 cpm (± 1)

  e (i) 29 min (± 0.50 min)
    (ii) 7.50 min (± 0.50 min)

6 c 19.50 min (± 0.50 min)

  d (i) 21.50 min (± 0.50 min)
    (ii) 32 min (± 0.50 min)

  e (i) 18.50 min (± 0.50 min)
    (ii) 30 min (± 0.50 min)

INDEX

PERIODIC TABLE

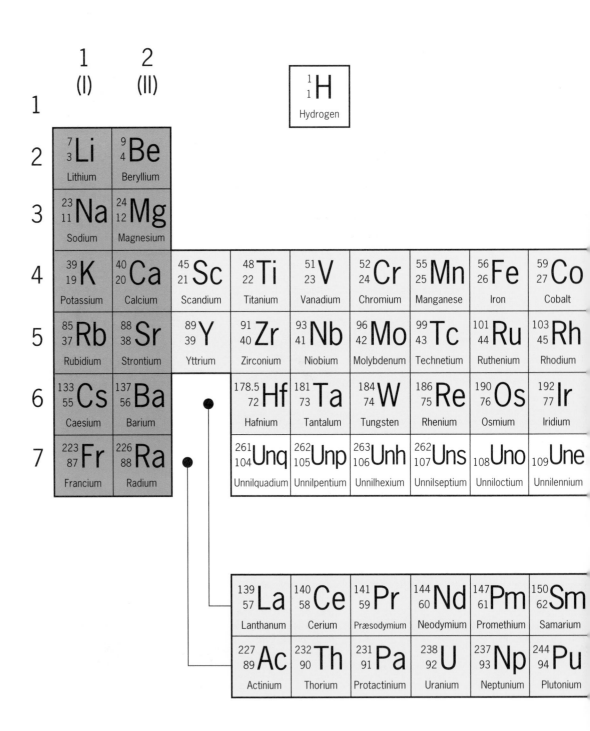

	1 (I)	2 (II)						
1								

$^{1}_{1}$H Hydrogen

| 2 | $^{7}_{3}$Li Lithium | $^{9}_{4}$Be Beryllium |
| 3 | $^{23}_{11}$Na Sodium | $^{24}_{12}$Mg Magnesium |

4	$^{39}_{19}$K Potassium	$^{40}_{20}$Ca Calcium	$^{45}_{21}$Sc Scandium	$^{48}_{22}$Ti Titanium	$^{51}_{23}$V Vanadium	$^{52}_{24}$Cr Chromium	$^{55}_{25}$Mn Manganese	$^{56}_{26}$Fe Iron	$^{59}_{27}$Co Cobalt
5	$^{85}_{37}$Rb Rubidium	$^{88}_{38}$Sr Strontium	$^{89}_{39}$Y Yttrium	$^{91}_{40}$Zr Zirconium	$^{93}_{41}$Nb Niobium	$^{96}_{42}$Mo Molybdenum	$^{99}_{43}$Tc Technetium	$^{101}_{44}$Ru Ruthenium	$^{103}_{45}$Rh Rhodium
6	$^{133}_{55}$Cs Caesium	$^{137}_{56}$Ba Barium		$^{178.5}_{72}$Hf Hafnium	$^{181}_{73}$Ta Tantalum	$^{184}_{74}$W Tungsten	$^{186}_{75}$Re Rhenium	$^{190}_{76}$Os Osmium	$^{192}_{77}$Ir Iridium
7	$^{223}_{87}$Fr Francium	$^{226}_{88}$Ra Radium		$^{261}_{104}$Unq Unnilquadium	$^{262}_{105}$Unp Unnilpentium	$^{263}_{106}$Unh Unnilhexium	$^{262}_{107}$Uns Unnilseptium	$_{108}$Uno Unniloctium	$_{109}$Une Unnilennium

| $^{139}_{57}$La Lanthanum | $^{140}_{58}$Ce Cerium | $^{141}_{59}$Pr Præsodymium | $^{144}_{60}$Nd Neodymium | $^{147}_{61}$Pm Promethium | $^{150}_{62}$Sm Samarium |
| $^{227}_{89}$Ac Actinium | $^{232}_{90}$Th Thorium | $^{231}_{91}$Pa Protactinium | $^{238}_{92}$U Uranium | $^{237}_{93}$Np Neptunium | $^{244}_{94}$Pu Plutonium |

Blocks

- Reactive metals
- Transition metals
- Poor metals
- Metalloids
- Non-metals
- Noble gases

3 (III)	4 (IV)	5 (V)	6 (VI)	7 (VII)	0
					4 2 **He** Helium
11 5 **B** Boron	12 6 **C** Carbon	14 7 **N** Nitrogen	16 8 **O** Oxygen	19 9 **F** Fluorine	20 10 **Ne** Neon
27 13 **Al** Aluminium	28 14 **Si** Silicon	31 15 **P** Phosphorus	32 16 **S** Sulphur	35.5 17 **Cl** Chlorine	40 18 **Ar** Argon

9 8 **Ni** Nickel	63.5 29 **Cu** Copper	65 30 **Zn** Zinc	70 31 **Ga** Gallium	73 32 **Ge** Germanium	75 33 **As** Arsenic	79 34 **Se** Selenium	80 35 **Br** Bromine	84 36 **Kr** Krypton
06 6 **Pd** Palladium	108 47 **Ag** Silver	112 48 **Cd** Cadmium	115 49 **In** Indium	119 50 **Sn** Tin	122 51 **Sb** Antimony	128 52 **Te** Tellurium	127 53 **I** Iodine	131 54 **Xe** Xenon
95 8 **Pt** Platinum	197 79 **Au** Gold	201 80 **Hg** Mercury	204 81 **Tl** Thallium	207 82 **Pb** Lead	209 83 **Bi** Bismuth	209 84 **Po** Polonium	210 85 **At** Astatine	222 86 **Rn** Radon
10 *	111 *	112 *						

*no names yet

| 52 63 **Eu** Europium | 157 64 **Gd** Gadolinium | 159 65 **Tb** Terbium | 162 66 **Dy** Dysprosium | 165 67 **Ho** Holmium | 167 68 **Er** Erbium | 169 69 **Tm** Thulium | 173 70 **Yb** Ytterbium | 175 71 **Lu** Lutetium |
| 43 95 **Am** Americium | 247 96 **Cm** Curium | 247 97 **Bk** Berkelium | 251 98 **Cf** Californium | 252 99 **Es** Einsteinium | 257 100 **Fm** Fermium | 258 101 **Md** Mendelevium | 259 102 **No** Nobelium | 260 103 **Lw** Lawrencium |

ACKNOWLEDGEMENTS

Acknowledgement is made to the following boards:
Midland Examining Group (MEG)
Northern Examinations and Assessment Board (NEAB)
Northern Ireland Council for the Curriculum Examinations and Assessment (NISEAC)
Southern Examining Group (SEG)
University of London Examination and Assessment Council (ULEAC)
Welsh Joint Education Committee (WJEC)

All the apparatus, equipment and chemicals for use with *GCSE Chemistry* can be obtained from: Beecroft & Partners, Northfield Road, Rotherham, South Yorkshire S60 1RR. Tel: 01709 377881 Fax: 01709 369264.

Thanks are due to the following for permission to reproduce copyright photographs*:

p.iii Andrew Lambert; **p.1** Fig. 1.1*t* George Holton/Science Photo Library; Fig. 1.1*m* Simon Fraser/Science Photo Library; Fig. 1.1*b* Doug Plummer/ Science Photo Library; **p.2** Fig. 1.2 Andrew Lambert; Fig. 1.3 David Scharf/ Science Photo Library; **p.3** Fig. 1.5 Michael W. Davidson/Science Photo Library; **p.4** Fig. 1.6 Andrew Lambert; Fig. 1.7 © Robert Ellis/Repfoto; **p.5** Fig. 1.11 Andrew Lambert; **p.6** Figs. 1.12, 1.13 Andrew Lambert; **p.7** Fig. 1.15 John Townson/Creation **p.8** Fig. 1.19 Mary Evans Picture Library; **p.11** Fig. 2.1 ESA/PLI/Science Photo Library; Fig. 2.2 Mary Evans Picture Library; **p.12** Figs. 2.3, 2.4*tl, bl* Andrew Lambert; Fig. 2.4*r* Damien Lovegrove/Science Photo Library; **p.14** Fig. 2.5*r* Andrew Lambert; **p.15** Fig. 2.7 Andrew Lambert; **p.16** Fig. 2.8 Pascal Rondeau/Allsport; **p.17** Figs. 2.9, 2.10 Andrew Lambert; **p.18** Figs. 2.11, 2.12b Andrew Lambert; **p.19** Fig. 2.13 Florence Durand/Rex Features; Fig. 2.14 Andrew Lambert; **p.20** Fig. 2.15 Nicholas de Vore/Bruce Coleman Ltd; Fig. 2.16*t* Andrew Lambert; Fig. 2.17 Zefa Pictures; **p.21** Fig. 2.18 Al Grillo/Still Pictures; Figs. 2.19, 2.20*t* Andrew Lambert; **p.22** Fig. 2.21a Rex Features; Fig. 2.21b BOC Gases; Fig. 2.22 Alex Bartel/Science Photo Library; Fig. 2.23 Andrew Lambert; **p.23** Fig. 2.24a Andrew Lambert; Fig. 2.25 J.C. Revy/Science Photo Library; **p.24** Fig. 2.26 Alain Compost/Bruce Coleman Ltd; Fig. 2.27 Geoff Tomkinson/Science Photo Library; **p.25** Figs. 2.28, 2.30*t* John Townson/Creation; Fig. 2.30*m, b* Andrew Lambert; **p.26** Fig. 2.31 The Laser Centre – Banbury; Fig. 2.32 Manfred Kage/Science Photo Library; **p.29** Fig. 3.1 Prof. Erwin Mueller/Science Photo Library; **p.35** Fig. 3.9 Mary Evans Picture Library; **p.36** Figs. 3.11, 3.12 Andrew Lambert; **p.37** Figs. 3.13, 3.14 Andrew Lambert; **p.38** Fig. 3.17 Andrew Lambert; **p.39** Fig. 3.18 Dr B. Booth/GSF Picture Library; Fig. 3.19 Zefa Pictures; Fig. 3.20 Andrew Lambert; **p.40** Fig. 3.21 Andrew Lambert; **p.41** Fig. 3.23*tl* Habitat; Fig. 3.23*tr* John Townson/Creation; Figs. 3.23*bl, br*, 3.24 Andrew Lambert; **p.44** Fig. 4.1 Zefa Pictures; **p.46** Fig. 4.6b Science Source/Science Photo Library; **p.49** Fig. 4.9a Andrew Lambert; **p.50** Figs. 4.11b, 4.13b Andrew Lambert; **p.51** Figs. 4.15b, 4.17b Andrew Lambert; **p.52** Fig. 4.20 Gary Gladstone/The Image Bank; **p.53** Fig. 4.21b Philippe Plailly/Science Photo Library; **p.54** Fig. 4.22b Galerie Apollon/Robert Harding Picture Library; Fig. 4.23*tl* Richard Megna/ Fundamental/Science Photo Library; Fig. 4.23*tr* Gray/Mortimore/Allsport; Fig. 4.23*bl* Sheila Terry/Science Photo Library; Fig. 4.23*br* Manfred Kage/Science Photo Library; **p.55** Fig. 4.25 The British Architectural Library, RIBA, London; **p.56** Fig. 4.27 John Townson/Creation; Fig. 4.28 Mr Andy Price/Bruce Coleman Ltd; **p.60** Fig. 5.1 Alan Murray, Tilcon Ltd; Fig. 5.2 Andrew Lambert; **p.61** Fig. 5.3 Andrew Lambert; **p.69** Fig. 6.1a Simon Fraser/Northumbria Circuits/Science Photo Library; Fig. 6.1b John Cancalosi/Bruce Coleman Ltd; Fig. 6.1c Ronald R. Read; **p.70** Fig. 6.3 C.B. & D.W. Frith/Bruce Coleman Ltd; **p.71** Fig. 6.5 Martin Bond/Science Photo Library; Fig. 6.6 Quadrant Picture Library; Fig. 6.7 Billiton; **p.72** Fig. 6.8 Zefa Pictures; **p.73** Fig. 6.10 with kind permission of ICI Chemicals and Polymers Ltd on Merseyside; Fig. 6.11*t* Andrew Lambert; **p.76** Fig. 6.14 Andrew Lambert; **p.77** Fig. 6.17 Adam Hart-Davis/Science Photo Library; **p.78** Figs. 6.19, 6.20 Andrew Lambert; Fig. 6.21 Mary Evans Picture Library; **p.82** Figs. 7.1, 7.2 John Townson/Creation; **p.83** Figs. 7.3, 7.4b, 7.5 Andrew Lambert; **p.85** Fig. 7.6*t* Martin Dohrn/Science Photo Library; Fig. 7.6*b* Lupe Cunha; **p.87** Figs. 7.9, 7.10, 7.11 Andrew Lambert; **p.88** Fig. 7.12 Andrew Lambert; **p.89** Fig. 7.13 Andrew Lambert; **p.90** Figs. 7.14, 7.15, 7.16 Andrew Lambert; **p.97** Fig. 8.1 Andrew Lambert; **p.98** Fig. 8.3 Primrose Peacock/Holt Studios International; Fig. 8.4 Rex Features; **p.99** Fig. 8.5 A.C. Waltham/Robert Harding Picture Library; **p.100** Fig. 8.8 Alan Murray, Tilcon Ltd; Fig. 8.9 Andrew Lambert; **p.101** Fig. 8.10*l* Roberto de Gugliemo/Science Photo Library; Fig. 8.10*r* Dr B. Booth/GSF Picture Library; **p.102** Fig. 8.11 Andrew Lambert; Fig. 8.12 Sheila Terry/ Science Photo Library; Fig. 8.13 Dr B. Booth/GSF Picture Library; **p.103** Fig. 8.15 Andrew Lambert; **p.104** Fig. 8.16b Andrew Lambert; **p.106** Fig. 8.19b Andrew Lambert; **p.107** Fig. 8.21 Andrew Lambert; **p.110** Figs. 9.1a, b Andrew Lambert; Fig. 9.1c Kevin Rushby/Bruce Coleman Ltd; **p.111** Fig. 9.2 Andrew Lambert; **p.112** Fig. 9.4a Morsø; Fig. 9.4b The Anthony Blake Photo Library; Fig. 9.5 British Airways; **p.113** Fig. 9.7 Andrew Lambert; Fig. 9.8 Robert Harding Picture Library; **p.114** Figs. 9.9, 9.10 Andrew Lambert; **p.115** Figs. 9.11, 9.12 Dr B. Booth/GSF Picture Library; **p.116** Fig. 9.13 British Steel plc; **p.117** Fig. 9.15 Dr Jeremy Burgess/Science Photo Library; **p.118** Fig. 9.16 Britannia Zinc Limited; Fig. 9.17 Kevin Burchett/Bruce Coleman Ltd; **p.119** Fig. 9.18 Gerald Cubitt/Bruce Coleman Ltd; Fig. 9.19 Hank Morgan/ Science Photo Library; Fig. 9.20 Zefa Pictures; **p.120** Fig. 9.22 Ford; Fig. 9.23 John Townson/Creation; **p.121** Figs. 9.24, 9.25 Paul Brierley; Fig. 9.26a Keith Scholey/Planet Earth Pictures; **p.122** Fig. 9.27 L. Ellison/Sheridan Photo Library; Fig. 9.28 John Townson/Creation; **p.123** Fig. 9.30a Paul Grundy/ Robert Harding Picture Library; Fig. 9.30b Manfred Kage/Science Photo Library; **p.127** Fig. 10.1*tr* Zefa Pictures; Fig. 10.1*bl* Kent Wood/Science Photo Library; 10.1*br* O.L. Miller/Science Photo Library; **p.128** Fig. 10.2a Zefa Pictures; Fig. 10.2b Zefa Pictures; **p.129** Fig. 10.5 Andrew Lambert; **p.130** Fig. 10.6 BOC Gases; **p.131** Fig. 10.7 Penny Tweedie/Science Photo Library; Fig. 10.8 Zefa Pictures; Fig. 10.9 John Townson/Creation; **p.132** Fig. 10.10 Alexander Tsiaras/Science Photo Library; Fig. 10.11 Vaughan Fleming/Science Photo Library; **p.133** Fig. 10.12*t* NASA/Science Photo Library; Fig. 10.12*bl* SmithKline Beecham Consumer Healthcare; Fig. 10.12*br* John Townson/ Creation; **p.135** Fig. 10.14 Sean Avery/Planet Earth Pictures; Fig. 10.15 David Guyon/Science Photo Library; **p.137** Fig. 10.18 Chris Howes/Planet Earth Pictures; **p.140** Fig. 11.1*tl* Nigel Cattlin/Holt Studios International; Fig. 11.1*tr, br* Zefa Pictures; Fig. 11.1*bl* The Anthony Blake Photo Library; **p.141** Figs. 11.2, 11.3 Andrew Lambert; **p.142** Fig. 11.6 Andrew Lambert; **p.145** Figs. 11.11 Ford; **p.146** Fig. 11.12 John Townson/Creation; Fig. 11.13 National Medical Slide Bank; **p.149** Fig. 12.1*t* Pascal Rondeau/Allsport; Fig. 12.2a Andrew Lambert; Fig. 12.2b John Townson/Creation; **p.150** Fig. 12.3 Sindo Farina/Science Photo Library; Fig. 12.4*l* Stevie Grand/Science Photo Library; **p.151** Fig. 12.5*r* Andrew Lambert; **p.153** Fig. 12.7a British Gas; Figs. 12.7b, 12.8 Andrew Lambert; Fig. 12.9 John Townson/Creation; Fig. 12.11 Robert Harding Picture Library; **p.155** 12.13*r* Andrew Lambert; **p.156** Fig. 12.15 Andrew Lambert; **p.157** Fig. 12.16 Andrew Lambert; **p.160** Fig. 13.1 Dr B. Booth/GSF Picture Library; Fig. 13.2 Zefa Pictures; **p.161** Fig. 13.3 Richard Folwell/Science Photo Library; **p.162** Figs. 13.5, 13.7 Martin Bond/Science Photo Library; **p.163** Fig. 13.9 Novosti/Science Photo Library; Fig. 13.10 John Mead/Science Photo Library; **p.164** Fig. 13.14 Andrew Lambert; **p.165** Fig. 13.16 Philippe Plailly/Science Photo Library; Fig. 13.17 John Murray/Bruce Coleman Ltd; Fig. 13.18 David Hall/Science Photo Library; **p.169** Fig. 13.24*b* Andrew Lambert; Fig. 13.25 NASA/Science Photo Library; **p173** Fig. 14.1 Andrew Lambert; **p.174** Figs. 14.2, 14.3, 14.4 Andrew Lambert; **p.175** Fig. 14.5 Andrew Lambert; **p.176** Fig. 14.6 James Holmes/Zedcor/Science Photo Library; Fig. 14.7 John Townson/Creation; **p.177** Fig. 14.10a Ecoscene/Sally Morgan; Fig. 14.10b Andrew Lambert; **p.178** Fig. 14.12*r* Andrew Lambert; **p.179** Figs. 14.13, 14.14 Andrew Lambert; Fig. 14.15a Rex Features; Fig. 14.15b Gerald Cubitt/Bruce Coleman Ltd; Fig. 14.16 Patrick Lucero/Rex Features; **p.181** Fig. 14.17 Alcohol Concern; Fig. 14.18a The Times/Rex Features; Fig. 14.18b Andrew Lambert; **p.182** Fig. 14.19*r* Andrew Lambert; Fig. 14.20 Hans-Peter Merten/Bruce Coleman Ltd; **p.183** Fig. 14.21 John Townson/Creation; **p.185** Figs. 14.25, 14.26 Andrew Lambert; **p.186** Fig. 14.27 Richardson/Custom Medical Stock Photo/Science Photo Library; Fig. 14.28 James King-Holmes/ICRF/Science Photo Library; **p.187** Fig. 14.29 Rex Features; **p.194** Fig. 15.6*r* Andrew Lambert; **p.196** Fig. 15.7 Andrew Lambert; **p.199** Fig. 16.1*t* Dr B. Booth/GSF Picture Library; Fig. 16.1*b* Andrew Lambert; **p.201** Fig. 16.7 Andre Maslennikov/Still Pictures; **p.202** Fig. 16.9 Andrew Lambert; Fig. 16.10 with kind permission of ICI Chemicals and Polymers Ltd on Merseyside; **p.204** Fig. 16.12 Andrew Lambert; **p.207** Fig. 17.1 Christian Zuber/Bruce Coleman Ltd; **p.209** Figs. 17.3, 17.5 Rick Falco/Rex Features; Fig. 17.4 Lewkowicz/Rex Features; Fig. 17.6 Zefa Pictures; **p.210** Fig. 17.8 Alexander Tsiaras/Science Photo Library; **p.211** Fig. 17.10 JET Joint Undertaking; **p.214** Fig. 18.1 Zefa Pictures; **p.215** Fig. 18.2b Dieter & Mary Plage/Bruce Coleman Ltd; **p.216** Fig. 18.4 Ecoscene/Nick Hawkes; Figs. 18.5, 18.6a Dr B. Booth/GSF Picture Library; Fig. 18.6b Zefa Pictures; **p.217** Fig. 18.7*t* Zefa Pictures; Fig. 18.7*b* Rex Features; **p.219** Fig. 18.9 Andrew Lambert; **p.220** Fig. 18.10a David A. Ponton/Planet Earth Pictures; Fig. 18.10b J.R. Bracegirdle/Planet Earth Pictures; Fig. 18.10c Zefa Pictures; Fig. 18.10d John Lythgoe/Planet Earth Pictures; **p.222** Fig. 18.14a Butler/Bauer/Rex Features; Fig. 18.14b Paul Morse/Rex Features; **p.223** Fig. 18.15 reproduced by permission of the Director, British Geological Survey. NERC copyright reserved.

*(*t* = top, *b* = bottom, *m* = middle, *l* = left, *r* = right)